1992

HEALTH CARE SYSTEMS

PHILOSOPHY AND MEDICINE

*Editors*

H. TRISTRAM ENGELHARDT, JR.

*Center for Ethics, Medicine, and Public Issues,*
*Baylor College of Medicine, Houston, Texas, U.S.A.*

STUART F. SPICKER

*School of Medicine, University of Connecticut Health Center,*
*Farmington, Connecticut, U.S.A.*

VOLUME 30

# HEALTH CARE SYSTEMS

## *Moral Conflicts in European and American Public Policy*

*Edited by*

HANS-MARTIN SASS

*Ruhr Universität, Institut für Philosophie, Bochum, F.R.G.; Senior Research Fellow, Kennedy Institute of Ethics, Georgetown University, Washington, D.C., U.S.A.*

and

ROBERT U. MASSEY

*School of Medicine, University of Connecticut Health Center, Farmington, Connecticut, U.S.A.*

KLUWER ACADEMIC PUBLISHERS

DORDRECHT / BOSTON / LONDON

**Library of Congress Cataloging-in-Publication Data** CIP
(Philosophy and medicine ; v. 30)
Based on the Twenty First Transdisciplinary
Symposium on Philosophy and Medicine, held July 23–26, 1985 at the Werner Reimers Stiftung in Bad Homburg vo der Höhe, Federal Republic of Germany; sponsored by Werner Reimers Stiftung and the School of Medicine of the University of Connecticut Health Center in Farmington.
Includes bibliographies and index.
1. Medical care–Europe–Congresses. 2. Medical care–United States–-Congresses. 3. Medical ethics–Europe–Congresses. 4. Medical ethics–United States–Congresses. 4. Ethics, Medical-United States–Congresses. 5. Resource allocation–Congresses.

I. Sass, Hans-Martin. II. Massey, Robert U., 1922– . III. Werner Reimers Stiftung. IV. University of Connecticut. School of Medicine. V. Transdisciplinary Symposium on Philosophy and Medicine (21st: 1985: Werner Reimers Stiftung) VI. Series. [DNLM: 1. Delivery of Health Care–Europe–congresses. 2. Delivery of Health Care–United States–congresses. 3. Ethics, Medical–Europe–congresses. 5. Health Policy–Europe–congresses. 6. Health Policy–United States–congresses. W3 PH609 v.30/W84 GA1 H4 1985]
RA395.E85H33 1988 174.2′094 88–37614
ISBN 1–55608–045–X

---

Published by Kluwer Academic Publishers,
P.O. Box 17, 3300 AA Dordrecht, The Netherlands.

Kluwer Academic Publishers incorporates
the publishing programmes of
D. Reidel, Martinus Nijhoff, Dr W. Junk and MTP Press.

Sold and distributed in the U.S.A. and Canada
by Kluwer Academic Publishers,
101 Philip Drive, Norwell, MA 02061, U.S.A.

In all other countries, sold and distributed
by Kluwer Academic Publishers Group,
P.O. Box 322, 3300 AH Dordrecht, The Netherlands.


Printed in The Netherlands.

# TABLE OF CONTENTS

# FOREWORD

The Twenty-First Trans-Disciplinary Symposium on Philosophy and Medicine convened July 23–26, 1985, at the Werner Reimers Stiftung in Bad Homburg vo der Höhe, Federal Republic of Germany. The symposium enabled scholars from different disciplines to convene; they came from the United States, the Federal Republic of Germany, France, The Netherlands, and the German Democratic Republic. The original theme, "Health Care Systems: An International Study in Moral Issues and Public Policy," was intended to illuminate the health care systems of the United States and the Federal Republic of Germany, but the organizers of the symposium soon realized that the perspectives of other European scholars would add significantly to the ability of the group to achieve the stated goals of the program. The results of this working group find expression in this volume, along with contributions from persons unable to attend the conference. The volume itself has subsequently developed from the symposium and bears the revised title: *Health Care Systems: Moral Conflicts in European and American Public Policy*.

The content of the symposium included comparisons among the health care systems of these countries, and served to draw attention to the decision-making processes which structure the various health care systems. Problems were raised and addressed, and some solutions were proposed.

This publication is the outcome of two years of work on the part of the contributors and the editors. I am pleased to have had the opportunity to work with the organizers and editors: Robert U. Massey, M.D. and Hans-Martin Sass.

The Werner Reimers Stiftung (in cooperation with the School of Medicine of the University of Connecticut Health Center in Farmington) served as host, and provided conference facilities as well as financial support for a project that is of extraordinary importance to the Foundation. The Werner Reimers Stiftung is also very grateful to the editors of this volume for editing the essays in such a way as to retain in the volume the general sense of the discussions at Bad Homburg.

The primary aim of the Werner Reimers Stiftung is to convene scholars from many nations who represent a wide variety of scientific and humanistic disciplines. The founder and factory owner, Werner

*Hans-Martin Sass and Robert U. Massey (eds.), Health Care Systems,* pp. vii–viii.

Reimers (1888–1965), charged the Foundation with the mission of promoting the most important ideas which emerge from scholars in the social and human sciences.

Thanks to the founder's foresight and the continuing efforts of the industrial company which he established almost a quarter of a century ago, it is possible to support and encourage research in the *Geisteswissenschaften* as well as the *Naturwissenschaften*.

The patronage of the founder – too seldom realized – began in Europe; it further developed in the United States, and once again returned to Europe. Werner Reimers wanted those who follow him to attend continuously to the enhancement of humanity. May this publication serve to realize his intentions.

*Bad Homburg v.d.H.*
*Bundesrepublik Deutschland*
*July, 1987*

KONRAD VON KROSIGK

# INTRODUCTION

Health care issues have gained a distinct visibility and prominence in *public policy* and in *economic debate*. Health care costs rose an average of 8.9% during 1985 in Organisation for Economic Cooperation and Development (OECD) countries, while the average inflation rate was only 3.2%, and the average increase in gross national products was 5.7% [9]. In the United States from 1980 through 1985, during a period of intensive efforts at cost containment and structural reforms, costs of health care rose from $248 billion to $425 billion, an increase of more than 11% per year ([1], p. 125]. During the last 15 years, total health expenditures divided by gross domestic product rose from 5.3% to 10.7% in the U.S., from 3.9% to 5.9% in the United Kingdom, from 4.7% to 8.1% in the Federal Republic of Germany, and from 4.3% to 9.1% in France ([11], p. 117). As a percentage of the national growth products of the OECD countries, the share of health care costs has risen from 4.2% in 1960 to 7.5% in 1984, while public expenditure for health care increased from an average of 61% to 79% in 1984. As Schieber and Poullier have shown ([11], p. 116), the public share of health spending was widely spread from 41% in the U.S. to 80% or more in the U.K., Sweden, and Italy. The (1) *public share* of total health costs in the U.K. was 85.2% in 1960 and 88.9% in 1984, while (2) the *health care costs as a percentage of GNP* increased from 3.9% to 5.9%; the figures for the U.S. were (1) 24.7% to 41.4% and (2) 5.3% to 10.7%, respectively; for the Federal Republic of Germany (1) 67.5% to 78.2% and (2) 4.7% to 8.1%; the figures for France were (1) 57.8% to 71.2% and (2) 4.3% to 9.1%; these data indicate that increases occurred similarly in very differently structured national health care systems with apparently marginal impact on the structure of the system.

These figures challenge the common approach to contain costs or to keep the percentage of health care spending within the gross national products low (1) by manipulating the structural elements of the various national systems in place, (2) by exchanging the players in the field, (3) by regulating costs, or (4) by selecting citizens according to diagnosis. If we wish to get a handle on the economic and medical problems we are facing as a good result of progress in medical knowledge and technology, we must address the underlying *normative and moral issues* in order

*Hans-Martin Sass and Robert U. Massey (eds.), Health Care Systems*, pp. ix–ixx.

to get a better assessment of the economic and political issues under debate in high tech and high cost medicine.

Also, as Schwartz has pointed out [12], current cost containment strategies, including the elimination of so-called "unnecessary care," reduction of in-patient treatment, cost control for drugs and various forms of competition can only reduce rising costs by a small, but nevertheless very important, percentage. The essential problems of spiraling costs as far as they are related to population growth of the elderly, rising input prices (the hospital market basket), and technological innovation, cannot be controlled by these and other traditional measures ([12], pp. 220, 224). What is needed is a fresh and innovative debate on more basic issues that includes our cultural understanding of health and health maintenance, our moral concepts of solidarity and distributive justice, the relationship of good personal health care responsibility, and a citizen's ethic in health matters to the medical profession's ethic and public policy issues concerning welfare distribution. Such a debate cannot simply continue to compare, exchange, or streamline existing systems of resource allocation, regulation and deregulation, in-patient or out-patient treatment, the percentages of taxpayers' money in health care spending, and the advantages and disadvantages of various forms of providing care in compulsory, mandatory, or voluntary systems of insurance or co-insurance, health maintenance organizations, and specialized cost provisions for the elderly and the needy.

Whether traditional health care systems based their cost sharing philosophy on 'solidarity' (as in German political culture) or on 'distributive justice' or 'autonomy' in combination with what has been called 'societal obligation' (as in the United States), there are very serious moral conflicts as the fair balance of 'societal obligation' and 'individual obligation' ([15], vol. 1, p. 4) for caring for health and for paying the costs – the economic ones and the non-economic ones. The prescription mentioned by Schwartz to remedy these spiraling health care costs, namely the rationing or restricting of progress in medical technology and beneficial services, is a painful prescription, indeed, but not a therapy at all. Curbing medical research, prospective payment according to Diagnosis-Related Groups (DRGs), stigmatizing certain forms of high tech medicine, or excluding certain citizens because of age or for other reasons would not only violate the Hippocratic ethic and tradition, it also cements an already ineffective mode of economizing, politicizing, regulating, and administrating medicine and health matters. What we need is to look into the *basic moral and cultural roots* which led to the existing forms of providing and administering health care in the various national settings, and to *reconstruct* the systems according to these and other original values. This will require a fresh moral and philosophical approach toward the special issues in health care and health policy and in the discussion of the entire concept of public health provision.

Among the OECD countries, only the United States has on a national level of political and intellectual prominence into these philosophical and moral presuppositions which constitute the basis of our actual economic and political problems. The report by the President's Commission and similar studies serve as a sample of what can be achieved in basic moral debate in a pluralistic society and what has to be achieved in order to avoid further bungled reforms that are medically unfeasible, politically unstable, morally inferior, and that lead nowhere.

The discussion regarding the cost and benefits of the market approach, which in the Federal Republic of Germany has just been initiated by the *Sachverständigenrat* (1985) report which calls for replacing or improving the statutory insurance systems by introducing more market factors [6], and the first report (1987) of the Concerted Action in Health Policy [7], are the first signs that a fundamentally different approach must now replace the recent patchwork regulations. It is important, as Vilmar [14] points out, to include the medical profession in such debates that serve health policy. It is not only important, it is mandatory to include those whose 'sphere of conduct' (as Aristotle remarks in the *Nicomachean Ethics* [1114 a 8]) "develops qualities corresponding to the activities they pursue." But it is not less mandatory to include in these assessments the ethicists and philosophers, i.e., the professionals in the arts and professions of reasoning and value assessment. All nations have a long way to go in order to achieve these interdisciplinary points of view and to make them a part of their culture and policy. The answer to spiraling costs in health care, as in other matters of high tech use and public financing, will not be found in economics and in politics; the answer can only be established by a cultural and moral reformulation and redefinition of the benefits of modern technology together with a renewed appreciation of the values of 'dignity,' 'responsibility,' and 'beneficence' in these new contexts. This is no easy task, as it will require change of public culture and attitudes, not merely change in regulation or cost provision.

What is needed is a just moral risk assessment of modern medical technology, the way it is organized, administered and paid for, and its impact on health, health attitudes, lifestyle, and the individual's responsibility for his or her health. This has also to be a moral assessment of the ways we apply values such as 'autonomy,' 'solidarity,' 'beneficence,' 'reciprocity,' 'social justice,' and 'dignity' to concrete situations of modern medical and health issues. If there is a triage in an age of abundant medical intervention capability and health information and health education for all, it is the scarcity of our moral and cultural resources to handle the increasing numbers of cases both morally and prudentially. Withholding treatment from defective newborns or from severely suffering chronically ill persons, for example, is not just an economic issue, it is first of all a moral and cultural issue; it is deeply related to our beliefs in the worthy goals of life, the meaning of death, and perhaps the very content of beneficence, autonomy, and ethics in high technology medicine. The patchwork approach to basic issues in the transition from stone age to information age is predominant in our political attitudes and public culture. It is a result of the prevailing *defensive ethic*, which tends to be technophobic in technology matters and paternalistic in policy matters.

Another reason for the contemporary dominance of defensive ethics is the not yet digested cultural transition from religiously and ideologically totalitarian societies to pluralistic and open societies, the latter being rich and diverse in values, attitudes, preferences, and goals among individuals and groups.

Both the plurality of values in a democratic society and the availability of most effective technology are not shortcomings; they are achievements we have to handle in a more aggressive and affirmative manner. We have to replace a defensive ethic by an *aggressive ethic*, i.e., by making the most aggressive, the most innovative, and the most responsible use of our moral capacities, our moral education, our individual responsibility, and our public moral awareness. Such a morally aggressive approach towards issues in public health care, health cost care, and personal health care will result in different outcomes in different cultural settings, the same way different stages of technology or forms of administration and government have different results in otherwise comparable countries, as the OECD study proves for the countries so relatively close to each other culturally as are these nations.

Collectively and individually, the so-called "developed" countries and their citizens seem to be underdeveloped in regard to their capacity to assess moral conflicts effectively enough as moral conflicts, not as economic, nor as political, nor as technical ones. But if we want the age of reason not to be just a part of the past, we have to come to grips with the moral challenges we face as a result of value pluralism and technological progress. No cost containment, no increase or decrease in taxpayers' monies, no new forms of medical regulation or self-regulation will solve these problems; such measures will only postpone the solution. They may well, to the contrary, actually increase financial and regulatory pressures, civil discontent, and contribute to the loss of personal responsibility, not only in health matters, but also in health care matters.

The essays in this volume address the *moral presuppositions and underlying ethical considerations* of the structures that exist to provide health care in different national settings: in the United States, West Germany and other European countries, and in the Soviet Union. Underlying all positions is the common belief that organizational structures and allocational strategies of national health care systems are deeply rooted in moral and cultural presuppositions and traditions. Health policy, whether comprising individual or concerted efforts, will be ineffective insofar as it fails directly to address the moral and cultural values and goals in existing systems, as well as their allocational procedures. The direct analysis and assessment of moral and cultural principles such as justice, rights, equality, equity, responsibility, solidarity, and beneficence is *preconditional* for an effective evolution and implementation or improvement of the structure of a public health care system.

In the past, medical ethics had been intertwined with broader ethical issues: the political questions of allocation of funds, government regulation, risk management, business ethics, administrative ethics, and the moral challenge to achieve a prudent balance between individual civic responsibility and public political regulation; the last has become a part of medical ethics. Questions of special importance in this volume are: (1) the *analysis of value conflicts* in designing politically feasible, economically affordable, and morally acceptable health care delivery systems; (2) the description and *assessment of existing American and European health care systems*, with special attention given to the moral principles that govern them and the organizational and financial means used to achieve the intended goals; (3) the *ethics of allocation*: *macro-level allocation* in terms of the gross national product; *high level micro-*

*allocation* between medical research; prevention and acute intervention; cost allocation in prospective payment systems between different Diagnosis Related Groups; *low-level allocational decision-making* in individual hospital or insurance settings; and (4) the special issues in contemporary medical ethics, such as the *ethics of preventing health risks* in recreation or occupational activities; the *future of the traditional ethos of professional medical conduct*; and the *impact of environment and regulation on health and health care in the future*.

The four basic goals in health care policy – "best of care," "equality of care," "cost containment," and "optimum choice" – cannot be achieved equally; they are not, as H.T. Engelhardt shows, all compatible. Further conflicts arise in allocation because the issues of allocation are fundamentally issues for *moral* argumentation, but the final decisions are made by the *political* process which mediates economic realities, political conflicts, and moral ideals. Complicating the situation even further is the cultural dimension of our concepts of 'health' and 'health care' [4]. This is especially obvious when we consider the "medicalization" of certain health and well-being issues, the fashionable language which claims "rights" to health care, and the debate on whether to regulate further or to deregulate medical care markets [13]. Basic moral convictions, with an emphasis on either public regulation or personal responsibility, guide the debate on public policy, including health care policy in welfare states [2, 3]. When establishing health policy we must consider that the concepts of health and disease vary from individual to individual and culture to culture; we must ask ourselves how we should value these differences in values and life styles. How can we justify imposing a uniform standard of medical intervention on a heterogeneous group of educated citizen? How can we avoid a decline in personal health risk management when we have a system that redistributes health care costs, including those which are the consequence of an individual's poor health practices and risky behavior?

Existing health care systems are, however, excellent paradigms to study how the resolution of competing moral goals has resulted in different administrative and regulatory settings. The 100-year-old German health insurance system, for example, which covers nearly the entire population by compulsory membership in sick funds, is in fact a system of heavyweight provider institutions with little competition and few incentives to provide medical services more efficiently, as von Schulenburg argues. In West Germany, established provider institutions make concerted efforts to balance the principles of "solidarity" and of "benefits in hand" against the pressing concerns of cost containment and medicalization of health issues [3]. The U.S. system, on the other hand, presently experiments with a variety of insurance and health maintenance organizations, both privately and publicly funded. The result, according to Buchanan, is a decrease in provider freedom that is only partially offset by an increase in consumer choice. DeWachter and Maynard describe the ingredients of the Dutch and British national health care systems, respectively, assigning the government the prime responsibility for health care, and making health a form of civil right, and health care a public obligation. They discuss the arguments for a morally just and prudent mix of private and public responsibility and funding under the assumption that health care is a governmental obligation. Lacronique and Maynard uncover nearly all the arguments for and against the realization of the solidarity principle through direct governmental execution of medical care in a retrospective analysis of the French and British health care systems.

Very much in contrast to the Western debates on equality and choice, the health care system of the Soviet Union focuses on health care as a national benefit, not only a benefit to each individual. Such an attitude permits preferential care for persons considered to be of greater national worth. Feshbach argues that the totally centralized system suffers from bureaucratic dominance and a lack of incentives for competition among providers. Reviewing the competing political, moral, and economic goals that must be addressed when establishing national health policy, Delkeskamp-Hayes reconstructs Rawls' principle of justice as a principle of produce, calling for a system of incentives for insurance and other means of reducing health risks for the individual, and measuring the success of realizing the principle of "fairness as prudence" by measuring the efficiency and reliability of the system.

In comparing health care in the United States to the German system, Rie accentuates the failure of cost control in Germany while pointing to some promising aspects of the emerging corporatization of health care delivery in the U.S. The development of 'corporate medical ethics' might relieve providers and patients from the shortcomings of governmental involvement, and governments from all direct involvement.

Mattheis, introducing the allocational debate, identifies government, insurance, and health care providers as the main players in the field, allowing only marginal input from the citizen as responsible patient.

This might be one of the reasons that allocational issues in health care policy, as in other areas of public resource allocation, are handled predominantly in a crisis management style, without the benefits of long-range planning and without encouraging educated citizens to safeguard their health. Brody points out that the choice principle – allowing the choice between preventive or primary care *versus* catastrophic intervention costs, or the choice for a voucher system *versus* a fixed percentage of heteronomously allocated funds within the G.N.P. – might permit us to abandon the triage situation by replacing triage and heteronomy with *individual allocation* and autonomous decision-making. Mattheis and Thust use the West German health policy system to demonstrate how, in fact, limited resources are managed, governed by the principles of *solidarity* and *equity* (not necessarily free choice), thus making further "rationing" unavoidable as technology progresses and demand increases. Spicker, in challenging Brody's argument in support of the voucher approach and self-rationing concept, calls not only for more *efficiency* in the existing system which must manage limited resources, but argues for the moral legitimacy of rationing. Spending less by fighting waste, *and spending less by providing less*, are the only significant ways to assure a system that will provide minimally adequate care; a wholesale reduction in inefficiency, inequity, and unfairness necessarily requires rationing.

Pfaff suggests reform, not revolution, of the existing West German quasi-market system. In addressing the economic and administrative aspects of health care allocation and cost overruns, he calls for cost-efficient structural, financial, and medical reform. Schöne-Seifert, like Delkeskamp-Hayes, discusses the looseness of the Rawlsian justice principle and focuses on the impact of more educated patient involvement in health care rationing decisions.

It is interesting that most of the contributors do not advocate accepting the framework of the triage and rationing situation. Strong moral arguments favor an increased involvement of the patient, or the prospective patient, i.e., the educated citizen, in health and medical care allocational decision-making. Regardless of the final direction of future structural or allocational reforms, preventive medicine, good health risk management in the workplace, and strengthening the essential traditional ethos of the healing professions have again become issues of great importance. Routine and low cost health care investments, such as annual stool analysis to detect blood, immunization against measles and other childhood discases, blood tests for detecting abnormal cholesterol and triglyceride levels, campaigning against drunk driving and smoking, are dull indeed, but these unexciting health investments are extremely effective even though they are not in the mainstream of the public's health concerns. Viefhues discusses the benefits of illness prevention, but also the moral and social costs which arise when there is public pressure to give up "unacceptable" life styles and the dissemination of incorrect or unverified financial and medical data which discriminate against persons who display non-preventive behavior. Brandts outlines the excellent track record of health-risk management in occupational health in the West German system as politically well established, morally accepted, and extremely efficient. This is the result of a happy and prudent cooperation of labor, capital, and management in safety research, early diagnosis, prevention, and therapy with heavy emphasis on risk prevention and management. It is urgent and challenging when Ballantine, who spent his lifelong career as a practicing surgeon, reminds us that in situations of allocational triage and economic cost containment there are still important traditional Hippocratic values in the medical profession, but they are endangered in a system which makes clinical judgement capitulate to majority vote, government regulators, or insurance executives.

This volume assesses different systems of public health care and allocation; it presents *controversial positions* and *conflicting recommendations*. It does not offer any single, watertight solution to the value conflicts or fairness and choice principles involved; indeed, there may not be such a solution. However, whatever medical, philosophical, political, economic, or administrative approach each author advances, they all seem to agree that focusing exclusively on political compromise or economic effectiveness will inhibit, not help, the promulgation of health and health care policy. When we re-open the moral debate on the value-preconditions of actual health care reality and its reform, we may not have to accept the concepts of allocation or triage for the future structuring of health delivery regulation and resource allocation. Moreover, we may not have to accept the thesis that the future of medical service will mean uniformity of life styles, endless rationing, and various forms of injustice as the result of insoluble problems. By incorporating reflections on the moral aspects of health care policy in contemporary and future discussions in the medical, economic, and political aspects of providing health care to all citizens, we shall have to reassess the risks and benefits of politizing medicine, of medicalizing non-health problems and life style behaviors, and of economizing in distributing health and medical resources. The educated citizen's involvement in health care decision making, in allocating resources, and in public solidarity and civil beneficence will have to be re-appreciated, not only for the sake of the health care system, but for the sake of better health and the provision of quality health care. In the end we shall require nothing less than a *healthy mix of public and private responsibility*.

If it is often true, as Pfaff points out, that economic health expenditures do not correlate positively, but often negatively with personal well-being, then the real triage we are facing is the shortage and shortcomings of our capability to handle morally abundant technological and relatively abundant medical and financial resources. The absence of professional ethicists from the work of health policy formation is indicative of such a "moral triage," which can only be overcome by an aggressive moral approach to matters of health and health care. Humanity always needs to be given priority over economy, as Vilmar puts it [14].

## ACKNOWLEDGEMENTS

This volume offers international perspectives in the analysis and assessment of moral values in different health care settings, which raise normative not just comparative issues. First drafts of the articles were presented and discussed during the Twenty-First Trans-Disciplinary Symposium on Philosophy and Medicine, hosted by the Werner Reimers Stiftung in Bad Homburg, Federal Republic of Germany, July 21–26, 1985. The Editors would like to thank the Reimers Stiftung through its Director, Konrad von Krosigk, for its financial support, exceptional hospitality, and stimulating environment. We also thank the University of Connecticut School of Medicine for supporting, in various ways, the participation of American philosophers and physicians in this engaging intellectual undertaking. We gratefully acknowledge the contributions at the conference of Professor Dr. med. Hans-Peter Wolff, Head of the Scientific Council of the German Medical Association (*Bundesärztekammer*) and of Dr. Rainer Flöhl of the *Frankfurter Allgemeine Zeitung*. We also extend our thanks to Lorraine Langlois for preparing the Index, and to Mrs. Susan Engelhardt in Houston, Texas, and Theo Ungewitter in Farmington, Connecticut, for their, unselfish participation from the first days of planning the symposium to the final process of seeing this volume through the press. Without the patience of the publisher, D. Reidel in Dordrecht, and the continuous counsel, encouragement, and editorial suggestions from the editors of the *Philosophy and Medicine* series, H.T. Engelhardt, Jr. and Stuart F. Spicker, this volume would not have come to fruition. My very personal appreciation is extended to Robert U. Massey, M.D., for his patience with my shortcomings and for having unhesitatingly accepted the heavier burden of this joint venture, leaving the easy tasks for me.

HANS-MARTIN SASS

BIBLIOGRAPHY


1. Anderson, G. F.: 1986, 'National Medical Care Spending', *Health Affairs* **5**(3), 123–130.
2. Aaron, H. J., Schwartz, W. B.: 1984, *The Painful Prescription. Rationing Hospital Care*, Brookings Institution, Washington, D.C.
3. Califano, J. A.: 1986, *America's Health Care Revolution*, Random House, New York.
4. Engelhardt, H. T.: 1986, *The Foundations of Bioethics*, Oxford University Press, New York.
5. Germany, Fed. Rep., Bundesminister für Forschung und Technologie: 1985, *Forschung und Entwicklung im Dienste der Gesundheit*, BMFT, Bonn, F.R.G.
6. Germany, Fed. Rep., Sachverständigenrat zur Begutachtung der gesamtwirtschaftlichen Entwicklung: 1985, Jaresgutachten 1985/86, *Deutscher Bundestag. 10. Wahlperiode. Drucksache* **10**/4295 (22.11.85) F.R.G.
7. Germany, Fed. Rep., Sachverständigenrat, Konzertierte Aktion im Gesundheitswesen: March, 1987, *Gutachten* (Bundesministerium für Arbeit und Soziales), Bonn, F.R.G.
8. v. Ferber, L. and v. Ferber, C.: 1986, 'Sozialmedizin – eine unbekannte Wissenschaft', *Theorie und Praxis der sozialen Arbeitswelt* **37**, 260–267.
9. O E C D: 1985, *Measuring Health Care*, OECD, Paris, France.
10. Russell, L. B.: 1986, *Is Prevention Better than Cure?*, The Brookings Institution, Washington, D.C.
11. Schieber, G. J., Poullier, J. P.: 1986, 'International Health Care Spending', *Health Affairs* **5** (3), 111–122.
12. Schwartz, W. B.: 1987, 'The Inevitable Failure of Current Cost-containment Strategies', *Journal of the American Medical Association* **257**, 220–224.
13. Thiemeyer, Th.: 1986, 'Gesundheitsleistungen – Steuerung durch Markt, Staat oder Verbände', *Sozialfortschritt* **35** (5/6), 97–104.
14. Vilmar, K.: 1987, 'Strukturreform im Gesundheitswesen', *Deutsches Arzteblatt* **84**, No. 22, 1544–1553.
15. United States, President's Commission for the Study of Ethical Problems in Medicine and Biomedical and Behavioral Research: 1983, *Securing Access to Health Care*, U.S. Government Printing Office, Washington, D.C.

# SECTION I

# VALUE CONFLICTS IN ALLOCATION AND CARE

H. TRISTRAM ENGELHARDT, JR.

# NATIONAL HEALTH CARE SYSTEMS: CONFLICTING VISIONS

## I. INTRODUCTION

Choosing the proper health care system[1] is difficult, for such decisions bear on life and death issues. The problem is complicated by Western culture's transition from a period of wide-spread belief in an after-life to wide-spread skepticism. For many it is unclear how one ought to choose among alternative levels of health care, if this life is the only life we shall possess. As belief in a Deity wanes, it becomes more difficult "to play God" and to shoulder choices that have implications for differential morbidity and mortality rates. It is hard to compare probabilities of marginal increases in life expectancy with different qualities of life, or to decide that a life of a particular quality does not merit further extension. Decisions regarding the proper character of a health care system depend on our views of the worth of a life. Different lives are saved with different qualities of outcome, depending on whether one invests resources in prenatal care or in intensive-care units for very defective, low-weight neonates. Different levels of investment in health care and different apportionments of health care budgets have implications not only for morbidity and mortality rates, but for the character and quality of living and dying.

One might wish to ask straightforwardly what amount of expenditure is worth what extension of life or decrease in morbidity. The difficulty is that often information regarding the comparative usefulness of particular investments is inadequate. In addition, high-intensity care interventions or those that involve new technologies tend to capture the attention of the public, and therefore of policy makers, more than the more mundane undertakings of preventive medicine. A quarter of a million dollars is more likely to be invested in providing for a liver transplant than in a campaign to dissuade teenagers from smoking, even if the latter may have a greater impact on morbidity and mortality rates. There is a distortion in our moral intuitions that causes us to weigh more highly saving identifiable lives rather than statistical lives [2]. Though such may be the appropriate response on an individual level, where one may be bound to aid friends before anonymous strangers, it would seem to be far from the prudent response to the challenge of fashioning public

*Hans-Martin Sass and Robert U. Massey (eds.), Health Care Systems,* pp. 3–13.

policy for health care. A rational choice would be to adopt that plan that will on average secure oneself and others against mortality and morbidity risks unless there are grounds for giving special weight to the provision of high technology care.

Given irresolvable disputes about the nature of justice, health care allocations are best made in terms of an insurance metaphor [5]. A particular health care system provides an insurance system against the untoward results of the natural and social lotteries: those untoward outcomes that lead to disease, deformity, and disability, or that diminish the ability to pay for health care. Because there are other untoward circumstances besides disease and disability against which a society may wish to prepare (e.g., provision of military defense as a security against attack by enemies), in addition to the interest in simply enjoying this brief life, and because our resources are limited, we must settle on a finite apportionment of funds to the health care system. Both individuals and societies cannot protect themselves against all the risks to which they are likely to be exposed and must choose to protect themselves against only some of them. To decide, for example, to discourage the provision of hemodialysis for individuals over a specific age reflects a decision not to invest funds against such losses at the natural lottery, but instead to invest them in other social undertakings or to have them available for private enjoyment.[2] Indeed, if one tries to invest all one's energies in protection against risks, there will be no resources available to enjoy the life so secured, and there will still be risks left unmitigated.

Though decisions regarding the investment of resources in health care have a sense of urgency because they bear on life and death, they are complicated by the fact that aggressive investments do not always bring significant benefits. Often, no matter how much one invests in time, resources, and money, some individuals cannot be restored to full health or provided the years of life they might find minimally satisfactory. In contrast, individuals can be provided an amount of nutrition that will be minimally satisfactory. Individuals dying of hunger in Ethiopia can in principle be given sufficient food and shelter to provide, all else being equal, a level of nutrition comparable to that available to Europeans and Americans. After a certain point, the further provision of food and shelter is no longer productive and, in fact, may be counter-productive, at least with respect to objective measures of morbidity and mortality. People can be more easily and effectively fed enough to achieve adequate nutrition than provided with health care to secure health. Still, there is the allure of more investment in nuclear magnetic resonators, intensive care units, or coronary bypass operations, which will incrementally increase the lifespan and decrease the morbidity of the population as a whole, and will make dramatic differences in the lives of particular individuals.

Because of the perceived importance of health care, many are less disturbed by the fact that some live in mansions and others in apartments, than they are by the fact that some may be able to purchase better care and therefore possibly live longer lives or lives with less morbidity. Concerns with health and health care are bound up with our very understanding of our finitude and our fear of death and circumscription. Inequalities seem dramatic and anything less than a full commitment of resources for health may appear to be inadequate. This is the case despite the fact that better apartments may more effectively increase the quality of lives of persons than the provision of intensive care units during terminal illness.

In summary, any discussion about health care policy is fraught with conflicting values and visions of the good life. Even where health care may not be as important for health and well-being as good food and adequate shelter, high technology medicine is more likely to be seen as a life-and-death issue than more prosaic concerns with adequate nutrition. This circumstance is complicated by the fact that the life secured by high-technology intervention may often be of a quality not worth the living or worth the resources that could have brought more pleasure elsewhere or earlier in life. The rather general problem of coming to terms with human finitude is thus raised by health care decisions as one determines what amount of financial and other resources ought to be invested in saving what number of lives, and of what quality. All of these difficulties are then compounded by the diversity of human understandings of the concrete character of the good life and of adequate health care as well as by the inequalities of resources, talents, and needs that mark the actual lives of individual men and women.

## II. THE MORAL TENSIONS IN HEALTH CARE POLICY

It is not just that a diversity of concerns cluster around health care, health care policy tends to involve incompatible goals:

(1) all should receive the best of care;
(2) there should be more equality in health care;
(3) the costs of health care should be contained; and
(4) it is good to maintain consumer and provider choice.

The first and third propositions cannot both be pursued without modification and compromise. One cannot have the best of care and still engage in serious cost containment without in fact losing at least marginal benefits from the health care foregone in the pursuit of cost savings. Similarly, propositions 2 and 4 cannot be pursued at the same time. One cannot attempt to achieve equality in the distribution of health care, while one allows patients and physicians freely to contract for services. Patients with greater resources will contract for better and more effective care. So, too, proportions 3 and 4 may be in tension. If one attempts to curtail health costs, one may come to intrude into the free agreements between patients and physicians. In short, one cannot have the best of health care for all at an inexpensive price while retaining the freedom of all the parties involved.

A common response is that we should not provide the best of care for all, but only minimally decent care or adequate care. In this model all individuals are provided with some level of care, which will be different from the care that can be purchased by those with ample resources. In most Western democracies differences exist primarily in the interpretation of what should count as decent care. There is a major difficulty in determining what will count as a decent or adequate minimum.[3] Some may have a more expansive version where the minimally decent amount provides nearly all of high technology medicine to all who could benefit. In this last case, the differences between the private and the public tiers would perhaps be defined primarily in terms of the amenities offered, e.g., a ward setting versus a semi private setting. In other societies, high technology medicine with only marginal promise of benefit is excluded from the public tier.

Even in socialist countries, such as Poland, a significant proportion of health care continues to be provided on a fee-for-service basis ([13], p. 204). It is probably very difficult, if not impossible; totally to extirpate a second, private tier. Indeed, it may not be feasible to fashion an enforceable notion of equality in health care. If that is the case, the only question will be the extent to which the private tier operates openly and the number of individuals who can have access to it. The more that tier is criticized and contained, the more it becomes available only to the privileged few.

The attempt to resolve the tension among the four desiderata of an ideal health care system will thus compel public policy makers to decide on the character and extent of inequalities in health care that their society will tolerate. It will also require an interpretation of minimally decent or adequate levels of health care. The more one concludes that such concepts of adequacy or minimally decent character cannot be discovered by the exercise of reason alone, the more the answers to such questions must be sought through public negotiation and democratic process. That is, one will be compelled to abandon a search for a correct answer in favor of a fair answer. The interactions among the four desiderata are thus dialectical in the sense of leading to mutual qualifications and completions of what would otherwise be onesided or incomplete understandings of health care.

As a result, it becomes clear that in order to compare the moral and public policy virtues of different approaches to the delivery of health care one will need to determine the significance of a number of key moral issues.

1. *Equality*

One will need to determine the extent to which it is morally desirable, obligatory, or useful to pursue equality in the distribution of health care resources. One will need in addition to determine whether requirements to distribute equally should apply to all forms of health care or only to some. One will also have to decide what will count as equality in health care services. The provision of the same services to all or of the same amount of money to purchase health care will not lead to a useful sense of equality, since health care needs are so different. At best one may be able to establish a limited list of health care services that would be available to all and that would be all that would be available.

2. *Freedom*

One will need to determine the extent to which the autonomy rights of care givers and receivers provide them with a claim to being free to arrange forms of treatment and/or reimbursement with which major segments of society may disagree.

3. *Private versus Communal Resources*

One will need to determine the extent to which there are both private and communal resources available for the purchase of health care. Insofar as there are any truly private resources, will these justify individuals in purchasing extra care in spite of a system being directed toward equal care?

4. *The Limits of Governmental and Societal Authority*

Unless one concludes that societies and governments may totally appropriate the resources and services of individuals, exclaves of private authority will remain where individuals with consenting others may make special arrangements for health care. Insofar as there are private resources and limits to governmental and societal authority, then black markets in health care may be special sanctuaries of freedom, despite attempts to create an all-encompassing national system of health care.

5. *The Importance of the Market*

As an empirical matter, one must determine in what areas the market is or is not useful in the distribution of health care resources. One will need to ascertain where profit motives best achieve the goals of health care. Some compromises may be acceptable, even on the part of those who ideally would have a single, all-encompassing governmental health care system. One might think of the arguments to tolerate prostitution on the grounds that attempts to prohibit prostitution would only make matters worse. For the committed egalitarian there may be an analogue in the question whether one should acquiesce in some patients' purchasing extra services openly, rather than encouraging clandestine corruptions of the health care system. One must also determine the extent to which the profit motive aids in specific areas, such as the development and/or distribution of pharmaceuticals or technological innovations, or the encouragement of physicians in the zealous discharge of their duties. Once such general empirical matters are known, they can be put in terms of theoretical moral concerns such as those raised by John Rawls [10] and his commentators, who have attempted to apply Rawlsian theory of health care.[4] For instance, should the distribution of health care be seen as an element of fair equality of opportunity that must therefore be distributed equally (whatever that might mean), or may it be distributed in ways that redound at least to the benefit of the least advantaged, so that one would expect some to receive better health care than others, as long as such is part of a system leading to the best health care possible for those receiving least?

6. *Discovering versus Inventing the Proper Patterns for Health Care Distributions*

To understand how one should interpret the data sought in point 5, one will need to decide whether the proper character of health care distributions is something that can be discovered through moral reasoning, or whether it is rather a pattern to be created through democratic negotiation. If one concludes that fair conclusions are more accessible than correct conclusions, one will then need to determine the proper processes for fairly negotiating a correct pattern of health care distribution [6].

7. *Comparing Well-beings: Health versus Other Goods, and Healths versus Healths*

Whether one creates or discovers the answer to the proper pattern for allocating health care resources, one will need to decide how important health is in comparison with other goods in order to determine the appropriate size of the health care budget. One will need in addition to compare particular forms of health care (treatment of leukemia versus treatment of arthritis versus the provision of cosmetic surgery) in order to determine how to apportion the health care budget itself. An answer in terms of "pure process" is provided by the market. In such a circumstance, the amount of the budget to be allocated to health care and the apportionment of the health care budget itself can only be determined by the final outcome of numerous decisions by men and women in the marketplace – or at least the market augmented by charity. A general solution in terms of a discoverable pattern may be sought in a centrally planned economy directed to providing resources to meet the "true needs" of citizens. The more one despairs of the possibility of discovering the "true" ranking of needs, the more one is forced to explore the issue of who has the authority to create a ranking and to impose it on others. The more centrally planned the health care economy, the more salient the question becomes of the grounds for the authority to impose a particular ranking by force.

A careful comparison of national health care systems will undoubtedly disclose that each relies on multiple and heterogeneous moral or philosophical justifications. Yet an analysis of the supposed and the actual justifications is essential if we are to make sense of such large-scale public policy choices. Moreover, any health care policy will require a solution to the tension among the four desiderata of health care and will require attention to the foregoing seven points.

## III. THE INTERPLAY AMONG LEVELS OF HEALTH CARE ALLOCATIONS.

One can distinguish four levels of discussion regarding the allocation of health care resources:

(1) high-level macro-allocational decisions through which it is determined what portion of the gross national product will be allocated to health care:

(2) low-level macro-allocational decisions through which it is determined what proportion of the gross national product will be allocated to particular health care endeavors;

(3) high-level micro-allocational decisions through which patterns for the allocation of health care resources to particular individuals are determined (e.g., by lottery versus social worth); and

(4) low-level micro-allocational decisions through which actual allocations to particular patients are made in light of (3).

These levels necessarily interact. The more one restricts the amount of funds allocated for health care, the more one will need to set policies on the micro-allocational levels. Not all patients who could possibly benefit will receive care. On the other hand, if at the lower-level micro-allocational level particular cases attract publicity so that funds are recruited for their care, as occurred with kidney transplantation in the United States, both low-level and high-level micro-allocational decisions will be altered [13]. The proper character of the interplay will depend on one's analysis of the seven questions raised in Section II, and will bring into question moral commitments at different allocational levels. The decision to provide equal care for all and to curtail costs in order better to maximize the social good will collide with moral arguments made on the basis of rights to autonomy and/or on the limits of governmental authority made at the micro-allocational level by individuals willing to forego other goods in order to have special access to high-technology care. Moral concerns to provide the best of care at a micro-allocational level will call into question commitments to serious cost containment, just as concerns for cost containment will raise the issue whether marginally useful treatment is in the best interests of patients, or whether it should be foregone and the resources invested elsewhere.

## IV. SUMMARY

Not unexpectedly, discussions of health care allocations bring together a number of fundamental issues about proper human relations in conditions of moderate scarcity. A comparison of the underlying values of national health systems affords an opportunity to compare moral visions. It is important, as a result, not simply to explore economic differences, but to understand how these economic differences are lodged within broader moral concerns and views of proper social relations. In fashioning health care policies, one encounters the essential interplay of economic realities and human ideals. As the world's societies face the problems of financing health care in a context of high technology and moderate scarcity of resources, a careful comparison and analysis of competing and conflicting possibilities must be undertaken, as occurs in this volume. We are only beginning to come to terms with the problem of balancing our Promethean visions of progress that draw us on to developing ever more effective health care with Hippocratic concepts of limits to reasonable care. Each will need to be given its place. There is the additional difficulty of deciding how to establish balance and proportion in democratic societies that recognize individual rights, including at least limited property rights. These are major challenges for moral and political theory.

*Center for Ethics, Medicine and Public Issues,*
*Baylor College of Medicine,*
*Houston, Texas, U.S.A.*

## NOTES

[1] In this essay I follow the President's Commission's Report in drawing no distinctions between 'health care' and 'medical care' because of the ambiguous ways in which both terms are used in the literature [9].

[2] For a description of the ways in which the British Health Care System functions through setting informal impediments to access to hemodialysis over age 50, see [1].

[3] There is a major difficulty in defining what should count as a decent amount of health

care. Apart from the problem of defining what 'equal care' would mean (e.g., an equal amount of money versus an equal amount of services versus an equal amount of success), there are problems in defining a 'decent minimum' or 'adequate level of health care'. If a decent minimum is defined in terms of what will provide a benefit or meet a need, the concept of decent minimum will expand dramatically. Consider, for example, that individuals may benefit from cosmetic surgery or the fact that providing ICU care may only extend the lives of individuals a few days or weeks. Should such costly care be a part of the decent minimum amount? For an interesting analysis of these issues and an attempt by the President's Commission for the Study of Ethical Problems in Medicine and Biomedical and Behavioral Research to define an adequate level of health care, see [9].
[4] There has been a considerable effort to apply the work of John Rawls and to analyze its significance for ethics, economics, and public policy. See, for example, [4]. There have been a number of attempts to apply Rawls's work to the allocation of resources to health care, even though Rawls eschews such applications. See [3, 7, 8, 12]. For a general overview regarding issues of justice and health care, see [11].

## BIBLIOGRAPHY


1. Aaron H. J. and Schwartz, W. B.: 1984, *The Painful Prescription: Rationing Hospital Care*, Brookings Institution, Washington, D.C.
2. Calabresi, G. and Bobbitt, P.: 1978, *Tragic Choices*, W. W. Norton and Company, New York.
3. Daniels, N.: 1983, 'A Reply to Some Stern Criticisms and a Remark on Health Care Rights', *Journal of Medicine and Philosophy* **8**, 363–371.
4. Daniels, N. (ed.): 1974, *Reading Rawls*, Basic Books, New York.
5. Engelhardt, H. T., Jr.: 1986, *The Foundations of Bioethics*, Oxford University Press, New York, esp. Chapter 8.
6. Engelhardt, H. T., Jr.: 1984, 'Shattuck Lecture – Allocating Scarce Medical Resources and the Availability of Organ Transplantation', *New England Journal of Medicine* **311**, 66–71.
7. Green, R. M.: 1976, 'Health Care and Justice in Contract Theory Perspective', in R. M. Veatch and R. Branson (eds.), *Ethics and Health Policy*, Ballinger Publishing, Cambridge, Mass., pp. 111–126.
8. Green, R. M.: 1983, 'The Priority of Health Care', *Journal of Medicine and Philosophy* **8**, 373–380.
9. President's Commission for the Study of Ethical Problems in Medicine and Biomedical and Behavioral Research: 1983, *Securing Access to Health Care*, U.S. Government Printing Office, Washington, D.C.
10. Rawls, J.: 1971, *A Theory of Justice*, Belknap Press, Cambridge, Mass.
11. Shelp, E. E.: 1981, *Justice and Health Care*, D. Reidel, Dordrecht, Holland.

12. Stern, L.: 1983, 'Opportunity and Health Care: Criticisms and Suggestions', *Journal of Medicine and Philosopy* **8**, 339–361.
13. Veatch, R. M.: 1985, 'The Ethics of Critical Care in Cross-Cultural Perspective', in J. C. Moskop and L. Kopelman (eds.), *Ethics and Critical Care Medicine*, D. Reidel, Dordrecht, Holland, pp. 191–205.

HANS-MARTIN SASS

# NATIONAL HEALTH CARE SYSTEMS: CONCURRING CONFLICTS

## I. INTRODUCTION

The moral dilemmas and politics of public health care are central to discussions of its economic and bureaucratic burdens. These burdens, and the assessment of appropriate, acceptable public response, become more serious and more challenging issues for two reasons: (1) Progress in medical technology and in the art and sciences of medicine have improved dramatically the successes of medical intervention. Past generations did not worry about economics or moral costs and benefits of helping people who suffered from infectious or deadly diseases, because, more often than not, no efficacious treatments were available. As in all cases of progress and the application of science and technology, questions are raised with regard to the moral and economic assessment and management of each particular technology [24]; (2) Conflicting values and goals appear in open societies that honor, on the one hand, value pluralism, political liberalism, free choice, and risk-taking, and, on the other hand, social solidarity. These societies promote equality of all citizens and justice for all, and "minimally decent" health care as a public good. Societal goods, however, do often compete with each other, and they need prudent and experienced assessment and management. This is particularly evident in the case of public health care, since any proposed alternatives inherently include great conflicts between the individual's free choice and his or her obligation towards the *res publica*, and between the societal goals and governmental mandate to protect citizens from risk, to establish solidarity networks, and to provide the widest possible opportunity for the individual to pursue freely whatever are his or her value preferences.

More efficient medical technology and more pluralistic life styles introduce conflict not only with regard to what we consider to be a minimally sufficient health status, and what degree of medical intervention this would require, but also with regard to the amount of public funding we want to allocate for this good. While the political and economic tensions in realizing the competing goals of our health care

*Hans-Martin Sass and Robert U. Massey (eds.), Health Care Systems,* pp. 15–36.

policy are more manifest, the intrinsically moral tensions are also obvious when the goals are listed: best of care, equality in care, cost containment, and consumer and provider choice [10]. None of these goals can be achieved totally; the reasonable realization of any one goal requires a compromise among all four. What we need, therefore, is a thorough *moral cost-benefit analysis* for setting priorities among competing goals. A moral assessment if required *prior* to a political analysis of public interest in the citizen's health, and prior to the question of whether health as a public good should be provided by governmental or quasi-governmental bodies and apportioned by the forces of the market, or by a mix of public-private regulations. Not only is this moral assessment crucial to planning an effective and viable delivery of public health care, but, inasmuch as health care delivery is one of the many urgent challenges to our traditional value system presented by technological progress and value pluralism, a moral assessment will provide a better understanding of the often competing goals of our personal and public convictions, patterns, and standards.

In what follows I will present a position which emphasizes the necessity of guarding the individual citizen as the prime beneficiary of technological progress and increased personal choice. To do so I will discuss the "costs" or responsibility which are necessary to achieve the benefits of freedom, i.e., the requirements and preconditions for making educated and experienced choices. I shall present my position in four stages by attempting to:

(1) define the moral and cultural dimensions of the concepts of "health" and "health care";

(2) distinguish between "health care" and "medical intervention";

(3) review the issue of "personal-public mix" in health care; and

(4) outline "priorities for governmental risk management" in health care.

Before addressing these concerns, and by way of introducing my arguments, I would like to offer two illustrations: The first case is intended to open up our understanding of "health." In most discussions, especially political discussions, the concept of health has been reduced to something clinically definable, the nondeviation from a standard norm and/or some status achievable by medical intervention.

While on a trip to China in 1979, I asked my interpreter if he knew the age of a very famous scholar. Upon hearing that the scholar was elderly, I expressed my concern about his ability to live long enough to educate another generation of Chinese academics. "Don't worry," was the interpreter's response, "he is a very learned and educated man, he will live very long." The point of this communication is that the young interpreter did not mention China's national health care system, scientific progress in medicine, improvements in housing or nutrition following the end of the Cultural Revolution, or preferential treatment of cadres and people of high social or national worth. Instead, the logic of his answer was this: quality of life and life expectancy result primarily from the individual's self-care, not his or her genetic heritage or access to quality health care and state of the art of medicine, adequate nutrition, and housing.

In Western society we tend to forget that good health is not simply the result of biological or medical manipulation; this condition can more often be properly attributed to culture – the *res non naturales*: good management of nutrition, sleep, emotional health, and psychic or spiritual harmony. Remembering the history of the long established but now largely forgotten Western tradition of dietetics (the *res non naturales*) demonstrates the extent to which public debates over cost control of health care financed by taxpayers are focused on statistical charts (measuring rate and cost of illness), rather than by the general principles of harmony and well-being. These principles operate in our lives by means of our personal accumulation of wisdom and personal responsibility, acquired through risk taking and experience. Unfortunately, our academic discussions and public debates construe health to be something which is provided by another: the state, the market, the doctor, the super-ego, or "big brother." It is against this background that we find discussions on the provision of health care confined to arguments for or against the "market approach" to health care provision, or even "socialized medicine."

The second illustration is the complex problems we face when discussing the "equal access issue" when we argue it in the traditional manner of the welfare state debate. In an article on equitable access to health care, Daniels [9] discusses the concept of equal access through an example of an office coffee machine. He illustrates the concept by describing the location of the machine and the employees' access to it in terms of gender. The complicated problem of access to the American corporate coffee machine has been solved by the Dutch office culture: a young lady, often hired exclusively for this service, stops every hour or so at every work station. This system provides equal and free access to tea care. The young lady might even offer a free cookie with the tea. But what does this equal access to tea mean for a coffee drinker? What about the rationing of the cookies – one per serving is hardly enough food for the day. What the Dutch secretaries and their supervisors get is equal and free access to one particular form of drink care, i.e., the traditional and culturally established and accepted beverage: tea, not coffee, not liquor, not soft drinks. They also have the opportunity to refuse the tea. They do not get food care, for the cookie is a courtesy rather than adequate nutrition. Also, they do not get all the forms of care for all their needs: no love care, neither during office hours nor after, no information care, which might include free access to the equally accessible telephone lines for private purposes. The free and equal access to the tea wagon is not the employer's nor any employee's "right," not is it a corporate "obligation." It is a free and cultivated expression of and appreciation for certain forms of civilized behavior and courtesy, a sharing of values and customs, equity, and equality. Where there are shared values, the sharing is not justified by weighty arguments, e.g., employing the "rights" language, or principles like justice or beneficence. In this case equality appears as a cultural expression, not a biological right or a political obligation. For those who do not share in these values, who would rather trade tea for coffee, this particular equal access setting appears paternalistic. It provides heteronomous care, introduces uniformity, suppresses free choice, individuality, and personal responsibility. Social equality here, as always, can only be achieved partially, or sectorially, never universally. We rightly ask, why do they not provide food care, love care, or health care on the premises of the Dutch office? Do human beings only need tea? In the Netherlands, as in all developed countries, the government provides (as a public good) "traffic care" by building functioning highways; but what are equal access benefits if you do not have a car and a driver's license? What benefits do you derive from a heavily subsidized mass transportation system if you lack the means to purchase a ticket? The tea wagon illustration indicates the moral and political difficulties we must address when issues of equal access are raised. Cultural or individual value preferences must be made salient, and tacit assumptions must be clarified.

## II. MORAL AND CULTURAL DIMENSIONS OF THE PUBLIC HEALTH CARE DEBATE

(1) We have at least two approaches to understanding the notions of "health" and "disease." One is the technical understanding employed by the medical profession which reflects progress or change in medical theory; technical understanding based on the interpretation of certain data, which is frequently a subject for public debate. The other approach is the individual's experience of being well, happy, and in harmony, which bases itself upon the mix of feelings, goals, and values of the individual person; this includes all the cases of complaints whose causes are undiscoverable diagnostically. There are cases in which both conceptions of health coincide, even though they might differ with respect to the seriousness of the case: the pain of a toothache can be devastating, but it is not directly life threatening; high levels of cholesterol are never painful, but may be a threat to life. Then there are the borderline cases – mild infection, certain forms of headache or arthritis – which are perceived by some individuals as serious illness, while not at all recognized as such by others. Further complicating the nosology of disease are the changes that occur in cultural preferences and modes over time. For example, masturbation was defined as a disease in 19th-century medical textbooks, while public weeping was viewed as a healthy expression of sensitivity and caring in educated men and women. In contemporary times, masturbation is no longer described as a disease in medical textbooks, and public weeping is not viewed as an expression of healthy behavior in adults in social or family life. In our century, this change in cultural preferences is further complicated when we have our own definitions of diseases or syndromes like "vegetative dystonia," or "getting old," which are not accepted as related to life style or life time, and which may or may not receive medical treatment.

The moral and cultural dimensions of medical art and history, medical science, personal awareness and sensitivity, and the ever-changing modes in values constitute the ingredients out of which emerge public debate and risk management in connection with public health and medical services.

(2) Health care has been considered as both a basic right which should consequently be protected by the government (cf. [19], p. 330), and as a right which each person may freely pursue without interference. We shall not develop this discussion here [24]; for our purposes it is enough

to claim that good or reasonable health is a prerequisite for the achievement of many other human goals, as is, for example, a reasonable amount of intelligence, prudence, education, and even good luck. But different from intelligence or prudence, health might be "consumed" or "traded" in the pursuit and enjoyment of other values. Smoking is a detriment to good health, yet smokers do trade a portion of their health or longevity for the pleasures of smoking. If, based on the general principle of solidarity, taxpayers or insurance agencies agree to pay for the increased health care needs of a smoker, are they agreeing to do so based on a principle of distributive justice or beneficence, or to avoid paying later for health restoration, or simply to avoid individual or public harm? According to Worldwatch, a Washington-based research group, the annual cost of the side-effects of smoking in the U.S. is $100 billion, i.e., three dollars per pack of cigarettes sold (*Washington Journal*, Jan. 24, 1986). But what life style does not include some detriment to health? Curtailing drinking or smoking has been mentioned by some as leading to frustrations which might result in serious illness for some individuals, and to costly social discontent. Drinking causes liver cirrhosis, but studying late, writing books, and giving papers at conferences may not be a healthy life style either; it weakens the body, ruins the cardiovascular system, and makes the body more vulnerable to various forms of infections or heart disease. Gourmet eating, taking drugs, even most forms of physical recreation carry their high probability of long-term or short-term health risks. Minimizing health risks related to life style, even cracking down on those behaviors with a high probability of risk, would require strict regulation of some life styles which are acceptable, as well as those considered non-acceptable and punishable. This regulation would undermine our claim to health as a precondition or prerequisite for enjoying other values, and more importantly, it would enforce uniform heteronomous personal values. A health care system, equally and uniformly financed, would be required to pay equally and uniformly for all costs and treatments claimed by free choice of the participants, or to limit costs and treatment, but thereby compromise individuals' values, life styles, and even health in order to minimize payment for potentially high risk activities.

(3) Given that individual and cultural perceptions of "health" and "disease" are constantly changing, and that health often has a "trade value" in decisions made by individuals with a whole range of pleasures, goals, and life styles to choose from, the public debate on health care

includes often neglected moral and cultural dimensions. Public and political analysis, assessment and management of health risks, and governmental involvement in risk management has to take these cultural and moral aspects into account. To date, debates on public health care issues concentrate almost entirely on the economics of health care and cost-benefit analysis based on medical technology. Ignoring an assessment of cultural and moral definitions of health, and omitting a discussion of personal values and responsibility renders public health care systems inadequate to respond to the more fundamental rights and responsibilities of the people. Public health care, and therefore any discussions on a public health care system, must take these cultural and moral dimensions into account. The risk of not doing so is a system that would be administratively unmanageable, economically unfeasible, and morally and culturally unacceptable.

## III. HEALTH CARE SYSTEM OR MEDICAL INTERVENTION SYSTEM

The weaknesses in our policies as a result of not taking cultural and moral values into account are not always readily apparent because the existing so called Health Care Systems are in actuality Medical Intervention Systems.

(1) The two values, discussed by Engelhardt ([10], p. 6) – "best care" and "equal care" – are only in part contained in a theory and practice of "adequate care." The 1911 German State Insurance Regulation states that public health care has to be "adequately sound" [7], and the 1982 President's Commission in the United States defines an adequate level of health care as "a floor no one ought to fall below, not a ceiling above which no one may rise." The Commission goes even further and holds that "by creating impossible demands on society's resources for health care, such formulations have risked negating the entire notion of a moral obligation to secure care for those who lack it" ([22], vol. 1, p. 4). Although in theory health care systems introduced or funded by the government do not exclude certain population groups or disallow certain treatments from public funding, in practice they are excluded by administrative or procedural means, e.g., the practice of not accepting people over a certain age for hemodialysis. Whatever the theory and practice of rationing health at the medical intervention level is, the individual case is treated as a sub-issue of the general and uniform regulation. The individual citizen has no influence in the ways and

means or costs of his or her treatment in totally public health care systems, and should not, as preferential treatments would violate equality (although such a treatment is given to individuals of high national value on the grounds that preferential treatment for those individuals and their friends and family is socially beneficial, thus justifying the extra costs).

(2) But even if we were economically able and morally willing to provide everyone with the best (in the sense of fullest possible) medical intervention, there might not be any significant increase in the health status, quality of life, or even life expectancy of our citizens. In reviewing the steady increase in health status and life expectancy in developed countries over the last 100 years, it is apparent that those positive results have not been the result of equal access to and increasing availability of medical intervention technology in specific situations. Better health and longer life is the result of a general increase in opportunities for better housing, nutrition, education, and occupational safety, and advances in medical and biological research, e.g., anesthesia, radiology, and antibiotics. We did not defeat tuberculosis by socializing health care and opening the plush, private Swiss sanatoria, which had catered to the rich and superrich, to the poor and needy; tuberculosis was eradicated as a result of major breakthroughs in research on antibacterial drugs, not by the moral issue of equal and best treatment for rich and poor.

In a 1975 World Health Organization study, Grundy and Mackintosh confirm this in their statement: "The improved health owes more to the changes in the external environment . . . we do not become ill because of the lack of a specific protective therapy but because we live in a healthier environment. In its preoccupation with the minutiae of diagnosis and pathogenesis of disease, medicine is in danger of neglecting what has hitherto provided its most powerful resource, the manipulation of the external environment" ([14], p. 450). A French study in the same year collected data on life expectancy and mortality among males aged 35 to 60; while the average probability of death in that age range was 174 in 1000, it was as high as 470 among the unemployed compared to 149 among the employed. Within the employed group it was as low as 71 to 91 among professors and professionals, 120 among farmers, 157 among office personnel, and 253 among unskilled laborers. This and other data led Schwartz [1] to the conclusion that even though there are plausible interrelationships between medical breakthroughs and life expectancy, there are nonmedical cultural factors which determine health, quality of life, and life expectancy essentially. The Federal Ministry for Labor and Social Welfare in the Federal Republic of Germany states, in its 1983 Report, that nutrition and social and economic factors outweigh medical care in protecting and improving health [8].

Recalling the case of the old Chinese scholar, the argument is relevant that factors of personal life style, education, awareness of health risk, and good life management, including health risk management, influence health and well-being more than medical intervention, regardless of how such intervention is paid for, whether in a market system or in a socialized setting. Even if access to health care were not constrained by measures to restrict the increasing costs of medical intervention, the expansion of medical intervention systems would not secure better health or better access to health care.

(3) On the contrary, an undiscriminating expansion of public medical intervention systems carries great risks to the health and well-being of citizens: (a) it could create an increase in unfair expectations that one can spread the costs and negative side-effects of life styles that carry great health risks, while at the same time enjoying the benefits and positive side-effects of such life styles; and (b) it might create behavioral and economic incentives to use "medical intervention" to treat emotions which are simply part of life: grief, anxiety, frustration, despair, old age ([25], p. 13), thus further weakening self-responsibility and risk management competence of the individual citizen.

A continuation of the public medical intervention system will only prolong the conflicts we now face: the discrepancy between the theoretical definition of "adequate health care" and the practical provision of health care while officially subscribing to the competing goals of "best" and "equal" care; the overemphasis on the role of medical intervention in health care to the exclusion of other preconditions for well-being from the economic and administrative schemes of public health care; the practice of socializing the health care costs of some particularly hazardous life styles without holding the individual responsible, a practice which diminishes the individual citizen's risk competence and his or her rights and obligations to care for personal health. These concurring conflicts are only intensified by the fact that what we all call public health care in fact is only public medical intervention care.

(d) There are, however, risks to the individual which society must

## IV. THE PERSONAL-PUBLIC MIX IN HEALTH CARE RESPONSIBILITIES AND RIGHTS

(1) The debates in health care policy basically revolve around a discussion of two different systems: the market system versus the governmentally regulated or administered system. The advocates of heavy government involvement in health care construe health to be a special good, something basic and preconditional to other values in life. They conclude that because health is a prerequisite for exercising other rights, it must be provided for in a welfare system, either directly by government health care institutions or indirectly through strict regulation. The latter would involve regulation of the economic and medical aspects of a national health care system, created and sustained by legislative and administrative authority. The advocates of the market system believe that the market is the best means of balancing costs and benefits, and that within the market one can achieve goals, acquire assets, and contract for personal benefits (which subsequently benefit the *res publica*). Some claim that health is simply another asset or goal among many that do not need uniform governmental protection. One can also make the point that because health is somewhat preconditional to achieving other assets or goals in life, it is imperative that care for it be left to the individual's personal responsibility without paternalistic intervention. Neither one of the health systems – market system or socialized medicine – has ever existed in a pure form. Socialized medicine has been hampered by preferential treatment of people having power or other access within the system or by the forces of the black market which accompanies any heteronomously regulated system. Free market systems in health care or medical intervention, even in the absence of governmental regulation, have always been influenced by forces other than economic market forces, e.g., medical professional ethics, beneficence of individual citizens, charitable involvement of churches and humanitarian organizations. Also, as the contributions in this volume demonstrate, official public health policy has seldom achieved the pure realization of one systematic approach. Maynard describes the British system as a public-private mix ([17], pp. 133–166); Herder-Dorneich calls the German system a quasi-market system in which limited medical and technical competition among organizations (not persons) is allowed for in statutory and private insurance and reimbursement insurance; and specialized bodies such as the industrial

injury insurance agencies are self-governed by labor and capital [16]. Public political discussions and decisions in regard to the adequacy of the public health care system have concentrated exclusively on the issues of economic equality. The *individual citizen* as the main beneficiary and player has yet to be brought into the debate as an active agent, capable of and entitled to accepting responsibility.

I want to stress that in our future discussions we have to concentrate on the essential and crucial role the individual citizen has to play and is entitled to play in caring for his or her health. I do so *for moral and cultural reasons*, not primarily for economic reasons. While I personally think the forces of the market in the long run are the only ones which can provide best, i.e., most competitive, care in economic as well as in medical terms, Pfaff [21] has demonstrated that in economic terms non-market systems and quasi-market systems can contain costs and allow competition, albeit limited. I am arguing on moral and cultural grounds favoring the principle of personal choice.

(2) The choice principle as a moral value of highest priority for any person does not just call for an optimum choice among different providers or insurers of health care or medical care; it calls rather for not paternalizing the individual citizen's educated judgement as to what a "best," "adequate," "cost-efficient" treatment is. This establishes a priority of the choice principle for the consumer or prospective consumer of health care or medical care services for the following reasons:

(a) Not only do definitions of health differ, the values of well-being and life span are in conflict with other values, goals, and assets in personal life. How could a national health agency heteronomously define best, adequate, or even equal health or health care?

(b) The art and science of medicine together with housing, nutrition, education, workplace, and genetic heritage establish, improve, or hurt my health and well-being. How could a national medical intervention system justify dealing only with certain aspects of my health status, protection, improvement, or implementation and not with others?

(c) Health and well-being are threatened and hurt not just by natural and unavoidable causes, but also by uneducated, incompetent, even noncaring, irresponsible handling of one's own body and self and that of others. Should not, therefore, the broadening of the opportunity range and the establishment of risk competence in health matters be the *prime* political goal for achieving best, equal, and cost-efficient health of the citizens?

address: common risks caused by technology, epidemic diseases, or terrorism and war. These are the areas where direct governmental risk management clearly is essential and required.

Most desirable, however, is public intervention in educating and encouraging the individual in competent risk management, including health risks. The optimal public-private mix in health care, most compatible with the understanding of rights and responsibilities of educated citizens in open societies, would be public investment in the establishment of responsibility, risk information, and risk experience for its citizens. This would entail public divestment of direct health care and medical intervention care in favor of privatization and encouragement of market forces. This would result in the right to care for one's health, support for medical research and medical treatment, and the encouragement of charitable and beneficent individuals and organizations.

(3) Public protection against collective risks like war, epidemics, and the malfunctioning of the market place, along with broadening opportunities through education and providing for security and liberty, would

DIAGRAM: COSTS

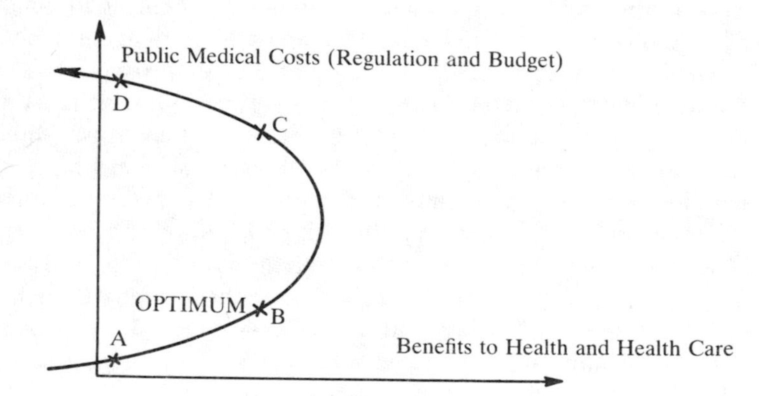


A – B = reasonable and fair public risk management
B – C = value ambivalence of public risk management
= increase in heteronomous value assesment
= decrease in personal risk capability and risk competence
C – D = suppression of free choice, security and liberty
= government as primary risk factor
Personal/public mix in health responsibilities

strengthen the preconditions for good health. Increasing public involvement in risk management on the other hand would include paternalism in defining unacceptable risks uniformly and be counterproductive in strengthening the individual's rights and responsibilities, including the right and the responsibility to care for his or her own health.

## V. OPTIMUM PERSONAL-PUBLIC MIX IN HEALTH CARE RESPONSIBILITIES

If we plot personal and societal benefits from risk care (which includes health care) on the horizontal axis of a chart, and the costs of governmental value and risk management (which includes health care) on the vertical axis, we can draw a reclining curve describing the interrelatedness of benefits and costs of governmental intervention. As in other areas of governmental risk management the degree of governmental involvement might vary. In general we can state that some sort of basic involvement, e.g., setting up frameworks for the market of values and valuables, encouraging certain forms of behavior while discouraging others, managing highest risk directly, is of great benefit to the individual; firm and reliable basic governmental risk management and opportunity management (A to B) is of immediate benefit. Further government involvement in risk management (B to C) involves increased regulation and, in all likelihood, more taxpayer's money. The costs include a smaller choice and responsibility range for the individual, while the increase in benefits is only marginal. Discussions of the cost-benefit range between B and C are most common in welfare states having a percent of above 30 percent of the Gross National Product given to social, redistributive, and transfer costs. Moral and economic issues existing in this realm of governmental involvement in risk management are generally unsolvable. Second class solutions by majority vote, administrative process, and lottery are at best only pragmatic. It is in this area that fashion, fads, or cultural trends in moral concern and risk perception overemphasize some risks while underemphasizing others ([24], p. 30 ff). Governmental risk management beyond point C is not an issue in our actual welfare state discussion or health policy; it marks the point beyond which governmental regulations in different areas are intertwined together to knit a network increasingly suppressing individual benefits, restricting personal rights, responsibilities, and opportunity ranges, and exploiting the individual for the benefit of the "big brother." There are totalitarian regimes, in which health policy is a part of the regime's self-care. Those extreme moral problems of public-private confrontation fortunately are not ours; the moral problems of the adequate and appropriate personal-public mix in health risk care, however, are.

(4) What are our moral arguments if we either advocate or deny equal access to intensive, costly, and marginally beneficial treatment for all? For example: free dialysis treatment regardless of location, age, or life style; lung cancer treatment for heavy smokers; liver and kidney care for heavy drinkers; cardiovascular care for workaholics; brain care for motorcyclists; preventive hydrotherapy for the overworked university professors; love care for the frustrated and unhappy; abortions, free contraceptives, etc. These and similar moral questions arise in free market as well as in welfare states. Whatever is decided and regulated, some moral arguments will be overridden; there will be moral majorities and moral minorities fighting for regulatory power, legislative influence, and maximum share of tax revenues. We thus make moral goals and personal convictions about life styles issues in a political fight for winning and dominance. This is very uncomfortable on any moral grounds as it takes ethics out of its cultural domain and places it in the arena of politics. It is also uncomfortable on utilitarian grounds as it creates winners and losers, raises unanswerable questions, and produces individual and social discontent. Primarily for these moral reasons, but also in the interest of promoting social peace and lowering taxes, we should *unburden* public regulatory involvement in areas where perception of risk varies among individuals, personal risks are related to personal benefits, and where full societal consent cannot be obtained and should not be fought for. It would be morally most beneficial if health policy would *restrict* its involvement in those questions which have different, even opposing answers, and where moral and economic costs are high and benefits marginal. Health policy, instead of creating concurring conflicts, should concentrate on *basic* health risk management. Reducing governmental responsibility in areas of conflicting values will increase personal responsibility qualitatively, allowing public health care systematically to establish, provide, and encourage the best health care for an informed citizenry. If these recommendations are followed, most of the problems that are the focus of health policy today will diminish; individuals and society will enjoy the fullest possible benefits of health care.

## VI. PRIORITIES IN PUBLIC RESPONSIBILITY FOR HEALTH CARE

In recommending new priorities in public responsibility for health care and in governmental health risk management, I do not focus primarily on the tension between market-oriented and nonmarket systems. A shift in priorities would be beneficial for all systems, no matter how they generate or allocate resources or make decisions. Reorganizing public activities in health care could relieve all systems from facing generally insoluble moral problems in allocation, and from the impossible task of generating ever increasing amounts of funds for medical intervention.

What would be provided would be *best and basic public health care*, as described in the initial stage of our curve (A to B), if health policy would focus exclusively on the following six services for health risk management:

(1) health education and health care education;
(2) health information, both general and to the individual;
(3) support for biomedical research;
(4) monitoring and encouraging optimal processes of health delivery and research;
(5) monitoring and encouraging insurance markets; and
(6) direct public health protection services only in selected areas.

(1) If good health is a precondition for many other goods, rights, and responsibilities, it follows that a high level of individual knowledge and expertise in health protection and improvement is a prerequisite for every educated citizen. Our public education systems, public culture, and adult education do not simply neglect health education; effective health education does not exist today. In the absence of public responsibility, health care education is provided by small private groups, sects, nutrition stores, and magazines, some of which provide product-oriented information. Even though I am generally very careful in using the rights language, I will use it here and stress the right to know and to become educated regarding how the body functions, how it can be maintained, protected, improved, and manipulated. This is as important for the educated and responsible citizen as knowing where we come from (education in history), what we stand for (moral education), how to calculate (mathematics education), and how to express oneself (language education). If we really want to open the opportunity range for all individuals, to enable them to fight the lottery of nature and their

milieu, to distribute the wisdom of living healthy lives on the basis of the *res non naturales*, we will have to educate our fellow human beings. We need to know about the relatedness of somatic states and states of happiness, the cost-benefit calculations of excessive life styles, the wisdom of the principle – *mens sana in corpore sano* – and all the treasures hidden in the unread books of the history of medicine, rarely taught in medical school curricula or mentioned in public debate, public education, or public health risk management. How much richer and happier, how much stronger, how much more competent in life management would the individual be if we would put more effort into this endeavor. How many more jobs would be created for dedicated and gifted health care educators and consultants; how much more would we enrich public moral and cultural debate by introducing health care issues on an adequate level of competence; how much money would we save, money formerly spent on symptoms resulting from excessive life styles or simply poor personal health care management?

(2) Only educated citizens can make good use of information, including health information and health care information. Providing adequate health information directly, or encouraging and protecting market forces in health information, should be the second most important priority in health policy. Access to health information, both general information about well-being, environment, food, and epidemic risk, as well as information on personal health status, are preconditions for good personal health management. *Access to general health information* is an integral part of continuous health education. Focusing primarily on availability and accessibility of information through books, libraries, and other printed and electronic media would relieve medical professionals of the unnecessary and unjustifiably exclusive possession of information, treated sometimes by both public and professionals as esoterica and arcana. It would increase health risk competence of individual citizens and reduce medical intervention to what it should be in a society of educated citizens: a specialized customer service for educated consumers.

While access to general health information establishes a new customer-provider relationship, *easy and equal access to special health status information and health care information* will broaden the individual's opportunity range by providing data necessary and preconditional for effective and prudent personal health care. It also separates information and analysis (diagnosis) from risk management by medical intervention (therapy) or other measures. Such a separation would be beneficial as it would provide opportunity for personal choice, especially in preventive and life style related health risk management. A Health Information Service (H.I.S.) run by competitive profit organizations on the local level could provide basic information on the status of cardiovascular or organ systems, together with understandable explanations, references, and advice on risk management options. H.I.S. services should be run on a competitive basis in order for the consumer to maximize the benefits from competition in information research, processing, and interpretation. The government could compete with the market forces on a profit basis if so desired; the government could even run the H.I.S. through specialized agencies, which then would reduce some of the risks of direct bureaucratic involvement in serving the citizen.

DIAGRAM: RISK INSURANCE

| BASIC INSURANCE | SUPPLEMENTARY INSURANCE |
|---|---|
| – mandatory | – voluntary |
| – covering 'basic needs' | – covering lifestyle related 'needs' |
| – privately funded | – privately funded |
| – publicly funded on credit base | – additional funding by solitary or beneficent private groups |

BASIC NEEDS are minimal needs, defined by democratic process
LIFESTYLE RELATED NEEDS are cultural needs, individually defined

BASIC INSURANCE TOTAL (B.I.T.)
covers
B.I.E. = Basic Insurance Education
B.I.M. = Basic Insurance Medicine
B.I.L. = Basic Insurance Labor
B.I.R. = Basic Insurance Retirement

INSURANCE PREMIUMS for basic covering may be paid for solidarity by public funds and will be registered in the books of NATIONAL INSURANCE LOAN (N.I.L.)

Personal/public mix of risk insurance

The question here is not primarily the choice between market forces in providing the best, most effective, and equally accessible diagnoses, but in providing information separate from treatment. We do that in other areas of life where government is involved in risk reduction.

Motor vehicle inspection does not repair my car but provides information regarding its functioning; some of this information includes mandatory requests for repair, etc. H.I.S. would provide similar information on functioning of the body together with advice for improvement and maintenance or for seeking more specialized counsel or treatment. Since human beings are different from cars, and since their "dysfunctioning" is a much more complex matter, recommendations would have to be much more flexible, offering health risk management option advice. Such advice would have to be more strict in systems socializing the costs of health care than in market-oriented systems. In systems of full solidarity costs are shared by all and arguments of fairness and justice would call for personal contribution, even sacrifice, in favor of the common good. But in both systems, access to specialized and customized health information is a civil prerequisite, necessary to the exercise of personal responsibilities and rights to protect health, as well as to achieve other goals in life. It would not only be in the best interests of every individual to have access to such personal health information; it is the citizen's right and responsibility to request optimal access to health information and health care information in personalized and understandable form. H.I.S. could be offered by a wide range of providers, including licensed professionals, walk-in computerized information centers, department stores, mail order businesses, and governmental agencies. They might offer the same or different software or diagnostic techniques, give simple or more detailed reports and health management option advice. A market approach would be best for inexpensive service and high quality of reporting and interpreting data. Protection of privacy would be a must.

(3) *Monitoring and encouraging* optimal forms of medical intervention, health education, health information, insurance services, and the inclination of institutions and individuals to act out of beneficence would all be essential to the public risk management of health for the benefit of all citizens. This would be done best through focusing on output, i.e., results, rather than input, i.e., cost-containment discussions. Also, governmental health care policy makers would be well advised to concentrate their efforts to favor and protect a framework within which a multitude of providers could offer all individuals optimal choices for service ([24], pp. 24–27).

(4) Among the monitoring and supporting services of governmental health care, the protection of Basic Insurance Medicine (B.I.M.) would

be the best means of implementing the equality principle by government. B.I.M. would be the basic financial support system for health care. In centralized states it would be paid for by government and administered by governmental agencies. In pluralistic societies an open market of providers and consumers of Basic Insurance for Retirement (B.I.R.), Education (B.I.E.), Labor and Employment (B.I.L.) would represent a flexible system of Basic Insurance Total (B.I.T.), improving services and decreasing costs by competition among risk-experienced insurers. Consumers would enjoy full benefits from market forces; also, they would pay premiums over time, relatively inexpensive as compared to full capital payment in crisis situations.

Having mandatory B.I.M. as a floor for medical intervention available to all citizens, there would then be provided free choice for Supplementary Insurance Medicine (S.I.M.) of various types and coverage; this would transform the actual problem of public cost containment in health care and medical care into a wide field of opportunity ranges for coverage. Supplementary medical services would be sought by free choice, depending on motivation, value preferences, need, and money. In centralized societies they would be granted according to social worth.

In differentiating between B.I.M. and S.I.M., the more controversial issues in cost intensive or life style related health care would be left generally unsolved. Supplementary costs would be paid for according to risk management preferences and resource allocation choices by individuals. The government would not be involved in making supplementary insurance mandatory or contributing taxpayers' money to it; it would, however, lend the premiums for B.I.M. to the individual in case he or she could not pay. The premiums prepaid by taxpayers' money for B.I.M. would cover the premiums paid for basic insurance for education, unemployment, and retirement, and be dedicated to the recipient as National Insurance Credit (N.I.C.) in the books of National Insurance Loans (N.I.L.). These loans would be paid back over the recipient's lifetime; some would get bonuses for speedy paybacks; others would never be able to repay the loan to this pool of solidarity. My proposal comes close to the concept of "negative income tax" as proposed by Milton Friedman and Baruch Brody; but for reasons of accountability and pride there should be separate accounting books for these loans. Some will prefer to buy expensive Chinese jade or racing cars, or to insure against property loss, rather than pay insurance premiums for basic medical expenses – so be it.

Recalling the curve noticed above, we should require governmental involvement in basic services which provide immediate and undeniable benefits to each and every individual (A to B). To establish a cost base for basic mandatory services and to lend the premiums to individuals is morally more justifiable and economically more efficient than either a system of voucher handouts or of heteronomously paternalistic medical intervention care. To differentiate between basic and supplementary health care, and to pay for the costs of basic care publicly, is superior to the existing market or non-market system of public health care primarily on moral grounds, because basic health care is more fundamental and less subject to individual interpretation. Subsequently it might also prove to be beneficial politically, as it strengthens personal risk competence, and economically, as it helps in avoiding risks and in competing for the best and most inexpensive risk management.

(5) There is a final point, where *strict political health risk management* is indispensable. Direct governmental responsibility cannot be denied in collective health and life risks like terrorism, war, and health threatening environments. Environments can be unhealthy because of epidemics or as a result of poisoning or polluting technological processes. It is a matter of political prudence whether government manages all or part of these health risks directly or in a mix of subcontracting and direct involvement.

"The days of our years are three-score years and ten . . . . So teach us to number our days that we may apply our hearts unto wisdom" (Psalm 90); our life is short, indeed. Good health is a part of good life and often prerequisite for a good and happy life; poor health may often be the result of an otherwise happy life. There are long lives, boring and poor; there are short lives, rich in value and experience. Well-being is a variable within the mix of life, never identical with medically measurable data. There is no conceivable way that public health care systems can deal with issues like these by directly assuming all health care responsibility. They never should, or they might be tempted to make arrogant judgments between adequate and inadequate, valuable and worthless. Health, life, well-being, life style are all personal issues. To expect answers to these issues, either from socialized medicine or from market-oriented medicine, is unreasonable, as answers can only be provided by each person. Public health care, however, can help assure that we are prepared to provide answers as informed and as competent as possible. Questions will always remain which are unanswerable: the questions of Job, the questions regarding unfortuitous pain and suffering, and all the questions related to human finitude.

*Institut für Philosophie*
*Ruhr Universität Bochum, F.R.G.*
*and*
*Kennedy Institute of Ethics,*
*Georgetown University,*
*Washington, D.C., U.S.A.*

## BIBLIOGRAPHY

1. Aaron, J. and Schwartz, W. B.: 1984, *The Painful Prescription: Rationing Hospital Care*, The Brookings Institute, Washington, D.C.
2. Badura, B. and v. Ferber, Chr. (eds.): 1981, *Selbsthilfe und Selbsthilfe-organisationen im Gesundheitswesen*, Oldenbourgh, München.
3. Beauchamp, D. E.: 1985, 'Community: The Neglected Tradition of Public Health', *Hastings Center Report* **15** (6), 28–36.
4. Braun, H. and Articus, St.: 1984, *Die haüsliche Versorgung pflegebedürftiger alter Menschen*, Melle, Knoth (Adenauer Stiftung, Forschungsbericht 40).
5. Brody, B.: 1988, 'The Macro-allocation of Health Care Resources', in this volume, pp. 213–237.
6. Brown, L. D.: 1983, *Politics and Health Care Organizations: HMO's as Federal Policy*, The Brookings Institute, Washington, D.C.
7. Bundesminister für Arbeit: 1981, *Insurance Code* (Reichsversicherungsordnung), International Labor Organization, Geneva.
8. Bundesminister für Arbeit: 1983, *Sozialreport*, BMA, Bonn.
9. Daniels, N.: 1983, 'Equity of Access to Health Care', see [22], vol. 2.
10. Engelhardt, H. T., Jr.: 1988, 'National Health Care Systems: Conflicting Visions', in this volume, pp. 3–13.
11. Ferber, Chr. v. *et. al.*: 1985, *Laienpotential, Patientenaktivierung und Gesundheitsselbsthilfe*.
12. Fink, U.: 1984, 'Neue Wege in der Sozial-und Gesellschaftspolitik: Modell Berlin', *Evangelische Verantwortung*, Dezember, 5–8.
13. Gaefgen, G.: 1984, *Die ethische Problematik von Allokationsentscheidungen am Beispiel des Resourceneinsatzes im Gesundheitswesen*, Hochschule St. Gallen, Forschungsblätter für Wirtschaftsethik Nr. 6.
14. Grundy, F. and Mackintosh, J. M.: 1957, *The Teaching of Hygiene and Public Health in Europe: A Review of Trends in Undergraduate and Postgraduate Education in Nineteen Countries*, World Health Organization, Geneva, Monograph series No. 34.
15. Harron, F., *et al.*: 1983, *Health and Human Values: A Guide to Making Your Own Decisions*, Yale University Press, New Haven, Connecticut.
16. Herder-Dorneich, Ph.: 1984, 'Zwischen Utopie und Pragmatik', *Bundesarbeitsblatt* **12** (Dez.), 5–8.

17. Maynard, A.: 1988, 'The Public and Private Regulation of Health Care Markets', in this volume, pp. 133–166.
18. Menzel, P.: 1983, *Medical Costs, Moral Choices*, Yale University Press, New Haven, Connecticut.
19. Moskop, J. C.: 1983, 'Rawlsian Justice and a Human Right to Health Care', *Journal of Medicine and Philosophy* **8**, 329–338.
20. Muennich, F. E.: 1984, 'Mehr Markt', *Bundesarbeitsblatt* **12** (Dez.), 8–11.
21. Pfaff, M.: 1984, 'Ordnungspolitik im Gesundheitswesen: Internationale Erfahrungen', *Bundesarbeitsblatt* **12** (Dez.), 13–18.
22. U.S. Government: 1983, 'Securing Access to Health Care', *President's Commission for the Study of Ethical Problems in Medicine and Biomedical and Behavioral Research*, U.S. Government Printing Office, Washington, D.C. (Vol. 1: Report; Vol. 2: Appendices: Sociocultural and Philosophical Studies; Vol. 3: Appendices: Empirical, Legal, and Conceptual Studies).
23. Rescher, N.: 1983, *Risk: A Philosophical Introduction to a Theory of Risk Evaluation and Management*, University Press of America, Lanham, New York.
24. Sass, H. M.: 1985, *Verantwortung unter Risiko*, Koellen Verlag, Alfter-Oedekoven.
25. Shelp, E. E. (ed.): 1981, *Justice and Health Care*, Reidel, Dordrecht, Holland.
26. Shelp, E. E. (ed.): 1982, *Beneficence and Health Care*, Reidel, Dordrecht, Holland.
27. Starr, P.: 1975, 'A National Health Program: Organizing Diversity', *Hastings Center Report* **5** (1), 11–13.

# SECTION II

# NATIONAL HEALTH CARE SYSTEMS

ALLEN BUCHANAN

# AN ETHICAL EVALUATION OF HEALTH CARE IN THE UNITED STATES

Four questions are widely thought to be crucial for evaluating any health care system from an ethical standpoint: (1) Does the system provide equal health care for all? (2) Does the system provide the best health care for all? (3) Does the system control health care cost effectively? (4) Does the system maximize free choice for both patients and providers? (see Engelhardt, this volume, p. 6). In this essay I first examine the meaning and presuppositions of each of these four criteria. I then argue that neither the first nor the second criterion is appropriate for the ethical evaluation of health care systems and suggest alternative criteria to replace them. Finally, I begin the complex task of applying the revised set of criteria to the ethical evaluation of the current U.S. health care system. One conclusion reached is that the extent of freedom of choice in the U.S. health care system, as well as the moral value of some forms of this freedom, have been greatly exaggerated.

## 1. PROVIDING EQUAL HEALTH CARE FOR ALL

Before it can be determined whether this is an appropriate ethical criterion for evaluating health care systems it is necessary to clarify the meaning of the phrase 'equal health care for all'.[1] It could mean either (a) all receive equal resources for health (i.e., equal amounts of money or services of equal monetary value), (b) all achieve, or have the opportunity to achieve, equal health status, or (c) all receive the same "level" of care (with respect to both "quantity" and "quality"[2], who are in equal need of care.

The first two interpretations of the goal of equal health care for all may be dismissed rather quickly. Since individuals have widely differing needs (and preferences) for health care, pursuing the goal of providing equal amounts of resources for health care for all would not only be inefficient but would also lack an ethical rationale. The goal of achieving equal health status for all must also be rejected simply because it is impossible to achieve, quite independently of whether it would be ethically appropriate if it were possible: there are some unfortunate

*Hans-Martin Sass and Robert U. Massey (eds.), Health Care Systems*, pp. 39–58.

individuals for whom even the most minimally tolerable health status is unattainable.

The third interpretation, according to which all those who are in equal need of health care are to receive the same level of care, is a coherent goal. However, whether or not this criterion is ethically plausible will depend upon how the notion of *need* is understood. On a broad interpretation of 'need', a health care need exists wherever the provision of health care services can reasonably be expected to produce net benefit for the patient: a need in this broad sense is simply a potential for net benefit. According to the broad interpretation of the criterion then, a system is ethically deficient if anyone lacks a health care service who could benefit from it as much as someone else who has that service, regardless of how large or small the benefit is. In spite of the fact that many people at least claim to subscribe to it, this broad interpretation of the goal of equal health care for those in equal need has little to recommend it from an ethical point of view, unless there are sound ethical grounds for attempting to arrange all social institutions so that benefits of all kinds are distributed in a strictly equal fashion among all citizens. Though I will make no attempt to rehearse them here, I believe there are a number of strong arguments against any such comprehensive egalitarian principle, not the least important of which is that even a modest attempt to implement such a principle would lead to intolerable restrictions on individual freedom, massive losses of productivity and hence of social welfare, and would obliterate not only individuality but also the social and personal recognition of all desert and merit.

Some who defend the broad interpretation of the goal of equal health care for those with equal health care needs might at this point attempt to dissociate it from a comprehensive equal benefits principle. The point, they might contend, is that *health care* is of such special importance that it requires egalitarianism even if other goods and services do not. Thus one may consistently espouse the goal of equal health care for those with equal potential for benefiting from it without embracing egalitarianism across the board with all of its undesirable consequences.

This reply, however, is inadequate as it stands. It is simply not true that *all* health care is of special importance, because it is not true that all benefits that can be gained from health care are of special importance. Some forms of health care (e.g., resuscitation of persons who will go on to live fulfilling lives) are extremely beneficial: other forms of health care (e.g., triple by-pass surgery or back surgery for some patients) are much less beneficial for those patients than alternative medical or non-medical uses of the same amounts of money would be. Once it is conceded that not all forms of health care are especially beneficial relative to all other goods and services, it is quite arbitrary to ascribe to a strictly egalitarian principle for the distribution of health care while shrinking from strict egalitarianism for other goods, such as food.

A more charitable interpretation of the goal of equal health care for persons with equal health care needs would be to jettison the broad notion of needs as potentials for benefit and restrict 'needs' to 'potentials for exceptionally great benefits.' In other words *needs* would be understood as *especially important interests*, to be contrasted with mere preferences and with the mere potential for being benefited to some extent or other. But to interpret the equality criterion in this way is to restrict its scope very significantly: it no longer calls for equal health care for all who could benefit from care, but only that all receive equal health care with respect to the satisfaction of those especially important interests that count as health care needs. The great difficulty then, of course, is to determine what counts as a health care need or basic health care interest. Until health care needs in this narrow sense are at least roughly specified, it is impossible to determine how closely a given health care system approximates or falls short of the criterion of equality.

Assuming that the class of health care needs (in the narrow sense) can be roughly specified, the ethical basis for using the equality criterion for evaluating a health care system could be of either of two sorts. According to some ethical theories a system which fails to provide equal satisfaction of health care needs (i.e., basic health care interests) is *unjust*.[3] Justice, however, is not the only ethical standard for institutions. According to some rather austere libertarian theories of justice, individuals have no general positive moral rights and hence, *a fortiori*, no general moral right to health care. Yet such a theory of justice is compatible with – and perhaps only plausible if conjoined with – a theory of ethical virtues, including the virtue of *beneficence* or *charity*, according to which a system which did not provide equal satisfaction of health care needs would be ethically defective, though not unjust.[4]

It is important to understand that whether equal satisfaction of health care needs (construed narrowly as basic health care interests) is understood as an ethical requirement of justice or of charity, its fulfillment is perfectly compatible with inequalities in the provision of health care that satisfies preferences or interests that exceed the minimum constituting the set of health care needs. For example, if Jones and Smith receive equal amounts of health care of equal quality for the satisfaction of their basic health care needs, and if the potential for benefiting from cosmetic nose surgery is not a health care need, then the criterion of equality is satisfied, even if Jones and Smith could both benefit equally from cosmetic nose surgery but only Smith gets it.

Since the criterion that all are to receive equal health care who have equal health care needs is so ambiguous, and since the distinction between needs as especially important interests and needs as preferences or potentials for benefit is so unclear and in need of theoretical support, it is advisable to abandon this formulation altogether. Instead, what is plausible in the equal need criterion can perhaps best be reformulated as follows: everyone is to receive an "adequate level" or "guaranteed minimum" of care.

This formulation has several advantages. First, by requiring equality only with respect to the provision of some limited level of care, a floor or guaranteed minimum of services, it captures the idea that not all health care is of equal value or even of sufficiently high value to make strictly equal provision of it plausible. Similarly, even if there are sound ethical reasons for everyone's receiving some minimal provision of food or shelter, it does not follow that there are equally good reasons for preventing inequalities in the amounts or quality of food and shelter above that minimum. Second, unlike the term 'need' the qualifier 'adequate' here is explicitly normative. Thus the necessity of providing a theory of which health care interests are of special ethical importance is frankly acknowledged and the temptation to assume an objective scientific interpretation of 'needs' (as in 'physiological needs' or 'caloric needs') is avoided. In particular, it is important to preclude the erroneous suggestion that needs in 'equal health care needs' refers to some set of conditions that can be ascertained by the exercise of medical judgment alone.

In this less ambitious form the equality criterion – more accurately called the universal guaranteed minimum criterion – is a highly plausible standard for the ethical evaluation of health care systems, if sufficient content can be given to the notion of a 'guaranteed minimum' or 'adequate level' of care. I will make no attempt here to provide such a specification in detail.[5] Fortunately, an exhaustive account of what is included in the minimum is not required for all uses of the criterion in the ethical evaluation of a health care system. If we are confident that whatever else the minimum includes it at least includes certain uncontroversially highly beneficial – and relatively inexpensive – forms of care, then we can infer that any system which fails to provide this 'minimum minimum' is ethically inadequate; this we can know without determining whether the ethical failure is a matter of injustice or of lack of charity. Among the most uncontroversial candidates for inclusion in the care of the minimum are professional obstetrical care during delivery, a minimum of one examination by a physician or mid-wife during pregnancy, modern and reasonably prompt emergency care for severe trauma victims, routine preventive care and physical examinations at least every three years, and low-risk, relatively inexpensive surgical repair of conditions that prevent or seriously hinder major functioning, at least where surgery carries a very high probability of preserving or restoring functioning (e.g., thyroidectomy and maintenance medication for cancer of the thyroid).

Before applying this skeleton of a universal minimum criterion to the U.S. health care system, it is important to see that this criterion must be supplemented in such a way as to distinguish between the physical availability of the minimum set of services and the barriers which individuals must overcome in order to utilize them. A system might make some basic set of services available to all who could manage to reach the location at which they were to be rendered, yet some might face much greater difficulties than others in reaching that location. Depending upon the particular features of the system, some might have to wait longer periods before being treated. In general, whatever reasons support the use of the criterion of equal availability of a minimum of care will provide some weight for the additional requirement that no one is to incur "excessive obstacles" in the form of time costs, travel costs, or direct outlays of funds in attaining the minimum.[6]

Even without further specifying either the content of the universal minimum, without fleshing out the intuitive but vague notion that obstacles to utilizing the minimum are not to be excessive for anyone, and without determining whether failure to meet this twin standard is a matter of injustice or a lack of charity, we can nevertheless conclude that the U.S. health care system is seriously ethically deficient. As a recent Presidential Commission on medical ethics has convincingly documented, there are at present between 22 and 25 million Americans who have neither private health care insurance nor coverage under Medicare or Medicaid.

TABLE I

| Survey | Number of people without insurance (in millions) | Share of U.S. population without insurance |
|---|---|---|
| National Medical Care Expenditure Survey (National Center for Health Sciences Research 1977) | 25 | 12.6% |
| National Opinion Research Center (U. of Chicago, 1976) | 22 | 12% |
| National Health Interview Survey (National Center for Health Statistics, 1978) | 23.5 | 11% |

Source: President's Commission. *Securing Access to Health Care*, p. 93.

Perhaps as many as another 20 to 22 million have coverage which fails to provide at least some of the least controversial elements of an adequate minimum. Moreover, the same Commission also documented the fact that many Americans, particularly in rural and inner city areas, face severe obstacles to utilizing care when it is available.[7]

These figures should come as no surprise once two facts are noted. First, while many individuals whose incomes put them far above the official poverty line are unable to afford anything like comprehensive health insurance (especially if they are not eligible for group plans through their employer), less than 50% of those at or below the poverty line are covered by Medicaid.[8] Second, since Medicaid reimbursement rates are in some cases as low as 40% of prevailing private insurance reimbursement rates, many physicians refuse to treat those "covered" by Medicaid.[9]

It is important to qualify this rather bleak picture of the access problem with two points. First, approximately ninety percent of the American population does have some form of public or private health care insurance. Second, some of those who lack coverage nonetheless have been able to secure at least some forms of care, either through the explicit charity of public hospitals or through the hidden welfare system of "cross-subsidization", whereby a part of the surplus revenue from paying patients has been used to fund care for some of those who cannot pay. However, growing cost-containment competition and new prospective payment arrangements designed to limit charges for Medicare hospital care are making it very difficult for hospitals to continue the practice of cross-subsidization.

TABLE II
Health care coverage, 1980 (in millions)

| | |
|---|---|
| Total | 231.0 |
| Private | 161.2 |
| Medicare | 21.7 |
| Medicaid | 15.2 |
| Medicare and Medicaid | 6.1 |
| Other Public[a] | 5.3 |
| Uninsured | 21.5 |

[a] Champus + VA programs.
Source: Survey of Income and Education, Dept. of Health & Health Services, in President's Commission, *Securing Access to Health Care*, p. 91.

The American health care system and medical profession have received well-deserved praise for technological innovation and research, and for providing very high quality care for the majority of citizens. My purpose here is not to deny or to minimize these impressive accomplishments. Instead, I wish only to point out that serious ethical problems remain, especially in the area of access to care.

## 2. PROVIDING THE BEST HEALTH CARE FOR ALL

Although this second proposed evaluative criterion is frequently voiced, not least in the pronouncements of organized medicine and politicians, it does not in fact express a plausible goal, either from the standpoint of ethics or of economic rationality. Presumably providing the best health care for all entails making available for everyone *all technically feasible health care services which can be expected to provide any net health benefit for the patient*. In other words, the goal of providing the best health care for all requires us to allocate all our resources to health care up to the point at which marginal health benefit falls to zero (subject only perhaps to the distributional constraint that each is to receive the best possible care compatible with everyone's else receiving it).

But simply pursuing the goal of providing the best health care for all is to focus exclusively upon maximizing health benefits, without regard for the *opportunity costs* of allocating the resources in question to health care services rather than to the production of other goods and services that may produce *more* benefit. There are decisive reasons of self-interest as well as ethical reasons not just for eliminating "unnecessary health care," i.e., care that produces *no* net health benefit, but also for not providing care that produces *less* net benefit than alternative uses of the same resources. To fail to see this is to overlook a very simple truth, namely, that health care is *not* of infinite value because there are other things in life that are worth pursuing besides minimizing the probability of death or illness. Each of us acknowledges this simple truth in the way we live. One could reduce significantly the probability of death or disability from trauma by never traveling in an automobile, but most would find this too costly, in terms of time, money, and enjoyment. Just as it would be irrational for an individual to place no limits on how much he or she spends to minimize the probability of death or illness, so it is irrational for a society to fail to limit how much it spends on health and to provide "the best health care for all."

Unrestrained pursuit of the goal of providing the best health care for all is not only irrational, it is ethically dubious as well, for at least two reasons. First, a society that relentlessly would pursue the goal of providing the best health care for all would be one whose citizens had exhibited a failure to appreciate other good things in life besides health – art, literature, music, scientific research unrelated to human health. Such a single-minded, indeed obsessive concern with capturing all possible health benefits, no matter how small and no matter at what costs, would amount to a kind of cultural mono-mania which, one would surmise, was rooted ultimately in an infantile denial of human mortality. Second, the idea of providing the best health care for all has in the U.S. at least been used as a rationale for failing to deal with the failure to provide even the more basic and less costly, but highly beneficial, forms of health care to those who now lack them. For example, spending millions of dollars a year on artificial heart implants and coronary by-pass surgery is often justified on the speculation that with practice and improvement these techniques will eventually contribute to the goal providing the best health care for all. Yet this vast outlay of resources clearly could be used to provide other services – either medical or non-medical in nature (such as better food, education, and housing) –

which would produce greater benefit and would better serve the goal of providing a universal minimum of the most important health care services.

## CONTROLLING HEALTH CARE COSTS

One of the most distinctive features of health care in the U.S. has been the predominance of the third-party payment, fee-for-service system, the perverse incentives of which are well known. First, the physician has no effective economic incentive to restrain utilization of treatments and tests because the higher the utilization the higher his revenues. Second, what has traditionally been a fundamental tenet of medical ethics, the principle that the physician is to do the best he can for each patient, puts the moral seal of approval on the physician's urgings of self-interest. Third, fear of malpractice liability – whether that fear is realistic or not – also encourages the physician to order extra tests or treatments, even when they are of dubious value relative to their costs. Fourth, the physician can easily reconcile the provision of care that promises meager benefits relative to costs with his commitment to acting in the patient's best interests, because he knows that the insurance premium is already paid and that the costs of treating this particular patient are being spread over the entire insurance pool.

Similarly, quite aside from the fact that he is likely to defer to what the doctor orders in any case, the insured patient has no effective incentive to restrain utilization so long as the services in question promise any net benefit to him, since he has already paid his premium (unless his policy requires him to make a substantial copayment).

The goal of curbing health care costs can be viewed as a *public good* in the technical sense and, like other public goods such as energy conservation, clean air, and national security, may not be achieved through strictly voluntary efforts because of the free-rider problem. Although each consumer and physician may recognize that we would all benefit from lower health care costs, the incentives of the fee-for-service, third party payment system make it rational for each individual to act in ways which will prevent this goal from being reached. Each individual may reason that his own contribution to cost control will be so minimal as to be negligible and that either enough others will exercise restraint to achieve the goal or they will not, regardless of what he does. If the individual also believes that he will be able to reap the benefits of cost control regardless of whether he contributes to its achievement and if he views his own efforts at cost control to be a cost to him, then, so far as he behaves as an individual utility maximizer, he will not contribute and will attempt to take a "free-ride" on the contributions of others. If enough people act in this way the good will not be achieved – health care costs will continue to rise in spite of the fact that we all agree that curbing costs is necessary.

In addition to the "free-rider" problem, the *assurance problem* threatens to block voluntary action to control health care costs. Even if both the physician and the patient would be willing to exercise restraint in utilizing services *if they could be assured that others would do so as well*, neither may do so where such assurance is lacking. A physician may reason that whether or not health care costs are curbed will be determined not by his own rather insignificant behavior but rather by whether enough other physicians exercise restraint. Unless he can be assured that other physicians will also limit utilization, why should he unilaterally expose himself to malpractice liability (by falling below the "community standards of care" that determine what counts as malpractice), to a reduction of income, and to possible complaints from patients that he is not doing the best he can for them? Similarly, you and I as consumers of health care may bemoan rising costs, but may be unwilling to forego that extra laboratory test which has only a 0.005 probability of detecting cancer, unless we have assurance that others will exercise similar restraint. Even if each test is very inexpensive, the cost of this small added protection against misdiagnosis may be very great if many thousands of tests are done each year. Yet the cost of *my* having the test may be negligible, especially since the potential benefit to me of early detection of cancer is enormous. Granted this remarkable convergence of incentives, both for the provider and the consumer, it is not surprising that such a system generates increasing costs.

The final major contributor to the cost crisis in health care is the manner in which private insurers and government programs such as Medicare and Medicaid have paid for services provided by physicians and hospitals. Payment has been according to " reasonable and customary rates," and in a health care system in which organized medicine has approached a professional monopoly on the provision of reimburseable services, what is "reasonable and customary" has been determined largely by the providers of health care, in the absence of the restraints which price-competition provides in a genuine market.

It is important to understand the basis of the charge that the U.S. health care system has failed to satisfy the criterion of adequate cost containment. The fact that the percentage of the U.S. gross national product devoted to health care has more than doubled in the past dozen years is often taken as proof that there is a "cost-crisis" in health care. Such an inference, however, is invalid. There is no magic number which can be assumed to represent *the* correct proportion of social resources to be used for health care. It might be quite reasonable, for example, for a society as affluent as the U.S. to spend 20% rather than 11% of its G.N.P. for health care, *if* this allocation accurately reflected informed social priorities and if the allocation for health care were used efficiently. The real cost crisis in U.S. health care is not so much that we are spending too much, but that we are not getting our money's worth.

Until quite recently, the same system that produced the cost crisis also supported a hidden welfare system for some of the millions of patients who lacked private insurance and who were not covered by government programs. By overcharging paying patients, hospitals were able to "cross-subsidize" some care for indigent patients. However, the main strategies for cost containment are making it increasingly difficult if not impossible for hospitals to continue the practice of cross-subsidization without exposing themselves to serious financial risk. In an attempt to curb the abuses of "reasonable and customary" reimbursement, the U.S. Federal government has now instituted the Diagnostic Related Grouping (DRG) system for Medicare hospital payments. This is a prospective payment scheme whereby hospitals taking care of Medicare patients are paid pre-established amounts for various diagnostic categories. If the hospital's costs for treating a patient with appendicitis exceed the amount prescribed for that diagnostic category, then the hospital must absorb the excess. Assuming that the diagnostic category rates are set low enough to achieve significant cost reduction, hospital administrators will have smaller surpluses, if they have surpluses at all, from which to subsidize care for indigent patients.

Increased competition in health care, the other major strategy for curbing costs, has had a similar impact on cross-subsidization. The most visible aspect of the increase in competition has been the rapid growth of *for-profit health care corporations*, including hospitals, nursing homes, psychiatric facilities, dialysis centers, and free-standing "urgent care," out-patient surgical, and emergency centers. Non-profit hospitals, and public hospitals in particular, have complained bitterly that competition from for-profit hospitals has made it necessary for them to reduce the volume of charity care they have traditionally provided. The problem is not restricted to for-profit corporations, however: competitive pressure is mounting throughout the health care sector and the behavior of both non-profit and for-profit hospitals seems to be moving toward a norm of vigorous marketing, professional management, and closer scrutiny of costs.[10] In such an environment cross-subsidization is hardly likely to flourish.

So not only has the U.S. health care system failed to satisfy either the criterion that all are to be provided with at least a minimal level of basic health care services or the criterion of adequate cost-containment; even worse, current efforts to curb costs are exacerbating the problem of lack of access to care for millions of American.

## 4. MAINTAINING THE MAXIMUM AMOUNT OF FREE CHOICE FOR PATIENTS AND PROVIDERS

This fourth criterion for evaluating health care systems, if taken literally, contains a very controversial presupposition, namely, that the freedom, or rather, various freedoms, of different persons, in a given system can be individually quantified and then aggregated into a magnitude to be compared to similar aggregates of freedom in other systems. This presupposition is open to criticism on a number of grounds, not the least of which is that it may be difficult to spell out the notion of an *amount* of freedom even for one person and for one kind of freedom. For example, do I have more freedom, in any morally interesting sense, in one situation than in another simply because I have one more option in the former than in the latter? Or does that depend upon how valuable the additional option is? Presumably we cannot always determine whether a person has greater freedom in one situation than another simply by counting his options, since to do so would be to overlook the fact that ultimately we are interested not merely in increasing opportunities for action (no matter how trivial or valueless) but in increasing the range of *significant options*.

Is the significance of options to be understood in a purely subjective way, as defined by the preferences of the agent, or is it to include an objective element? Further, in which situation do I have a great amount of freedom: in a situation of which I have fewer options, each of which I can avail myself of without great effort, or in a situation with more options, each of which would require more effort or costs on my part to realize? Instead of pondering further these and a number of other perplexities that lay behind the notion of 'maximum freedom', I shall simply ask what kinds of freedoms are enjoyed by consumers and providers in the U.S. health care system, and then ask which freedoms are of greatest moral importance.

We can begin by considering freedoms which health care consumers have and do not have in the U.S. It is often said that one of the strong points of the U.S. system is that consumers have *the freedom to choose a physician* that meets their own standards. This, of course, is at best a very misleading half-truth. First of all, only those who are fortunate enough to have health care coverage have the freedom to choose a physician; those without coverage lack not only the freedom to choose among physicians but the ability to get any professional medical care at all for many of their problems, even in some cases of life-threatening emergencies. Second, many of those who are covered by government programs, in particular Medicaid, have little or no freedom to choose among physicians, and as was noted earlier may not even be able to find one physician in their region who will accept them, because of low Medicaid reimbursement rates. A more accurate statement, then, is that the U.S. health care system provides considerable freedom to choose a physician only for most of those who have private insurance of Medicare coverage but not for the more than 22 million who lack any form of coverage and not for many of the millions who have Medicaid coverage.

Nevertheless, the majority of people in the U.S., who are covered by Medicare or private insurance, have a good deal of freedom to choose a physician. And many seem to value this freedom highly.

There is one kind of freedom of choice concerning care, however, which largely has been lacking for those who have adequate health care coverage as well as those who do not: the freedom to show their preferences about the limited importance of health care (to them) relative to other valuable goods. In other words, the third-party payment, fee-for-service system that has dominated U.S. health care, as we have seen, has made it virtually impossible for consumers to control the amount of social and private resources that have been flowing into the health care system.

Sometimes it is said that a small class of terminally ill patients is like the black holes of deep space – capable of sucking in almost limitless quantities of health care resources, with dubious benefit. There is, however, a much more general health care *black holes* problem in the U.S. The health care sector, chiefly because of the high utilization incentives of the third-party payment, fee-for-service system, threatens to drain ever increasing amounts of private and public resources away from all other sectors of the economy. This is largely because the incentives that limit expenditure in other sectors of the economy are generally absent in the health care sector. In such a system, a system in which adequate cost containment is a collective good whose achievement is thwarted because of the free-rider and assurance problems described earlier, individuals lack control over their resources. This, surely, is a significant limitation on freedom. Only if the incentives characteristic of the current system are changed will we be able to allocate resources in a way which is rational.

Even if we set aside this lack of control over private and social resources as one important kind of freedom, it is difficult to justify the U.S. system on the grounds that the freedom to choose a physician enjoyed by the fortunate compensates for the lack of this freedom on the part of the unfortunate. The results of hypothetical choice or ideal contractarian or *ex ante* insurance decisions seem to support this conclusion. Given that he is behind a "veil of ignorance" that precludes him from knowing that he is a member of the fortunate group, a rational decision-maker whose aversion to risk falls within what can be considered the normal range would presumably choose a system in which he would be guaranteed access to progressional care, at least for serious problems, even if this system placed restrictions on his freedom to choose a physician, over a system which gave him greater freedom in the choice of a physician if he turned out to have coverage of the appropriate sort, but in which he might have no access to important forms of care at all.

Further, it is very implausible to maintain that we in the U.S. must choose between having a system in which some enjoy a wide freedom of choice among physicians while millions lack any access at all to some important health care services and a system in which everyone has access to important services but most or all have no freedom of choice among physicians. To assume that these are the only choices is to assume that the only alternatives are the current system or an extreme version of the National Health Service model in which the government not only provides all health care but does so in such a way that patients have little or no freedom to choose among physicians employed by the government. It is not at all clear, however, that even a National Health

Service need be so restrictive. More importanly, there are several plausible options for increasing access to care while maintaining a wide freedom in the choice of physicians for all without introducing a National Health Service in the U.S. One of the most attractive of these is the idea of promoting greater competition in health care while providing the poor with health insurance vouchers subsidized from general tax revenues.[11]

There is another facet of the issue of consumer freedom in the U.S. health care system which ought not to be overlooked. So far I have only examined the question of whether American consumers enjoy the *freedom to choose a physician*. This is not to be identified with the broader question of *health care consumer freedom in general*. A wide freedom to choose among physicians is quite compatible with significant restrictions on the freedom to choose between physicians and non-physician health care workers. As a matter of fact, a good deal of the activity of organized medicine in the United States traditionally has been as much if not more concerned with limiting or even eliminating the consumer's freedom to choose non-physician health care as with preserving the consumer's freedom to choose among physicians.

Examples abound: physician-dominated medical licensing boards have prevented nurses, midwives, and nurse-practitioners from providing services which they could in some cases have provided just as effectively (and much more cheaply) than physicians; physician-dominated bodies for the accreditation of medical schools and hospitals have been instrumental in eliminating approaches to medicine that did not fit the orthodox mold. As in cases of monopolistic licensure in many other professions, these attempts to eliminate or subordinate non-physician health care have been made in the name of "providing the best health care."

It is now widely recognized that actual medical licensing requirements in the U.S. traditionally have been structured in such a way as to be more effective in inhibiting potential competitors from entering the field than in ensuring that those already licensed continue to practice competently, although relatively recent efforts by specialty organizations are beginning to reverse this trend by imposing recertification requirements. Quite apart from the doubt which this casts on the motivation of those who have supported medical licensure, the very idea of limiting the consumer's choice to physicians is itself vulnerable to the criticism that even if physicians do provide the *best* health care, people should be allowed to choose health care that is less than the best, at least so long as it is safe and meets certain minimal standards of quality.

This point can be made clear by an analogy. It is as if the corporate executives of the Cadillac company had persuaded the government to prohibit the sale of Fords and Chevrolets on the grounds that only the best cars should be offered to consumers. To those who cannot afford a Cadillac but who could afford a Ford, the fact that a Cadillac is a *better* car provides little consolation. Health care consumer freedom in the U.S. in recent decades has to a large extent been limited to the freedom to choose among different "Cadillac" health care dealers.[12]

In addition to restricting the freedom of consumers to choose among alternatives forms of health care and restricting the freedom of nurses and others to offer services independently of physicians, organized medicine in the United States has, of course, limited the freedom of all those persons who might have developed alternative forms of care had they been allowed to. Any attempt to characterize "the freedom of health care providers" in the U.S., then, must first distinguish among various sorts of providers and recognize that physicians have often used their freedom to curtail not only the freedom of consumers but also that of other, non-physician providers.

Physicians in the U.S. traditionally have enjoyed a great deal of freedom of several different sorts. Proposals aimed at providing wider access to health care, from the development of health maintenance organizations that offer more affordable services, to Medicare, National Health Insurance schemes, and more extreme plans for a National Health Service, have all met strong opposition from organized medicine in the U.S. in the name of preserving physicians' freedoms. Too often suggestions for improving access and for curbing costs have been dismissed on the grounds that they would unduly interfere with physicians' freedoms. What is needed is a careful account of what the relevant physician freedoms are, how various reforms would affect them, and, above all, what the moral status of those freedoms is.[13] The latter question is of great importance because not all freedoms are of equal, or even of very significant, moral value. If an institutional change would violate a freedom or liberty which is so basic as to warrant talk of a *right* to that freedom, then that is a potent objection to making the institutional change in question. On the other hand, some freedoms do not deserve the strong protection provided by rights and others are themselves morally suspect, especially if they are enjoyed by some through interfering with the freedom of other persons.

It has been observed that physicians in the U.S. traditionally have enjoyed considerable freedom from interference by others in four areas: (1) the diagnosis and treatment of patients; (2) the entrance requirements to the profession; (3) the organization of medical practice; and (4) the financing arrangements for medical services.[14] Another freedom enjoyed by American physicians – and frequently declared to be worth preserving at almost any price – is (5) the freedom to choose one's specialty and practice where one wishes.

I have noted that (2), physicians' freedom in controlling the entrance requirements to the medical profession has sometimes been exercised in ways which limit the freedom of consumers and non-physician health care providers or would-be health care providers. I have also suggested that at least in some cases these restrictions on freedom have not been justified, especially when they were imposed in the name of ensuring "the best possible health care," a goal which, I have argued, is neither rational nor ethical for a society to pursue at the cost of other important values. The special expertise to physicians does, of course, support a strong presumption that they are to play a major role in determining the requirements for entry into their professions. But the long record of self-serving behavior in this regard supports the conclusion that this freedom should be limited by regard for the public interest and the freedom of others.

Similarly, a great deal of physician freedom in the diagnosis and treatment of patients is desirable and justifiable on grounds of the physician's special expertise – and because of the vital need to prevent medical practice from becoming a tool (or rather a weapon) of state power and social control, as it has become in the Soviet Union. Nevertheless, as I have argued, the need to control costs is of sufficient gravity from both a rational-prudential and an ethical point of view to warrant some restrictions on this freedom as well. The trick, of course, will be to devise ways of limiting this physician freedom without undermining the patient's trust in the physician as his advocate, as someone whose primary concern is with the patient's well-being.

For the same reasons, a good deal of physician freedom in the organization of medical practice may be desirable, but this freedom too can hardly be regarded as sacrosanct. This is especially clear once we recall that physicians benefit from massive public subsidies, both in the form of educational and research support and public funding of Medicare and Medicaid. Indeed, the organization of medical practice in the U.S. already varies widely from community to community and is not now controlled exclusively by physicians. Moreover, there are many degrees and modes of control possible between the extremes of complete physician autonomy and various forms of socialized medicine.

The case for broad physician freedom in shaping the financing of health care is far weaker, both because of the need to control obvious conflicts of interest and because physicians have no special expertise in financial matters. As Dan Brock has argued, the proper scope and limits of each of the first four physician freedoms must be determined by a complex process of weighing social benefits and costs; it is simply not plausible to claim that any of them is a basic moral right which it would be impermissible to limit by appeals to general utility or the rights and interests of other than physicians.[15]

The fifth physician freedom, the freedom to choose one's specialty and place of practice, is ambiguous. It could refer simply to the freedom of occupational choice which all Americans enjoy as a matter of legal right, the freedom to try one's hand at any legal occupation, but with no guarantee, of course, that one will succeed in the market. On this interpretation, the freedom to choose one's specialty and place of practice is compatible even with a National Health Service, so long as physicians are permitted to offer their services wherever they wish in a market which is allowed to exist alongside the government system, as in Great Britain. On the other hand, the fifth physician freedom could be understood, in a quite different way, to be the actual range of opportunities for choosing specialties and practice location *that will be economically viable or lucrative* which American physicians now enjoy. There is little doubt that the establishment of a National Health Service, and, to a lesser extent, more modest proposals for improving access to care by limiting the number of training spots in over-supplied specialties such as general surgery, as well as other measures to encourage the movement of physicians into undeserved locales, would limit *this* physician freedom.

The wide range of opportunities for selecting economically viable and even lucrative specialties and practice locations that American physicians now enjoy is, however, in part a result of lack of competition in health care, and in particular stems from questionable exercises of physicians' freedom in determining the entry requirements for the profession, the financing arrangements for medical practice, and the allocation of public monies among different specialty training programs. If a case could be made for arranging our social institutions in such a way as to guarantee physicians not only the freedom to choose an occupation (and to succeed or fail in the competitive market) which all Americans enjoy, but in addition the freedom to have an exceptionally broad range of economically viable occupational choices as a result of being protected, it could not rest upon the wildly implausible assumption that physicians have a basic moral right to such a privileged position. Instead, it would be necessary to show that the attainment of some especially important social good depended upon according physicians this special benefit. It is hard to imagine, however, that whatever social benefits might accrue to preserving the exceptional occupational opportunities which U.S. physicians have come to have as a result of a combination of massive public subsidies for medical education, research, and patient care and the monopolistic exercise of control over licensure and training could outweigh the fundamental goals of improving access and controlling costs.

In sum, any serious attempt to evaluate the U.S. health care system from the standpoint of consumer and provider freedom must yield mixed results. That system is not so bountiful in consumer freedom as its more enthusiastic celebrants have proclaimed. Moreover, the much-touted freedoms of American physicians are not plausibly viewed as basic moral rights and in any case have sometimes been secured and exercised in indefensible ways.

*Department of Philosophy,*
*University of Arizona,*
*Tucson, Arizona, U.S.A.*

## NOTES

[1] My discussion of the goal of equal health care for all draws on pp. 17–22 of President's Commission for the Study of Ethical Problems in Medicine and Biomedical and Behavioral Research, 1983, Report, Volume One: *Securing Access to Health Care*, Government Printing Office, Washington, D.C.

[2] The distinction between quantity and quality of care, though often taken for granted, is far from unproblematic. Since this problem is not central to the concerns of this essay it will not be addressed.

[3] For discussion of justice in the distribution of health care, see President's Commission, 1983, Appendices, Volume Two, *Securing Access to Health Care*, Government Printing Office, Washington, D.C.; Daniels, N., *Just Health Care*, Cambridge University Press, Cambridge, England; Shelp, E. (ed.), *Justice and Health Care*, Reidel, Dordrecht, Holland, 1981.
[4] [5].
[5] For the beginning of an attempt to work out the content of the "decent minimum" or "adequate level," see President's Commission, 1983, Report, Volume One: *Securing Access to Health Care*, pp. 35–43.
[6] ([1], I, pp. 42–43).
[7] ([1], I, pp. 54–108).
[8] ([1], I, p. 95).
[9] ([1], I, pp. 86–87).
[10] Brock, D., and Buchanan, A.,:1986, 'Ethical Issues in For-Profit Health Care' *For-Profit Enterprises in Health Care*, National Academy Press, Washington, D.C., pp. 224–249.
[11] [3].
[12] See the discussion in the chapter on medical licensure in Friedman, M., 1962, *Capitalism and Freedom*, University of Chicago Press, Chicago, Ill.
[13] For what I believe to be the best available discussion of this neglected topic, see Brock, D., "The Distribution of Health Care and Individual Liberty", in President's Commission, 1983, Appendices, Volume Two, *Securing Access to Health Care*, pp. 239–63.
[14] ([6], p. 98).
[15] ([1], p. 253).

## BIBLIOGRAPHY

1. Brock, D.: 1983, 'Distribution of Health Care and Individual Liberty', President's Commission, *Securing Access to Health Care*, Government Printing Office, Washington, D.C., Vol. II, pp. 239–264.
2. Daniels, N.: 1981, *Just Health Care*, Cambridge University Press, Cambridge, England.
3. Enthoven, A.: 1980, *Health Plan: The Only Practical Solution to the Soaring Cost of Medical Care*, Addison-Wesley, Reading, Mass.
4. Friedman, M.: 1962, *Capitalism and Freedom*, University of Chicago Press, Chicago, Ill.
5. Shelp, E. (ed.): 1982, *Beneficence in Health Care*, D. Reidel, Dordrecht, Holland
6. Stroman, D.: 1976, *The Medical Establishment and Social Responsibility*, Kennikat Press, Port Washington, New York, p. 98.

J.-MATTHIAS GRAF VON DER SCHULENBURG

# THE HEALTH CARE SYSTEM OF THE FEDERAL REPUBLIC OF GERMANY: MORAL ISSUES AND PUBLIC POLICY

The Nobel prize winner Milton Friedman once wrote: "National health insurance is an . . . example of misleading labeling. In such a system, there would be no connection between what you would pay and the actuarial value of what you would be entitled to receive, as there is in private insurance. In addition it is not directed at insuring 'national health' – a meaningless phrase – but at providing medical services to the residents of the country. What its proponents are in fact proposing is a system of socialized medicine" ([2], p. 104). Although Milton Friedman did not name the German health care system, his statement is an excellent short description of the German statutory health insurance.

More than 90 percent of the population in the Federal Republic of Germany are covered by one of the 1,200 sick-funds. These sick-funds provide a very comprehensive coverage of ambulatory, hospital, and dental care with very little or no co-payments. In addition the health insurants receive cash benefits for maternity leave, taking a "cure leave" (*Kur*), or to compensate for lost wages in case of illness. These benefits are financed by a pay-as-you-go system. The contributions depend only on the insurants' income and family status because the major goal of the German statutory system is to provide equal access to high quality health care for all citizens. This goal is called the *Solidaritätsprinzip* (principle of solidarity). The second principle most characteristic of the German system is called the *Sachleistungsprinzip* (principle of benefits in kind). Like Health Maintenance Organizations in the U.S., German sick-funds reimburse physicians, dentists, hospitals, and drugstores directly, so patients do not have to prefinance their health services.

In this paper we will explore the basic principles of the German health care system in more detail. In addition, we will show the economic consequences and problems of the principle of solidarity. The paper is organized as follows: The first section contains a general view of the German health care scheme; in the second section the supply of health care services is discussed in more detail. The third section deals with the goal of solidarity, cost trends, and cost-containment policy.

*Hans-Martin Sass and Robert U. Massey (eds.), Health Care Systems*, pp. 59–74.

## I. THE GERMAN HEALTH CARE SYSTEM

The German statutory health care scheme dates back to the Health Insurance Act of June 1883, the oldest part of the social security system created by Chancellor Bismarck to stem the socialist movement among German industrial workers. The Health Insurance Act was followed by the acts concerning accident insurance in 1884, retirement funds in 1889, public assistance in 1924, and unemployment insurance in 1927. In 1911 the Health Insurance Act was revised and became the second book of the *Reichsversicherungsordnung* (social security act) which is still in effect. Since then the German health care became the most regulated sector of the economy. After World War II, the West German legislature enacted numerous statutes concerning public health care; the most important of these resulted in the founding of the *Bundesministerium für Jugend, Familie und Gesundheit* (Federal Department of Youth, Family Welfare and Health) in 1961. The Government established statutory fee-schedules for physician and dentist services in 1965, which were revised in 1982, specified the employer's liability to continued wage payments in case of illness (*Lohnfortzahlung*), and created the *Approbationsordnung* (medical profession education act) in 1970. Of importance were further the 1972 act concerning the financing of hospitals (*Krankenhausfinanzierungsgesetz*) and the disputed health insurance cost-containment act (*Krankenversicherungskostendämpfungsgesetz*) of 1977 which was followed by the second health insurance cost-containment act in 1982. With these regulations the German health care system became the most regulated sector of the entire economy (for a detailed description see [1, 4, 9, 10, 12, 16, 17]).

The basic elements of the German health care scheme are: all employees or retired persons who earn less than a fixed minimum per month have to join a statutory sick-fund. Others may join voluntarily (see Figure 1). However, there is only limited competition among the sick-funds operating in the same region because blue collar workers have to join the local sick-fund or the vocational sick-fund if one exists. The sick-funds are organized in associations both at state and federal level to negotiate the fees with hospitals and the insurance-doctors' association (*Kassenärztliche Vereinigung*). In addition, representatives of the sick-funds' associations join the *Konzertierte Aktion im Gesundheitswesen* (national health conference) where overall health care cost-ceilings are negotiated. In the past, however, this national health

| Total population 61 million = 100% | | | | | |
|---|---|---|---|---|---|
| others 9% | | | Statutory sick-fund 91% | | |
| | | | Voluntary membership 12% | Compulsory membership 79% | |
| Without any coverage 0.2% | Public assistance 1.8% | Private insurance 7% | Substitute sick-funds 26% | Local sick-funds 42% | Vocational sick-funds 23% |
| | | Insurants of statutory sick-funds with supplementary private insurance 8.7% | | | |

Source: [15].

Fig. 1. Health insurance coverage in the F.R.G.

conference was seen to have only limited power when it came to enforcing such ceilings (see [5]).

Each sick-fund is a self-governing and self-sustaining body of public law. Self-governing means that the board of directors of the sick-fund is elected by the insurants and the employers. Self-sustaining means that a sick-fund has to calculate the insurants' contributions so that the total of current revenues covers the total of current expenditures (pay-as-you-go financing). To meet the goal of solidarity the contributions must be stated as a percentage of the insurant's gross income. The contribution rate, therefore, varies from 9 to 15 percent. When determining the contribution, the sick-funds are not allowed to use other criteria (such as age, sex, or state of health) than the member's income and family status. Children and unemployed spouses pay no contribution, retired persons pay a lower percentage than employees. For employees the contribution is shared equally between themselves and their employers. Sick-fund insurants receive a very comprehensive insurance coverage so that co-payments – as compared to the U.S. – are relatively small. However, a law had already been passed under Helmut Schmidt (Social-Liberal Coalition), but only put into practice under Helmut Kohl (Christian-Liberal Coalition) which imposed a charge of DM 5 per day up to a maximum of DM 70 for hospital care. In addition, the latter Government raised the co-payments for dentures from 20 to 40 percent and the fee per prescribed drug from DM 1.50 to DM 2.

## II. SUPPLY OF HEALTH SERVICES

### 1. *Administrated Prices*

If the goal is solidarity when financing medical care, one has to restrict the consumer's freedom of choice and/or impose price control measures, otherwise costs will explode. This is clearly shown by the German experience and will be discussed in Section III. In this paragraph we will describe in more detail the price-control mechanism within the German health insurance scheme. To understand the German system of price control, one has to remember three things: (1) Since more than 90 percent of the population is covered by the statutory health insurance scheme, it plays the dominant role. For patients covered by private health insurers the Federal Government has imposed price regulations for physician and hospital services. (2) For ambulatory care the patients are free to choose among all doctors who are members of the insurance-doctors' association (*Kassenärztliche Vereinigung*) which includes 94.7 percent of all office-based physicians. By court decision of 1960 every office-based physician has a right to membership. At present the physicians are fighting for a reduction in the number of medical students and new entries of young doctors into the statutory health care scheme. (3) There is a sharp distinction between ambulatory physician care and hospital care. Office-based physicians do not normally have hospital privileges. Hospital-based physicians work on a salary basis, and their salaries are covered by the per diem rates negotiated between the sick-funds and the hospitals. Only private patients (i.e., patients who are not covered by a sick-fund or public assistance) have to make additional payments for physician services in a hospital. Table I provides an overview of the pricing mechanism in the German health care system.

### 2. *Physician Services*

In West Germany, office-based physicians play the dominant role in the health care sector as a whole. Every patient who is covered by a statutory sick-fund must first consult a physician in order to receive any type of medical care. Only office-based doctors may provide ambulatory care, prescribe drugs and medical appliances, and decide who is to be hospitalized. If the case is not an emergency, the hospital requires a referral order written by an office-based physician before the patient may be treated.

Between 1960 and 1985 the number of physicians, and especially the

TABLE I
Pricing mechanism of the German health care system

| Medical service | Pricing mechanism | Parties involved |
| --- | --- | --- |
| Hospital care | Per diems fixed by state administration, covering 'approved' operating costs; voluntary negotiations between hospitals and sick-funds in advance; if they come to terms, the State Government authorizes the results. | State Government, hospitals, sick-funds. |
| Ambulatory medical and dental services | (a) Statutory health insurance: fee-schedule negotiated between sick-funds' association and insurance-doctors' association at federal level; mark-up negotiated at state level.<br>(b) private health insurance: 'market pricing' within margins of official fee-schedule | Sick-funds' association and insurance-doctors' association. Physician, patient, Federal Government. |
| Pharmaceuticals | (a) producer price: 'market pricing'<br>(b) trade mark-up: fixed by Federal Government<br>(c) pharmacy mark-up: fixed by Federal Government. | Producer, physician, Federal Government. |
| Public health services | No pricing mechanism. | |

Source: [3].

TABLE II
Number of physicians in the F.R.G.

| | 1960 | 1970 | 1980 | 1985 |
| --- | --- | --- | --- | --- |
| Physicians: | 74,486 | 92,773 | 135,586 | 156,593 |
| – office-based | 45,320 | 48,830 | 58,908 | 65,780 |
| – in hospitals | 21,544 | 35,060 | 65,714 | 75,730 |
| – others | 7,622 | 8,877 | 10,964 | 15,083 |

number of hospital-based physicians, increased rapidly. In 1985 about 42 percent of all physicians were office-based (see Table II). To stem the rapid increase in physicians, the so-called *Ärzteschwemme* ('physician

TABLE III
Average income[a] of office-based physicians and dentists in the F.R.G. and the U.S.

| | F.R.G. | | U.S. |
|---|---|---|---|
| | 1971 | 1980 | 1979 |
| (1) physicians | DM 116,727 | 180,858 | US$ 78,400 |
| (2) dentists | DM 109,502 | 239,463 | |
| (3) average income of all employees | DM 18,097 | 36,633 | US$ 15,052 |
| (4) ratio of (1) to (3) | 6.5 | 4.9 | 5.2 |

[a] Before tax, professional expenses deducted.
Source: [14].

glut'), the insurance-doctors' association as well as the sick-funds' association have proposed a reduction in the number of students entering medical school, a longer training for young physicians, and a more difficult licensing procedure for foreign physicians (see [6]). Although physicians have, to some extent, the possibility of creating demand (see [8]), their relative position in the income hierarchy has slightly decreased (see Table III), but they still enjoy a very high income, as do their colleagues in the U.S. Until 1981 the number of foreign physicians working in Germany increased but has been declining since. In 1985, eight percent of the physicians working in Germany had foreign passports.

Whereas in 1952 only 34 percent of the office-based physicians were specialists, this proportion had increased to 56 percent by 1982. A major reason for this increase is financial, since general practitioners or family doctors are not permitted to provide patients with many of the services which have relatively high fees (such as electrocardiograms, electroencephalograms, x-rays). In spite of this increase, the percentage of specialists in the U.S. is still much higher than in West Germany. Unlike in the U.S., group practices are rare in Germany: in 1975 only, 2,301 physicians (about four percent of all office-based physicians) worked in the 1,131 existing group practices.

In the 'twenties, the statutory health-insurance scheme was in great disorder. Therefore, Chancellor Brüning enacted an emergency decree in 1932 shortly before Hitler came into power. By this decree insurance-doctors' associations were founded in each state, and every office-based

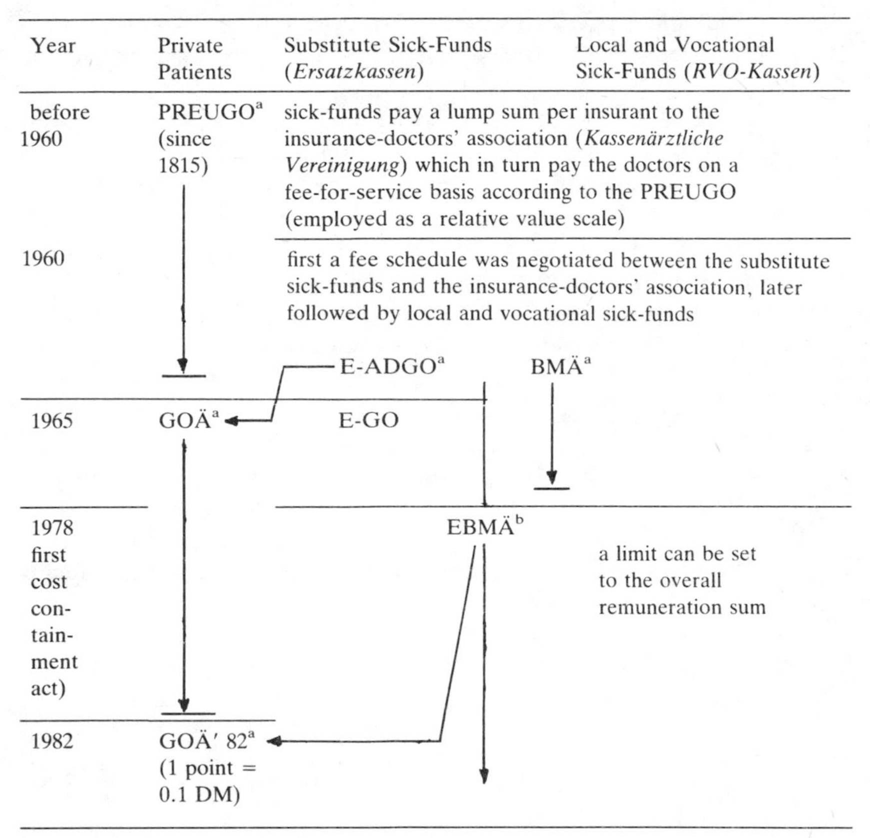


[a] Fee schedules with absolute values (fees in DM):
PREUGO = Preussische Gebührenordnung
E-ADGO = Ersatzkassen – Allgemeine Deutsche Gebührenordnung
BMÄ = Bewertungsmaßstab Ärzte
GOÄ = Gebührenmaßstab Ärzte.
[b] Fee schedules with relative values (point values) EBMÄ.

Fig. 2. Fee schedules and mode of remuneration in Germany.

physician treating sick-fund patients had to join the one existing in his state. Since then, the remuneration of physician services has been subject to negotiation between the sick-funds and the insurance-doctors' association. Figure 2 shows the form of remuneration since 1932. It is important to note that no direct negotiations or contacts take place between individual physicians and sick-funds. Physicians receive payment from their association, which in turn is compensated by the sick-funds.

The insurance-doctors' associations and the insurance-dentists' associations were very successful in their negotiations with the sick-funds, therefore physicians and dentists belong to the highest income classes of society. Interestingly, this is also true of the relatively liberal U.S. health care system (see Table III). It is also interesting that dentists outstrip physicians and are now in the highest position of the income pyramid. This is mainly due to the relatively constant number of dentists during the past two decades and the increase in the expenditures for dental care.

### 3. *Hospital Care*

There are three characteristics of hospital care in the F.R.G. which dominate the current discussion: (1) the relatively long length of stay, (2) the financing of hospital care and the explosive increase in hospital-care expenditures, and (3) the governmental planning of hospital facilities (*Krankenhausbedarfsplanung*).

Problem (1): As Table IV shows, the average length of stay in a hospital in Germany is twice that observed in the U.S. This is also true for certain diagnostic groups like heart disease or cancer (see [12], p. 339). A major reason for the relatively long length of stay in German hospitals is the sharp delimitation between the ambulatory care provided by office-based physicians and hospital care. This is because most hospitals are not allowed to offer outpatient care, even if the patient had previously been hospitalized. A referral order written by a private physician is required in order for a patient to receive hospital care. Given the poor flow of information between private and hospital physicians and the fact that office-based physicians do not have access to hospital equipment, hospitals repeat many diagnostic examinations already performed by private physicians prior to hospitalization . Other medical examinations which could be done before hospitalization are a standard part of hospital treatment itself in Germany, adding to the length of stay and, ultimately, to the cost of stay. In addition, the sharp delimitation between ambulatory and hospital care has another effect: patients stay longer in hospitals because hospital doctors, unsure about the medical care patients will receive after discharge from hospital, prefer to treat them in hospital until the recovery is complete.

TABLE IV
Comparative data on the F.R.G. and the U.S.

| | F.R.G. | U.S. |
|---|---|---|
| Population (in 1000) | 61.200 | 236.700 |
| age under 15 years | 16% | 20% |
| 15 – 30 | 24% | 27% |
| 30 – 45 | 21% | 20% |
| 45 – 65 | 24% | 19% |
| 65 and older | 15% | 12% |
| life expectancy: male | 70.5 | 70.7 |
| : female | 77.1 | 78.2 |
| deaths per 100,000 inhabitants | 1,169.5 | 875.8 |
| cancer (ICD 140–208) | 262.3 | 183.3 |
| circulatory collapse (ICD 390–459) | 589.2 | 437.2 |
| infant mortality, deaths per 100,000 under one year of age (ICD 760–779) | 430.8 | 629.7 |
| hospitals | 3,119 | 7,051 |
| beds per 10,000 of population | 111 | 59 |
| bed-occupancy rate | 72% | 74% |
| mean length of stay (in days) | 16.4 | 7.7 |
| physicians per 100,000 of population | 244.7 | 200 |
| specialists | 33% | 85% |
| female physicians | 27% | |

The solution to the above problem seems to be very simple: hospitals should have the possibility of providing outpatient care, and private physicians should be granted access to hospital facilities and be allowed to provide hospital care. In 1977, however, the various doctors' associations in Germany were powerful enough to hinder the passage of legislation that would have made possible the taking of necessary measures (such as an amendment to the social insurance act).

Problem (2): The financing of German hospitals is regulated by federal law, i.e. the *Krankenhausfinanzierungsgesetz* (hospital financing act) of 1972 and 1985 and the *Bundespflegesatzverordnung* (hospital-care rating decree) of 1973. These acts replaced and standardized former regulations which were inadequate in various respects.

The principle components of current arrangements for hospital financing are: capital expenditures are totally financed by state budget funds; the hospital's current costs (i.e., operating costs) must be financed by per diem payments; per diem rates must be fixed in such a way as to enable the hospital to cover all current costs of management. These rates are uniform for all patients, but differ among hospitals. University hospitals providing very sophisticated care have the highest per diems, small local hospitals have the lowest rates. A precondition for hospitals to receive budget funds for their capital expenditures is their meeting specific standards and conforming with the *Krankenhausbedarfsplan* (hospital need plan) which is discussed below.

The hospital-financing scheme described above has an inherent problem: clearly there is no economic incentive for hospitals to shorten the length of stay. Patients are frequently hospitalized on Fridays and often discharged on Mondays, even though no specific medical treatment is provided on weekends. The hospital financing system offers no incentive to reduce cost and to use resources economically because the per diem rates are calculated on the basis of past cost data. The higher the cost of hospital care was in the past, the higher the present rates will be. Recently the Federal Government has changed the hospital-financing scheme. Beginning in 1986, hospitals will be allowed to make profits and will have no legal claim to receiving per diems covering total cost. Thereby, the sick-funds will gain power, and for some hospitals the per diem negotiations will be tougher.

Problem (3): The hospital-financing act restricts federal funds for capital expenditures to those hospitals which have received accreditation (similar to a Certificate of Need in the U.S.). Combining hospital accreditation and financing was intended to achieve two major goals: a more equitable regional distribution of hospital facilities and available hospital beds, and a reduction in the number of hospital beds with a concomitant increase in occupancy rates.

The German hospital-need planning implies some obvious problems. First, the act contains no clear criteria or standards for hospital planning by state governments. Therefore, the criteria underlying the hospital-need plans differ from state to state or are not indicated explicitly at all. Second, it turns out that, in practice, hospital-need planning affects only construction of new hospitals or additions to already existing ones. Once a hospital has been listed in a 'hospital-need plan' it can hardly be excluded again. This is the reason why hospital-need planning can change the structure and distribution of hospital care only in the long run. Finally, it is now very difficult for private hospitals to become accredited. Only four percent of all hospital beds are under private ownership. Hospital-need planning seems to have paralyzed the private sector.

4. *Pharmaceutical Services*

The sick-funds provide about 15 percent of their entire expenditure on drugs and medical appliances. This is more than for dental care and dentures and nearly as much as for ambulatory physician services (18.3 percent). Two reasons could be given for these extraordinarily high expenditures: excessive consumption and high prices. German surveys indicate that German doctors prescribe on average about 11 drugs per person per year – almost three times as much as their U.S. colleagues. Surprisingly, about 60 percent of all drugs prescribed in West Germany are primarily prescribed at the patients' request; furthermore, doctors declare that 40 percent of the prescriptions are unnecessary, ineffective, or even noxious. One reason for the relatively large consumption of medication by German patients might be the coverage of prescribed drugs by the sick-funds and by most private health insurers.

The contribution of the German legislature towards lowering drug expenditures was the passage of 'Article 368 of the Social Insurance Act' in 1977 on the basis of which a black-list (*Negativliste*) of drugs was compiled which are normally used in the treatment of 'trifling illness' (*Bagatellerkrankungen*). These drugs have now to be paid out-of-pocket by the patient and are no longer covered by the sick-funds. In addition, co-payment for prescribed drugs has been increased several times and amounts to DM 2.00 per prescribed drug since 1983.

In comparison with other countries, prices for drugs are extremely high (see [12], p. 344) because of a lack of price competition on the pharmaceutical market. The German pharmaceutical market shows that solidarity and full coverage of health services require price regulation, otherwise prices will explode. During recent years the Government took two steps to revitalize price competition on the pharmaceutical market. First, the parliament created the *Transparenzkommission am Bundesgesundheitsamt* (drug 'transparency' commission). Its task is the publication and distribution of lists (*Transparenzliste*) comparing prices of medical compounds with the same use and/or ingredients. Second, the insurance-doctors' associations and the sick-funds are compelled by law to negotiate yearly an overall ceiling on drug expenditures (*Arzneimittelhöchstbetrag*). The insurance-doctors' associations must then influence their members by moral suasion, prescription guidelines, and economic incentives to implement the negotiated ceiling.

## III. 'SOLIDARITY' COST TRENDS AND COST CONTAINMENT

The major goal of the German health care system is solidarity, i.e., everybody should have equal access to health services free of charge, and the health care expenditures are financed by contributions according to the insurant's ability to pay (i.e., income and family status). The previous paragraph shows that a solidarity type system has its price, namely, limitations on the freedom of choice and regulations on prices and production techniques. The Social Security Act and the sick-funds' associations together with the insurance-doctors' associations define the maximum level of care someone is entitled to receive under the statutory health-insurance scheme. This is quite different from the market-oriented system in the U.S. where everybody may buy as much health service as he or she is willing to pay for, and the public discussion tends to focus on what would be a decent minimum package of health care the society has to offer to everyone. Anyone joining a Health Maintenance Organization in the U.S. that restricts his or her freedom of choice does so voluntarily.

Both health care schemes have been very cost-driven, so that in both countries the health-policy discussion is dominated by controversy over reasons for the explosive increase in health care expenditures and proposals to stem spiraling cost. In Figure 3 we have plotted the health care costs as a percentage of the gross national product for both countries. The most striking characteristic of the German expenditure growth is the rapid increase from 6.4 percent of the GNP in 1970 to 9.4 percent in 1975, and the leveling-off since then. The cost explosion in the early 'seventies might have been due to the social reform policies of the Social-Democratic Chancellor Willy Brandt (who came into power in 1969), to the relative increase in the number of old people, and to the high cost of investments that were necessary to build up health care facilities after the war.

In 1973 an intensive public discussion on the cost of medical care was initiated by Heiner Geissler ([3]) who was until recently the Secretary of Youth, Family Welfare, and Health. The formerly indolent sick-funds discovered the power given to them by the Social Security Act and insisted on conditions that would allow stable contribution rates (in

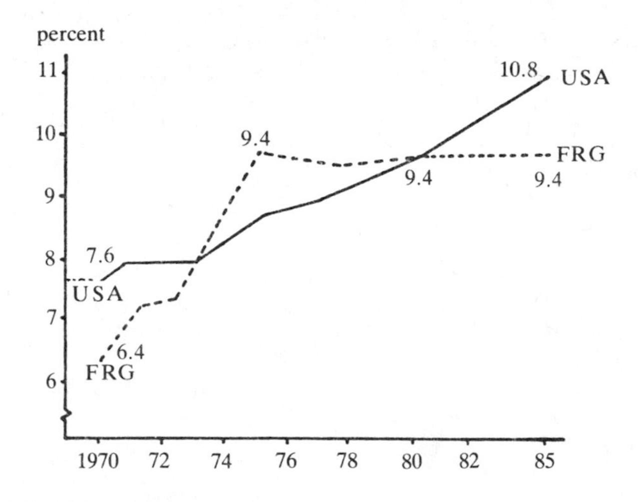


Fig. 3. National health expenditures as a percentage of gross national product in the U.S. and the F.R.G.

percentage of income). Support came from a public campaign as well as from studies by the Scientific Institute of the Local Sickness Funds (Wissenschaftliches Institut der Ortskrankenkassen) in Bonn, a research institute financed by sick-funds. As a result, regulatory measures taken by the State and the Federal Governments produced the health insurance cost-containment act of 1977 which introduced expenditure limits on ambulatory medical care as well as on prescribed drugs (*Arzneimittelhöchstbetrag*). However, it is quite possible that the public discussion of the problem and the reorientation of the sick-funds' own concept of what type of service they should provide and for whom, were actually more important than the cost-containment act (and earlier regulatory measures) in stopping the escalation of medical costs.

The comparison of cost trends in both countries shows that it is easier to contain health care cost in a highly regulated system where the money comes from one 'pipeline', than in a heterogeneous system like in the U.S. where health care costs are financed from numerous sources. Table V shows that the prices for health services as well as the national health care expenditures in real money units increased more rapidly in the U.S. than in the F.R.G. during the past years.

One of the major future problems of the social security scheme in general, and the health care system in particular, is a result of

TABLE V
Average increase in national health care expenditure in the period of 1975–1981

| | F.R.G. | U.S. |
|---|---|---|
| National health-care expenditure | 7.5% | 13.6% |
| prices of health services | 4.9% | 9.7% |
| National health-care expenditure in prices of 1975 | 2.5% | 3.6% |

Source: [11].

TABLE VI
Average contributions to the statutory health insurance and life-cycle net transfers of a person born in that year (in German marks)

| Year | Contribution per year | Expected value of life-cycle net transfers |
|---|---|---|
| 1900 | 377 | 215 |
| 1910 | 410 | – 298 |
| 1930 | 468 | –2,955 |
| 1950 | 629 | –1,272 |
| 1970 | 1,518 | –1,903 |
| 1990 | 4,399 | 16,805 |
| 2010 | 10,469 | 62,164 |
| 2030 | 26,727 | 132,905 |
| 2050 | 64,702 | 166,162 |
| 2070 | 143,141 | 79,459 |
| 2090 | 294,629 | 1,403 |

demographic changes taking place. During the last two decades the German population had declined slightly so that the number of old people increased rapidly (see the comparison of age structures in Table IV). The health care cost increase and the change in the population structure will cause an increase in the average pay-as-you-go contribution. This will lead to an unequal and unequitable intergenerational distribution of the burden of financing the scheme. Paul Kleindorfer and myself (see [7] and [13]) have studied these intergenerational transfer problems by employing a life-cycle model of overlapping generations. Below only an extract of the results of one projection is discussed to illustrate the problem. Table VI shows an estimation of the average contributions to the sick-funds from 1900 to 2090. From this one can calculate the value of all contributions, less the benefits, which one person of the generation born in the considered year can expect to pay to the sick-fund system. As shown in Table VI, all generations born till 1970 break even or were net winners. Future generations, however, will lose by joining the system. In the studies mentioned above, intergenerational equity measures are developed to change the financing mode of the statutory health insurance so that more intergenerational equity can be achieved. Otherwise one has to fear that future generations will vote against the solidarity type system. This causes a major threat to the future and stability of the German statutory health insurance scheme.

*International Institute of Management,*
*Wissenschaftszentrum,*
*Berlin, F.R.G.*

## BIBLIOGRAPHY

1. Abel-Smith, B. and Maynard, A.: 1978, *The Organization, Financing and Cost of Health Care in the European Community*, Commission of the European Community, Luxembourg.
2. Friedman, B. M. and Friedman, R.: 1981, *Free to Choose*, Avon Books, New York.
3. Geissler, H.: 1981, Fee for Service Compensation of Physicians in the Federal Republic of Germany, WIdO (mimeo), Bonn, 6.
4. Glaser, W. A.: 1978, *Health Insurance Bargaining: Foreign Lessons for Americans*, Gardner Press, New York, pp. 95–116.
5. Hamm, W.: 1980, *Irrwege der Gesundheitspolitik*, Mohr-Siebeck, Tübingen.
6. Herder-Dorneich, Ph. and Schuller, A. (eds.): 1984, *Die Ärzteschwemme*, Nomos, Baden-Baden.
7. Kleindorfer, P. R. and Schulenburg, J. -M. Graf v.d.: 1986, 'Intergenerational Equity and Fund Balances for Statutory Health Insurance', in Schulenburg, J. -M. Graf v.d. (ed.), *Essays in Social Security Economics*, Springer, Berlin, Heidelberg, New York pp. 108–129.
8. Kraft, K. and Schulenburg, J. -M. Graf v.d.: 1986, 'Co-Insurance and Supplier-Induced Demand: What Do We Have to Expect as the Physicians' Response to

Increased Out-of-Pocket Payments?', *Zeitschrift für die gesamt Staatswissenschaft* 142, pp. 360–379.

9. Landsberger, H. A.: 1981, *The Control of Cost in the Federal Republic of Germany: Lessons for America?* Department of Health and Human Services, Publ. No. (HRA) 81–14003.
10. Reinhardt, U. W.: 1981, 'Health Insurance and Health Policy in the Federal Republic of Germany', *Health Care Financing Review* **3**, 1–14.
11. Schieber, G. J. and Poullier, J. P.: 1985, *Comparison of Health Care Expenditures*, OECD, Paris.
12. Schulenburg, J. -M. Graf v.d.: 1983, 'Report from Germany: Current Conditions and Controversies in the Health Care System', *Journal of Health Politics, Policy and Law*
13. Schulenburg, J. -M. Graf v.d. and Kleindorfer, P. R.: 1986, 'Wie stabil ist der Generationenbertrag in der sozialen Krankenversicherung? Zum Problem der Gerechtigkeit und Akzeptanz intergenerativer Umverteilung', Güfgen, G. (ed.): 1987, Ökonomie des Gesundheitswesens, Duncker und Humblot, Berlin, pp. 413–434.
14. Statistisches Bundesamt: 1984, Fachserie 14, Reihe 7.1., Wiesbaden.
15 Statistisches Bundesamt: 1985, *Statistisches Jahrbuch 1985*, Bundesdruckerei, Berlin.
16. Stone, D. A.: 1979, 'Health Care Cost Containment in West Germany', *Journal of Health Politics, Policy and Law* **4**, 176–199.
17. Stone, D. A.: 1980, *The Limits of Professional Power: National Health Care in the Federal Republic of Germany*, University of Chicago Press, Chicago.

MICHAEL A. RIE

# THE AMERICAN AND WEST GERMAN HEALTH CARE SYSTEMS: A PHYSICIAN'S REFLECTIONS

Health care systems evolve in nation states and mirror the cultural value structure of the societies they serve. Professors Buchanan, von Schulenburg and Thust outline the evidence to support such a claim with regard to the American and West German systems. These three essays provide the reader with two nations that approach the funding distribution of personal health care services in seemingly different ways. Yet, both systems are caught in similarly situated dilemmas that are constant for health care in post industrial, high technology health care systems. Whether one is a privately insured American worker or a "compulsory health insured" citizen of the Federal Republic of Germany, high costs of health care are being balanced by complex social and cultural values and individual claims to service consumption and competing visions of allocatable justice. The human instinct of self-preservation drives the majority of citizens to desire health care as a valued commodity to enhance the quantity and quality of mortal existence even when such services may be of very limited benefit. But as Buchanan correctly points out, there will always be other desires in human populations that are competing for the finite resources in private and public sectors of the economy that enhance life and give individual and collective meaning to it.

The German system is stable and mature, having been in existence for 100 years. Von Schulenburg and Thust depict a system of social insurance that dominates the nation and originates primarily by legislation. This has led to a hierarchical system of institutional and individual physician providers purveying their commodities to a complex phalanx of insurance funds that are mandated by national insurance laws. A doctor must belong to the German Medical Association which negotiates fees and disperses them as "pay master" for the funds. The strengths of the German system as defined by Von Schulenburg and Thust have been that economic demands of providers are negotiated by national or quasi-national negotiated fee schedules for services on a fee-for-service basis. When the national economy grows, fees are

*Hans-Martin Sass and Robert U. Massey (eds.), Health Care Systems,* pp. 75–83.

"allowed" to rise by the process of benign political non-intrusion into negotiations. If provider demands exceed economic growth, then the negotiators know that they must temper the rise of fees to avoid political intrusion in the negotiating process. Our German medical colleagues have grown up in a more structured society that thrives on obedience to state authority and permits greater intrusion into matters of personal and professional life than in America. It is a system in which both doctors and patients are similarly nurtured from birth to understand and accept the order of society. It is a beneficent system and includes many social amenities such as spa vacations from work that are unknown in America. It is a system that lays claim to near universal access to health care as put forth in the notion of solidarity. Solidarity binds citizens and providers to "play the game" because it is the only game in town.

Yet it is regulated welfare state that has failed to provide incentives for efficiency of resource utilization and has been used as a political "pork barrel" to subsidize the West German pharmaceutical industry. Doctors who treat non-hospitalized patients are not integrated into the institutional care structure and have major incentives to act in entrepreneurial ways that may generate excess volume of services of ill-defined value and quality to patient consumers. Likewise, the absence of any utilization review or standards of utilization of hospital care results in heavy expenditures and lengthy hospital stays. One might therefore hypothesize that such a system will undergo transformation to a less regulated largess in coming years, because the imbalance between individual claims to care and available resources will become more of a political liability to government officials as the population ages and the productive younger generation's capacity to pay for the older generation declines. With an excess of physician manpower and institutional bed capacity, the system must lower unit fees which drives providers to "unbundle" services into smaller and less meaningful services, or the system must then decrease volume of services by attacking providers for inefficiency (as has been the case in America). The latter has one time value as pointed out by Aaron and Schwartz [1], but only puts off the day when one, out of necessity, must be driven to monetarize individual entitlement of consumption. The latter enjoys no political forum at present in either the F.R.G. or the United States, but one might well speculate that it is coming in both countries but from different beginnings. The notion of "unbundling" was graphically though anecdotally confirmed by a German physician who pointed out how one can increase office visit volume by conducting an in-depth physical examination in several visits as opposed to a single comprehensive office visit at one session. The heart and lungs are examined at one session while pelvic, rectal and sigmoidoscopic exams are conducted at the next visit, and so on. Likewise, patients come to the physician demanding drugs or medical work exemption certificates. These generate office visit fees, irrespective of the medical worth of the service rendered, while the service remains of personal value to the patient who is receiving secondary social gain. Such incentives are seductive ways for underworked physicians to "earn a living." But how can Germany deal with excess provider capacity by policies that would close medical schools, unemploy existing physicians, and unemploy hospital workers, thus leading to stagnation in the health sector of the economy? The answer, I allege, will be by increasing competition which I shall call "the providers' right to die" when such providers' services are no longer desired by the system. Unquestionably other avenues of retraining and redeployment will surface as public policy initiatives, but the German system in its own fashion will come to terms with "the painful perscription" [1]. In so doing, Germany will come to discover the virtues of increased copayments for office visits and other potentially marginal utility services as a way to inspire individual patient responsibility away from "free rider" consumption of communal resources. Are Germans more patriotic than Americans? Will Germans, by virtue of their culture, be more accepting of explicit rationing criteria than Americans? To what extent does a democracy with an authoritarian hierarchical culture permit a government to diminish individual entitlements to health care without political repercussion?

Professor Buchanan attempts to describe the contemporary American health care system but omits its historic origins and the relationship of its history to the social revolution and dynamic changes that are occurring almost daily in the mid-1980s. His introductory passages review the philosophic quandary of defining a just distribution of health care services so as to come to terms with what he calls "equal health care for all." He claims strict egalitarian distributions are ethically unsound, which may be interpreted to mean that they represent one among many incomplete descriptions of distributive justice. As he notes, the quest to give all to everybody in a strictly egalitarian system creates intolerable intrusions on individual liberty, productivity, and the diversion of resources from other valued human enterprises. Accepting egalitarian distribution systems in capitalist nations assumes the validity of the claim that health care is of special importance even if other services and goods are not. Even if such a claim were judged correct by democratic plebiscite, one is still left with the problem that the aggregate productivity of the non-health care economy must support the universal beneficence of the egalitarian system within limits that do not threaten the future viability of the non-health care economy. When these limits are reached, government is left with the burden of creating the moral authority to prioritize human entitlements in the health care sector.

It is perhaps useful to review a few salient points of American history, government, and economics so as to highlight the cultural differences between the American and West German health care systems. The United States is not a single culture but a melting pot of successive waves of multinational, multiracial, and multireligious immigration integrated into a dynamic, secular, pluralist democracy. People came to the United States over the last 210 years to make their own lives in a free land that was relatively underpopulated, rich in natural resources, and permissive in its style of government. To be sure, greed, economic monopoly, and predatory economic practices were in evidence. Yet the society grew, prospered, and the average standard of living and general health improved dramatically. In American society the physician lived at the margin in the economic pecking order prior to the wide-spread use of health insurance after World War II [5]. The work ethic "produced," and individuals could see the results of that ethic. In this environment the social commitment of the working generation to the older generation was at the local level and the Social Security Act was a partial subsidy to the elderly that was easy to fund. In 1935 there were more than 10 workers for every retiree. In 1985 there are less than four. During the last 50 years, when these demographic changes occurred, medicine has moved from a labor intensive cottage industry to a capital intensive technology-based service industry with heavy labor input. The results are the well-known circumstances of high health care costs associated with increasing numbers of older citizens who "consume" vast quantities of health care lavished on them (with or without their consent) by an equal access payment system. Moreover, high technology medicine has created the specter of extremely expensive and questionably useful care not only in the last six to twelve months of life of the elderly but also in the younger population. (Fig. 1)

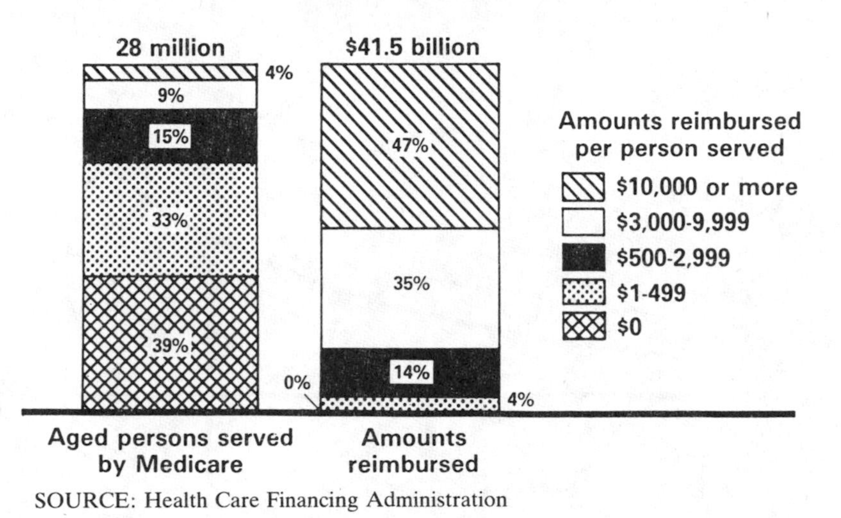


Fig. 1. Graphic demonstration of American Medicare disbursement enrollee cohorts. Note that the top four percent of enrollees consume 47 percent of the budget with two thirds of such expenditures occurring in the last 6–12 months of life (data after Dorothy Rice [4]).

The Federal government created the Medicare program on July 1, 1966 (following the Blue Cross/Blue Shield private health insurance model), which guaranteed health care "entitlement" to those over 65 years of age. This new excess demand fueled the expansion of hospitals as well as physician and other manpower pools, between 1966 and 1977. The proportion of the nation's resources devoted to health care increased as a proportion of the Gross National Product. Unlike Germany, the American health care system failed to curtail its expansion as the economy stagnated during the period 1977 to 1981 (Fig. 2), leading to the rise of government and employer demands to impose price controls on the existing structure. The latter appears to be a major power shift to bulk purchasers of health care, e.g., federal government and large employers via private insurance companies. However, as each payer group argues persuasively that each should pay its fair share of costs for only its constituents, the access problem alluded to by Buchanan becomes increasingly evident and leads to inexorable demands for

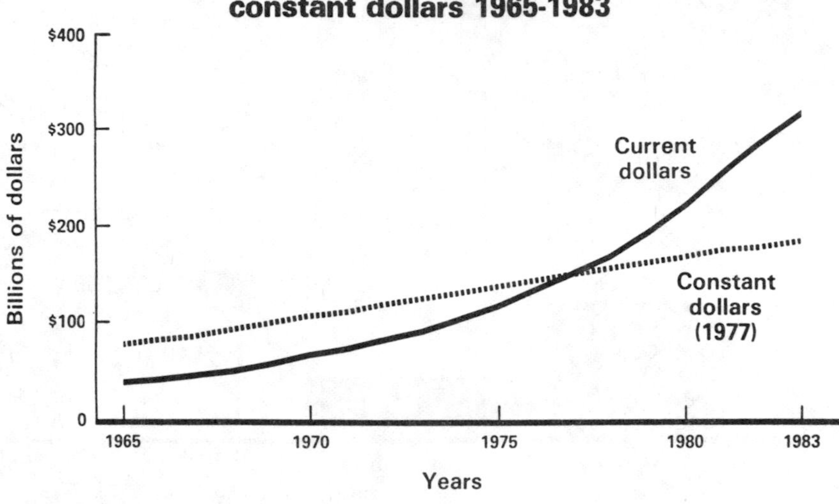


SOURCE: Health Care Financing Administration

Fig. 2. Data from Dorothy Rice [4].

"social justice" and access to health care for all. To be sure, the American system is working for almost 90 percent of the population and has provided the "highest standard of care" in the world to date. The allusion to 30 uninsured million Americans by Professor Buchanan (roughly 12.5 percent of the population) is politically and economically timely in the public debate because the classical methods of cost shifting to insured accounts in hospitals and the traditional "Robin Hood" ethic of office practitioners of old are being challenged by the corporate control that is sweeping the American health care system. The corporate-government purchaser mentality proceeds at this writing on the assumption that the business ethic of competition, free markets, and health care as a commodity purchase are the emerging new order in the U.S. Observers from both the political left and right agree that a government run and financed health care monopoly is unlikely to emerge. The reasons for this situation are left unstated by Buchanan who, like many academic American observers, looks with overpowering criticism at the system's failure to accommodate the unfortunate 12.5 percent of the population who lack a "decent minimum" access to health care because of organizational "cracks in the social welfare safety net." For America in economic transition, I would submit that the issue is: "Should one throw out a very successful, duplicative, and expensive system for a 12.5 percent 'failure' rate or should one increase efficiency by corporatization thus making care more affordable for the least well off, even if this implies greater public funding of health care in a bankrupt nation?" There is already a suggestion that corporatization may be achieving regionalization of high technology medicine, whereas government planning without allocatory power has been a well-documented disaster for nearly 20 years.

Buchanan appears to have overlooked the powerful celebration of individual autonomy that Americans demonstrate in all aspects of their individual lives. Even a socialistically inclined people like those in Massachusetts (who recently voted 72% to 28% in favor of a national health plan) oppose laws mandating the use of seatbelts in automobiles as an unwarranted intrusion by the state into private matters. This theme of autonomy and public zeal in the pursuit of individual health is perhaps best captured in the following remark which spans two centuries of sociologic observation: "To be sure, many observers, beginning with de Tocqueville, have remarked that Americans are singularly concerned with their individual well-being. Since the 1830s, when de Tocqueville visited America, the United States has been swept by a series of popular movements concerned with improving health variously through diet, exercise, moral purity, positive thinking and religious faith. Today, were a revived de Tocqueville to observe Americans jogging in parks, shopping in health foods store, talking psychobabble, and reading endless guides to keeping fit, eating right and staying healthy, he would probably conclude that, if anything, the obsession is now more pronounced" ([5], p 7).

What has thus far occurred in the American health care revolution is preliminary and difficult for either Buchanan or this commentator to analyze within an ethical framework. It does however appear that the power transformation to corporate structures of health care can only raise the question of what ethos physicians should abide by when they are employed and controlled by corporate structures [4]. Corporations may well achieve efficiency at the price of physicians' freedom that Buchanan might find unacceptable in terms of the value of preserving patient-doctor Hippocratic trust relationships. Yet there is nothing in corporate philosophy that precludes a corporation from choosing to circumscribe its business prerogatives around a new vision of "corporate medical ethics" so as to demonstrate to workers and other citizen consumers that it is a corporation that is concerned with individual health care; in short, a *virtuous* health care corporation. Finally, I will argue that corporatization of health care is politically advantageous to the U.S. Congress that desires to "get out of the health insurance business" for the elderly. Medicare is not an annuity but rather a transfer payment program from young to old. By telling older Americans to go out in the marketplace and spend wisely with what is available as a transfer payment, elected officials are able to avoid the moral issue of societal resource transfers from young to old, given the presently unfavorable demographics. America has always celebrated capitalism, youth, and productivity and relatively downplayed intergenerational duties at the public level [3]. This historical fact was best captured by the Honorable Richard Lamm, Governor of Colorado, who spoke of the elderly's "duty to die." Though he was castigated by many, the historical and philosophic roots of this issue are usually neglected by commentators but have been well summarized by Battin [2].

It is the historical context of both nations as well as the transformation of the post-industrial world from isolated nations to world economy participants that has led us to the present. A global pandemic recession with decreasing financial velocity and decreasing occidental birth rate will surely lead us to uncharted waters in the bioethical issues which surround the economics of health care. How Germany and the U.S. respond will depend on the cultural roots and values of their health care systems and on the realization that high-technology medicine will require not only the energies of finite human beings with finite communal resources but also demand agonizing new choices about individual entitlements to health care as one among many goods that fulfill human existence.

*Massachusetts General Hospital,*
*Boston, Massachusetts, U.S.A.*

## BIBLIOGRAPHY

1. Aaron H. J., and Schwartz, W. B.: 1984, *Rationing Hospital Care*, Brookings Institution Press, Washington, D.C.

2. Battin, M.: 1987, 'Age Rationing and the Just Distribution of Health Care: Is There a Duty to Die?' *Ethics* **97** (2), 317–340.
3. Daniels, N.: 'Am I My Parents' Keeper?', in *Securing Access to Health Care*, President's Commission for the Study of Ethical Problems in Medicine and Biomedical and Behavior Research. Appendix K, pp. 265–291.
4. Rice, D.: 1986, 'The Medical Care System: Past Trends and Future Projections', *New York Medical Quarterly* **6** (1), 39–70.
5. Starr, P.: 1982, *The Social Transformation of American Medicine*, Basic Books Inc., New York.

J. F. LACRONIQUE

# SOCIALISM, EQUITY, AND COST CONTAINMENT IN HEALTH: THE FRENCH EXPERIENCE

## I. INTRODUCTION

In May 1981, 25 years of "Gaullist" politics ended following a presidential election that gave to the socialist candidate, François Mitterrand, a slight but decisive majority. The parliament was then dissolved, and a new chamber was elected which gave to the Socialist party an absolute majority, and the possibility of an unprecedented total power to foster a real change in French politics.

During his campaign, the candidate, François Mitterrand, had proposed 110 "resolutions" that comprised a full political program. Among these propositions, about ten were in the field of health care, which provided answers to several questions that generated controversies, especially the linkage between the rate of growth of health expenditures in comparison to that of the G.N.P. imposed by the government after 1979.

The general philosophy of these answers was to consider that the economy could be stimulated by incentives to consume, rather than by any rationing mechanisms. The "left" was thus permitted to conduct a policy that would completely change the pervasive idea that cost-containment was inevitable in any industrialized country [1].

This essay is an attempt to explain the basis for this policy and its evolution during the five years of socialist experience in France.

## II. THE FAILURE OF A NATIONAL AND A UNIQUE HEALTH INSURANCE SYSTEM

At the end of World War II, France established a system of social security that was partly inspired by the socialist experience under the Leon Blum government in 1936, and partly derived from the Beveridge concept of a national health service that was at that time applied in Great Britain [11]. At its founding, the system was intended to be a general, national, and comprehensive system that could conciliate private medical practice and national health insurance [17, 14].

This principle has been respected. Although the scheme for coverage

*Hans-Martin Sass and Robert U. Massey (eds.), Health Care Systems,* pp. 85–96.

is the same for about 98 percent of the population (everyone having a definite role in the society – including clergypeople, artists, and even prostitutes – are entitled to social security as long as they pay for membership), there are still several "programs" of coverage in different professional branches (e.g., Electricité de France, railroad workers, civil servants) with some special privileges.

But the system has never been unified, and remains divided into one main system called "Régime général" – which covers 75 percent of the population (salaried employees), and several minor systems, according to various types of work (e.g., agriculture, tradesmen, etc.) [8].

The scheme for coverage is simple. The patient directly pays the health care provider, but is secondarily reimbursed for his or her medical expenses [18]. The reimbursement is not total: there is a co-payment requirement that can be reimbursed by another private system called "mutuelle," which only 20 percent of the population uses. The amount of co-payment varies greatly according to the type of care one receives, the institution used, and the total amount to be paid. Another feature of the French health care system is *pluralism*: The patient can select the physician and the institution he prefers and is guaranteed to have equal access to any aspect of the health care system under similar conditions. The major determinant for choice, then, is not the price of care, but the reputation of the provider and institution for quality care [4, 9].

This system is certainly highly appreciated by the French population, since it combines freedom of choice and is based upon egalitarianism [17]. But it is also a relatively expensive system: in 1984 the health care expenditures represented about 8.5 percent of the G.N.P., and the administrative cost of running the social security system is now about 10 percent of the total amount [10].

The basic intention of the socialists was to simplify the system, making it one unified scheme, thus dissolving numerous smaller programs. In reality, there was no attempt to achieve that goal, for that would have challenged the privileges of a number of social groups. In November, 1985, on the 50th anniversary of Social Security, the President declared that there would be no change in structure for Social Security. This declaration put an end to several propositions proffered by different groups – largely inspired by the United States' H.M.O. experience – that would stimulate competition between different health insurance plans.

## III. THE INDIVIDUAL/COMMUNITY PARTICIPATION IN HEALTH EXPENDITURES

The French social security system, then, is a private institution that provides a public service. The resources are directly collected from the wages of all workers through a complex scheme that has often been changed during the past ten years in terms of the direction of redistribution: those with higher salaries pay more for the same protection than those with lower salaries. The funds derived from participation in social security are allocated to a number of different special funds: the most important is a sickness fund, which is totally independent of the public budget. However, the traditional gap between the budget and the real expenses is covered by the national budget. The socialists have often criticized this system claiming that it only exploits the work force, and draws on no other financial resources. The strategy was to "fiscalize" part of the health budget, which would then allow the Government to derive revenues from all sources of benefits, including profits on investments and taxes on alcohol and tobacco.

One year after the implementation of these taxes, the French Government was ordered by the European Community Court to level them off and to pay a fine, since these taxes discriminated against imports of foreign products and did not restrict the production of French wine and tobacco in any way.

The plan to create a "health budget" as part of the national budget was studied in the first two years of the socialist government. An "exceptional social participation of one percent" was imposed on all taxpayers in 1983, but was then abandoned.

In 1984, a new attitude toward reducing the fiscal burden prevailed, and no real change is now noticeable in the repartition of the charges. Thanks to a massive and successful effort to reduce the rate of growth of health expenditures, the budgets in 1984 and 1985 for social security showed no deficit for the first time in 25 years. This unnoticed result, although it was not announced in the "socialist program" before 1981, will certainly be considered in the future as the most positive reform in terms of general economic policy [2, 12, 13].

## IV. SUPPRESSION OF THE SYMBOLS OF CONSERVATISM

At the onset of the establishment of French social security, the idea that the consumer should always be aware of his expenditures was built into the bylaws of the system. According to this principle, no insurance system could entirely cover the totality of the bills, and a modest co-payment should thus be borne by every individual. This principle, called *ticket moderateur* (since it was aimed at moderating the amount of consumption of care), was not applied until 1980. At that time, a program imposed by the last Gaullist government to curb the health care cost was revitalized and this old and virtually forgotten principle was made applicable at once.

This single measure had unforeseeable consequences: the private insurance schemes, requiring very low cost (called *mutuelle*), believed that they would be out of business if they could no longer offer the full coverage for the co-payment [11].

Before 1980, the *mutuelles* performed only marginal services, with relatively modest public influence. But these *mutuelles* existed before the social security, and thus were considered merely historical landmarks. The mere attempt to reduce their role stirred an unprecedented move of solidarity among the members, and the President of the Republic received millions of letters of protest against this "aggression." The government quickly retreated, and never tried to implement the *ticket moderateur d'ordre public* (T.M.O.P.); but this simple affair clearly had a negative influence on the following Presidential election. The T.M.O.P. was one of the symbols of conservatism that was announced would be abolished should the left emerge victorious. Of course, the suppression of the T.M.O.P. was easy to support, and marked the evident "re-birth" of the mutualist movement that is now considered a respectable political force.

Other symbols of conservatism were not as easy to overthrow. The most important example is the *Ordre des médecins*. This institution is a body of representatives of the medical profession, which reviews all ethical and disciplinary aspects of medical care. It was created during World War II by the government that collaborated with the German occupation, and the "left" believed that it served only a corporatist ideology [3]. In the past the *Ordre des médecins* stood for conservatism in the discussion of the law of 1975 that made abortion possible in

France under a Gaullist government (at the time, the socialists supported the government against its own majority).

One of the propositions of the socialist candidate for the Presidency was to dissolve the *Ordre*, without offering any alternative, since the mere idea of "orderliness" was unacceptable. When the socialists came to power everyone, including the leader of the institution, thought that the *Ordre des médecins* was living its last days. But there was no attempt to rescind it, since no one wanted to take the risk of controversy over an issue that did not really matter. During the five years of socialism, the *Ordre des médecins* continued to live, and there is no sign that its survival is in question.

A final symbol of conservatism is found in what remains of the private sector in public hospitals. Since 1965, full-time physicians in public hospitals possessed the right to a limited private practice in their specialty services. This "right" was largely exploited by a small number of physicians, and there was evidence that regulations had to be established and implemented [3]. The government (1980) tried to create a system that would allow fees-for-service in the private sector to be collected by hospital administrations, but President Giscard d'Estaing issued personal orders to postpone the effective date of the regulation.

Under the socialist government, then, the proposition of suppression of any private practice in public hospitals has been adopted by law, but it will not be applied before 1987, which leaves time for a new government to defeat the bill. The political right has always demonstrated its total opposition to any suppression of the private sector in public hospitals, and there is a strong chance that the resurrection of it will be one of the first measures that a rightist government will undertake in case of a change.

## V. THE IDEOLOGY OF "COMMUNITY HEALTH SERVICES"

A proposal to create "community health centers" has been one of the most controversial issues during the past five years. The idea was conceived from an experience in the city of Grenoble, where several salaried physicians offered free health services as well as educational programs in health maintenance and prevention. Financially, it was not obvious that the model would be acceptable; the majority of the physicians in the region of Grenoble criticized its underlying ideology on the grounds that it created a competition biased against private practice with virtually no concern for the quality of care.

The socialist party, through its health branch (*Association socialisme et santé*) pressed its concept of community health services which was greatly inspired by the Québec experience of *Centre locaux de services communautaires* (C.L.S.C.) in Canada.

The notion was to create community centers where all kinds of community services could be linked – including social workers, dentists, nurses, and psychologists – under a financial scheme of contracts for "programs" based on the principle of participative management by objective. The idea was, in theory, most attractive and could have become the new spinal cord for a complete change in the ideology of the French health care system, based as it had been on pure liberalism [5].

In the first year of the socialist administration, one or two "health centers" were created which produced very bitter local and national controversies. Thus, the plan to extend this model experience was postponed year after year on the "grounds" that it was too important to be implemented all at once. In official speeches, different ministers spoke about "experiences" that should be completed and carefully evaluated [16]. Yet everyone is convinced that there will be no longer any attempt to establish additional community health centers.

There is no need in France for the creation of new health institutions, since the system suffers much more from excess development than from underdevelopment: there are too many doctors and too many hospital beds. The real need is to reallocate the resources rather than to create new institutions. Moreover, the image of public institutions has declined in the mind of the general public, as in most Western nations, and the French government would not dare to suggest an unpopular reform that would surely yield very minor benefits. There is little likelihood that there will be any additional community health centers in the foreseeable future.

## VI. THE AMBITION OF A HOLISTIC HEALTH REFORM

To understand what happened during the first two years, one must appreciate that the Minister of Health, Jack Ralite, was a communist, one of four communist ministers in the socialist cabinet. Ralite began by visiting all parts of France, giving speeches on all health issues, and making apparent that he would become a very active and audacious minister of health. The first six months revealed no difficulties, and Ralite was apparently quite popular until he proposed a very ambitious reform that would drastically change the whole health sector, including medical research, industry, mental health, and medical education.

### A. *The Reform of Medical Education*

During the past ten years, the French government had been committed to reform its medical education in order to make it compatible with other European educational systems. The Rome free exchange treaty stipulates that freedom of circulation is guaranteed to all E.E.C. member states, and this freedom applies to medical doctors as long as they have comparable educational backgrounds. But the French educational system had a special feature that made it incompatible with other systems: in France there were *two* ways to become a specialist:

The first way was to compete for public *hospital* positions – for one to become an "extern" then an "intern." This competition was opened in the largest French cities, but there was no national equivalence: once a student obtained a position in Marseilles, for example, there was no way for him or her to obtain one in another city, except to initiate the selection process all over again. By tradition, this system called *Concours de l'Internat*, installed by Napoleon in the early 19th century, is considered an excellent one, and the French aristocracy of clinical medicine has generally been selected without exception by this process.

The second way was via the university. In France, one can pursue the way of the scholar, based on theory rather than on practice. The selection is not as difficult as it is in the hospital *Concours*, and the quality of the diploma is not as respected as is the first one. Traditionally, in clinical specialties, an "intern" was automatically granted qualification after having completed only the internship, even though he would never have had to pass any examination at the university.

This two-way system is unique among European nations, and it was therefore impossible to make it equivalent or truly comparable to any other European diploma. A reform was needed that could make the French educational system comparable to the others. Hospital training and university courses had to be combined into a single system based upon a selection process that would be national in order to facilitate exchanges between cities and countries. Before the change in government in 1981, the project was completed, but it was widely criticized by both the communists and the socialists on the basis that the newly-proposed system did not really pay sufficient attention to upgrading the general practitioner.

After the political change of 1981, the entire set of official laws was turned down, and a new reform was proposed. But at the time the two ministries of education and health were in competition for the management of medical-education reform. Very rapidly, good manners led to open hostility, and the two cabinet ministers revealed that they could not agree on an overall philosophy for the new reform.

This open dispute served to disrupt the prior confidence of the students, especially the current "interns" who were anxious to see their positions recognized as privileged. In October 1983, a strike began that lasted about two months, culminating in the resignation of the communist minister, Ralite; subsequently, a "mediator" – a personality who was not expected to be influenced by any political force – was appointed to negotiate the conflict; the reform that was then adopted was even more conservative than the one inherited from the previous government. After this rather bitter episode, there was no longer any difficulty, and the new education reform is still functioning without any major problems.

It should be mentioned that the major change in medical education is the drastic drop in the number of medical graduates from about 10,000 in 1978, to less than 5,000 in 1984 [15]. Thus the university is in a very unstable situation: the gradual decline in the total number of medical students makes it difficult to fill positions in the universities, and the development of new programs, especially in public health, is virtually at a standstill with the general freeze on new public positions.

## B. REFORMS IN THE HOSPITAL SECTOR

In 1965 and in 1970, two major laws concerning hospitals were passed by the French government:

(1) The 1965 law (*Loi Debre*) served to install full time positions in public hospitals – highly attractive positions that gave to the public sector the possibility of a transformation into "university hospitals," where patient care, research, and teaching could be performed in all services on a mandatory basis.

(2) The 1970 law (*Loi Hospitaliere*) helped to transform old, ill-equipped hospitals into modern institutions, and provided as well real

power to the directors of the hospitals. But the law also provided that the government would have control over the creation or designation of any hospital bed or any costly technical equipment.

Standards and quotas would be established to achieve a more equitable distribution of the facilities and equipment throughout the country. *Loi Hospitaliere* worked quite well in playing the role of a rationing device. But it also had its perverse effects, and the socialists and the communists were very much opposed to the consequences of this law on the distribution and employment of certain medical equipment, such as whole body scanners. France was, in 1980, the only European country where the "adequate ratio" was achieved: one whole body scanner for one million people.

In 1981, it became evident that the government would leave to the future a new law that could serve as a new historical landmark [7], which would encompass all aspects of hospital life, including the careers of doctors, all budgeting principles, the internal organization of hospitals, and methods of evaluation [6]. In 1985, a number of decrees were passed that dealt with all these subjects:

(1) The *global budget* in the public hospital has been generalized since January, 1986. Accordingly, each month social security provides an allocation that represents 1/12 of the annual budget. Each hospital is entirely responsible within this budget for its expenses for all Social Security members in the geographical area designated for hospital. The traditional fee-for-service system still applies, however, for all patients who come from outside the area.

(2) A *mandatory co-payment* for ancillary care has recently been imposed on a flat-rate basis. This measure could never have been accepted under a conservative regime, and would have generated furious opposition; moreover, there was almost no comment on it except from the communists.

(3) One single *career development scheme* is available for all doctors in public hospitals, substituting the previous system that made a difference between numerous categories of institutions. The system is now applied without any trouble.

(4) A system based on *Diagnosis-Related Groups* (as established in the U.S.) is presently installed under the initials PMSI (*Programme de médicalisation des systemes information*), with the objective of linking the information system in a hospital more closely to all its medical

activities. However, this system is not linked to the social security system and global budget, for it is solely intended to be an information system, and it does not direct a prospective reimbursement scheme.

(5) A *department organization* is now possible that will give greater flexibility to the management of services by an elective process that dissociates the functions and the grades of physicians. Under this organization a chief of service would no longer keep his mandate for a lifetime, but would be elected by his peers and other personnel, but on a temporary basis. However, this organization is not mandatory and is proposed as an "option" on a voluntary basis. A boycott of the provisions of this law has been organized by a majority of hospital doctors, and hence the government has not had to enforce it.

In short, this set of reforms is not complete. It is, however, an attempt to achieve greater coherence between different components, but it is much more management oriented than politically sensible. It could not be compared to previous reforms that had limited but precise direction. The image of a "monument" (which was the initial intention) is hardly established, even among those interest groups that are closest to the Socialist party.

## VII. CONCLUSION

After five years of socialism in France, no longer can anyone seriously advocate the policy that has been followed since 1983 as being inspired by socialist ideology. All the objectives initially set forth in the domain of health have been "postponed", in order to give priority to the general economy. From the expanded 23 percent annual rate of growth in health expenditures in the late 1970s, the French health care system has had to adjust to a reduction that does not exceed a rate of growth of eight percent per year. The fees of private physicians, the price of drugs, and hospital budgets are all severely controlled; furthermore, the rate of their growth does not coincide with the inflation rate, either. This policy is, of course, not popular, and the polls indicate that in 1986, 75 percent of physicians would vote for the rightist opposition.

The real consequence of this regulatory experience is a shift of public opinion against the socialist ideology regarding health does not any longer have any significance. It would be hard to find a single article advocating health policy in terms other than those which reflect a spirit of "real-politik" where any tribute to cost-containment is passively accepted. Even the communists cannot criticize this policy, since the general director of health (an equivalent position to the surgeon general of the U.S.) is a member of the Central Committee of the French Communist Party and has been, from 1982 to the end of 1985, a full member of the group that held political power. In the forthcoming years this radical dissociation between ideology and application will leave the health sector without any argument from the left. If there is a change of government, the right would have an almost free hand to conduct a conservative policy without taking any risk of criticism from the left.

*Institut Curie,*
*Paris, France*

## BIBLIOGRAPHY

1. Altenstetter, C.: 1980, 'Hospital Planning in France and the Federal Republic of Germany', *Journal of Health Politics, Policy and Law* **5** (2), 309–332.
2. Culyer, A. J. and Horisberger, B.: 1982, *Economic and Medical Evaluation of Health Care Technologies*, Springer Verlag, Berlin.
3. Escoffier-Lambiotte, C.: 1979, 'La santé des français', *Le Monde* (dossiers et documents), Paris.
4. Gallois, P. and Taib, A.: 1981, *De l'organisation du système de soins*, La Documentation Française, Paris.
5. Gros, F., Jacob, F., and Royer, P.: 1979 *Sciences de la vie et société*, La Documentation Française, Paris.
6. Hofman, P.: 1983, *La réforme hospitalière: bilan et perspectives*, Rapport au Conseil Economique et Social, Paris.
7. de Kervasdoué, J.: 1982, *Evolution des dépenses de santé: Communication devant le VII Commission du Plan*, le 25 Novembre, VII Plan, La Documentation Française, Paris.
8. Lacronique, J. -F.: 1984, 'France', in *Comparative Health Systems*, Raffel, M. W. (ed.), Pennsylvania State University Press, University Park, Pennsylvania, pp. 258–285.
9. Lacronique, J. -F. 1982 'The French Health Care System' in *The Public/Private Mix for Health*, McLachlan, G. (ed.), Nuffield Provincial Hospitals Trust, London, pp. 267–288.
10. *La santé en France: 1984, rapport au Ministre des Affaires Sociales*, La Documentation, Française, Paris.
11. Launois, R. and Le Touze, D.: 1982, 'The Role and Responsibilities of the Various Agents of the Health Care Delivery System: The French Scene' (paper presented to the H.E.S.G. meeting at Brunel University), Uxbridge, England.
12. Levy, E.: 1982, *La santé fait ses comptes: Une perspective internationale*, Economica, Paris.
13. Maxwell, R. J.: 1981, *Health and Wealth: An International Study of Health Care*

*Spending*, Lexington Books, Lexington, Massachusetts.
14. Maynard, A.: 1975, *Health Care in the European Community*, Croom Held, London.
15. Poullier, J. P.: 1984, 'Démographie des professions de santé dans les pays de L'O.C.D.E.', *Cahier de sociologie et de démographie medicales* **4** (24), 321–336.
16. Ralite, J.: 1982, 'Discours sur les réformes hospitalières', *La santé pour chacun* (Decembre), No. 8, pp. 1–8.
17. Rodwin, V. G.: 1981, 'The Marriage of National Health Insurance and Liberal Medicine in France: A Costly Union', *Milbank Memorial Fund Quarterly* **1** (59), 16–42.
18. Thorsen, L. C.: 1974, 'How can the U.S. Government Control Physicians' Fees under National Health Insurance? A Lesson from the French System', *International Journal of Health Services* **1** (4), 49–57.

MAURICE A. M. DE WACHTER

# ETHICS AND HEALTH POLICY IN THE NETHERLANDS

The Dutch health care system is in for substantial changes; the people, the professionals, and the politicians know it and have participated in it. Dutch policy-making in general has a long tradition of intertwining collective and individual responsibility. This paper attempts to freeze the process of health policy-making in the Netherlands. It halts the moving pictures in order that a close look may be taken at its components, particularly the ethical ingredient in health care policy.

For the reader to appreciate this snapshot it is essential first to have some idea of what happened earlier [8, 21]. Cost containment, for instance, is not a new issue of the 1980s. Right after the Second World War the Dutch government asked for restraint of all collective expenses, especially in the field of medicine. The aim was to accelerate investments for industry. While the Dutch were watching their pennies, Germany was taking off on its *Wirtschaftswunder* and France, as many other countries in Europe, lived through *les trentes glorieuses*, the thirty years of the postwar boom. By 1960 Dutch society had reached a sufficient level of affluence to allow for a badly needed explosion of health care facilities and medical technology. This led to building hospitals, introducing the latest equipment, and training professionals. Then came the recession with scarcity of resources and a need for their reallocation. This state of affairs in the mid-1970s was reinforced by the concern about diminishing returns; the moment when greater financial investments failed to produce more health was decisive.

Health economists of the Dutch government had chosen by 1975 to follow the Canadian model for the development of a new perspective on health care [9]. This well-known four-tiered model complements the traditional provision of health care with exogenous and endogenous factors such as environment, genetic and biological factors, and life style. Health is the focal point of this model, and, therefore, it allows policy-makers to set goals in terms of health improvement ([21], p. 1523).

Dutch policy-makers must reckon with the fact that health care in the Netherlands is an amalgam of a large number of bodies and persons

*Hans-Martin Sass and Robert U. Massey (eds.), Health Care Systems,* pp. 97–116.

performing relatively independent functions. It cannot be said that the country has a specific system of health care. Generally, the role of the government is limited to supervision and legislation to assure the quality of care. Private enterprise plays a dominant role in the area of therapy, and most health care institutions are in private hands. Thus, the Dutch health care system supports the theory that regulatory and marketing strategies are not necessarily like water and fire. Legislation, however, offers the Dutch government a firm grip on the distribution of health care, especially when costly and unusual equipment is required. In 1971 the Hospital Facilities Law was introduced. This law states that construction of buildings for in-patient care is to be allowed only with the approval of the Minister of Health. The same law gives the government the power to control the functions of hospitals. More important yet is Article 18 of this law giving the Minister control over specialist facilities and equipment. He may stipulate "the procedure for determining the need and the granting of licenses" (Article 18, §2C). Under this law a number of technologies have been made subject to Article 18, including hemodialysis and renal transplantation, megavolt therapy, and computed tomography ([5], pp. 17–44).

The cost of these technologies, together with the heavy expenses of heart surgery, accounted – in the Netherlands as in other European countries – for less than two percent of the total cost of health care during 1980. According to de Groot and Visinescu, Dutch health care costs should not be cut by "big ticket" technologies but rather by examining the "ordinary" ones. The growth in the use of ordinary technology is probably a consequence of the increase of medical specialists and their more frequent diagnostic and therapeutic interventions ([5], p. 61). At the same time, public attitudes and expectations have soared in the Dutch population, which wants no cuts even in expensive treatment unless it fails to improve the quality of life. The 1982 Report for the European Economic Community includes a survey of public attitudes; some of the findings were as follows:

(1) A majority of the Dutch population considers that cuts in health care should not be made. If economies are necessary, then the salaries of medical specialists should be cut.

(2) It is considered that the individual has a right to high-cost technologies, even if the sum involved is a million guilders for a heart transplant (68 percent of respondents).

(3) Expensive operations which allow the patient to live for only a

short time or lead to a vegetative existence are rejected by 50 percent of respondents ([5], p. 59).

This paper presumes to capture the image of the Dutch health policy debate in that it records the government's request for advice about limits to care, then reflects the outcome of the public debate, particularly its ethical components, and finally reports some initial efforts to respond to the requested ethical advice in policy making.

Statistical information taken from the Central Bureau for Statistics 1984 [3] can be found in the appendix to this paper.

## I. THE LIMITS OF HEALTH CARE: A REQUEST FOR ADVICE

The government in its effort to create appropriate conditions for health care must ask for advice and await the recommendations of advisory bodies [23]. A case in point is the 1984 request made by the Secretary of Health on "Health Care Limits." This request was made to the two major advisory bodies, the Health Council and the National Council for Health.

The Secretary of Health distinguished five types of limits to health care:

(1) The determination of what does and what does not belong to health care: this aspect concerns the definition of health care, e.g., when is health care no longer medical care but rather social service or pastoral care.

(2) The limit of the ethically permissible: this aspect tries to define the area between the upper limit of optimal care to be aimed at and the minimum of health care which is due to all, e.g., in questions of life prolongation with seemingly no benefit. Additional problems arise when expensive treatment of one patient means that others may be denied their less expensive therapy. The minister suggests that decision-making in these matters must reflect norms and opinions existing in society. These, however, may differ among groups or change within the same group over a period of time.

(3) The determination of what is or is not effective therapy: this report compares scientifically demonstrable effects with social expectations (needs) and with social cost-effectiveness, e.g., what is the limit for introducing advanced medical technology? The minister explicitly asks how quality of life would play a role in these matters.

(4) The limit of what can and cannot be done financially, e.g., should

high technology be given to a few patients at the cost of many others who will not receive the care they need?

(5) The limit of the state authorities intervening in matters of health care: the objective here is to identify policy issues such as quality control, accessibility, and financial coverage.

The underlying assumption of all five points is that budget cuts are necessary because of a lack of funds. The five points themselves represent ways that may lead to cutting down the current and foreseeable health care costs. While all of them are also legitimate questions in times of plenty, there is an additional bite in times of scarcity. There may be a danger that these points receive attention only because of the time of dearth we now live in. Indeed, some ethicists worry that this may lead to the narrow viewpoint which sees financial resources as the proper basis and justification for limiting health care ([16], p. 32).

The public, as well as various professional groups concerned with the secretary's request for advice, have made considerable efforts to help in finding a solution. In the fall of 1984 a national convention was held on "The Limits to Health Care." Both medical and ethical criteria raised in this convention deserve our attention.

*Limits and Definitions of Health Care*

The introductory speech of the convention pointed out that the debate about limits to health care in the Netherlands had become a public issue long before the financial squeeze was on [24]. Psychiatric hospitals and institutions for the mentally retarded had come under fire during the early 1970s for overtreating patients, sometimes against their will. Then general hospitals had been criticized for unlimited therapeutic efforts. Finally, the nursing homes were called to task. The growing dissatisfaction with health care had already led to questions about the limits of health care. Then came the financial limitations and budget cuts, making the choices harder, at times even tragic ([1], p. 5).

The question about what does and does not belong to health care was approached by a medical sociologist [25]. This speaker tried to distinguish on the one hand the area of formal health care for which society should be responsible both financially and organizationally, and on the other hand the area of informal care where the individual citizen should carry the financial burden.

Four criteria, so it seems, would help to delineate the areas of formal *vs*. informal health care: They are: (1) the dividing line between the quantitative and the qualitative aspects of life; (2) the dividing line between biotechnology and psychotechnology; (3) the dividing line between what the individual and his immediate resources would be able to handle or not; (4) the dividing line between functional and structural changes.

The moral implications of the four criteria are obvious. The first criterion, for instance, rates survival higher than quality of life and, consequently, has society pay for treatment of organic dysfunction, but not for therapy of psychological difficulties, e.g., communication, marital problems, labor conflicts, etc.

The second criterion discerns types of intervention which correspond to the areas outlined as either quantitative or qualitative. Biotechnology covers surgery, drugs, radiation, etc., whereas psychotechnology concerns counselling. The former is paid for by the community, the latter is not.

The third criterion distinguishes self-help from help which only others can provide. Whenever a valid alternative to formal help can be found through self-help, one could argue that this alternative should be chosen if one wants to re-allocate scarce resources. Contraception, abortion, and numerous other forms of fertility-intervention might then be barred from the communal health provision list.

The fourth criterion would solve the embarrassing problem for the physician who "finds nothing wrong." He may then suggest that the patient re-examine his environment and the way he relates to people.

*What is Cure?*

Reflecting more fundamentally, one of the speakers of the fall 1984 convention addressed the question of how to define "cure" within the frame of reference known as "healing" [17]. He first defined five such frames of reference, each of them described by such characteristics as what is attainable, or what is permissible, or necessary, or thinkable, or feasible. The best known example of this theory is Thomas Kuhn's "paradigm of scientific thinking" where a given characteristic, viz., what is thinkable sets the framework within which all acceptable science must be bound. How does this apply to medicine?

For decades now we have learned that pathogenesis, the development of disease, is undoubtedly influenced by a person's failure to succeed or adapt at various levels of human relations.

Today's curative health care services, however, are still dominated by the Cartesian frame of reference. Within this context all of reality can be located, measured, and predicted. Nothing less than a radical change of this frame of reference is needed. General systems theories might be of help in bringing about this new frame. The Cartesian mind-set in medicine has resulted in diminishing returns: more prevention and more cure do not necessarily lead to further gains in health. Only a change in the way we think about health and disease may increase our grip upon reality.

Yet another theme of the public debate is the limited right to health care, and linked to this right is the government's role. Health care rights are limited by availability. The right to health care is a *social* right which entitles citizens to share in services to the extent that they are available; scarcity of resources, therefore, reduces the social right to care. On the other hand, the right to health care is also an *individual* right which does not fluctuate with available resources; government and authorities may not override this right by arbitrary restrictions.

Health care rights are also related to the objectives of health care. Since health care aims at solving problems of health, one must question the value of many interventions employed around the beginning and the end of life. Elsewhere too much diagnosis and even therapy become morally questionable because these practices may diminish services which others need.

What then is the government's role in helping to achieve health care rights? The Dutch Constitution (Article 22) expects the authority to play an overarching role of supervision and coordination. In the welfare state of the past several decades this role grew considerably, but with the limits of the welfare state in sight we must ask whether and how government intervention too should be limited.

In the Netherlands the expenses for all collective services went from 31.5 percent of the GNP in 1950 to 65.4 percent in 1980. Disability and invalidity compensations increased from 300,000 cases in 1972 to 670,000 in 1980; unemployment went from 70,000 in 1969 to 320,000 in 1981 ([15], p. 612).

Under these circumstances a trend to self-regulation in matters of health care expenditure has been strongly encouraged: patients and providers must restrain their demands. Moreover, the government's policy has made considerable efforts to control expenses. Government regulations should focus on three specific aspects of the right to health care: competence, territorial accessibility, and financial accessibility. The first is quality control which seeks to assure equal levels of health care provision throughout the country as well as at international levels of cooperation. The other two dimensions concern the role of the government regarding social justice and the distribution of health care through such mechanisms as health insurance. Even private systems of health care payments will need careful regulation of individual claims.

## II. ETHICS AND THE LIMITS TO HEALTH CARE

What is ethically justified in setting limits to health care? Even ethicists agree that, under the present circumstances, the need to limit health care costs is clear and is in itself no topic for moral debate. Rather, the ethical issue lies in the critical and responsible assessment of human values and interests which are at stake when re-allocation decisions must be made. The specific question then is about the consequences which such decisions may have for equity and solidarity among citizens ([16], pp. 33–34).

A few years ago a more extensive elaboration of this approach was given by Schuyt [15]. This ethicist asserted that a just distribution will occur when our welfare society strikes a new balance between individual and collective responsibility, as well as between individual and communal (family) solidarity. The main reason why this balance has been lost in the Netherlands and elsewhere is to be found in the unrealistic expectations which citizens have had about what the welfare state would provide. These expectations grew far beyond the original purposes of the welfare state; thus, for instance:

(1) the original struggle against injustice turned into expectations of a fully just society;

(2) the struggle against disease and poverty turned into the expectations of a totally healthy life and of riches for all;

(3) the struggle against lack of well-being became the need for optimal well-being;

(4) the struggle against lack of self-esteem turned into deciding for others what this self-esteem should mean.

These outgrowths of the welfare state can be corrected. We must prevent collective care from leading inevitably to the dominance of the state's responsibility and to the neglect of personal responsibility.

Relative deprivation should not lead necessarily to excessive preoccupation with one's own rights. It is conceivable that people will come to realize how much worse off others are and, therefore, accept more readily their own misfortunes. It is also conceivable that people will learn to see the link between their own choices and their financial consequences when part of their health bills are left for them to pay.

Along the lines of Galston's critique of those modern societies [4] which have lost the sense of appropriate limits, Schuyt recognizes immoderation of expectations as both morally questionable and practically dangerous. A new awareness of limits, set by society and nature alike, must express itself in drastic changes of the claims which governments should acknowledge. In order to prevent tragic bankruptcy of the whole welfare state system we must develop combinations which will allow the burden to be shared: part to be covered by the authority and part to be left to freely chosen complementary insurance. This shifting of the burden is not necessarily and of itself a lowering of the rights to health care.

Finally, Schuyt discusses possible consequences of shifting the burdens such as a general loss of the overall well-being. Schuyt believes, however, that the exact opposite may occur: overall moderation may prove to produce improved well-being. The experience – individually and together – of a free choice as well as of overcoming misfortune may be another way to improve self-esteem and the general well-being. Other consequences, however, may prove to be more serious. First, social inequality will affect disproportionately minorities and marginal groups in society. Second, there will be difficulty in defining the limits of health care rights and in expressing these rights in some minimal package available to rich and poor alike.

Arbitrariness and political opportunism would be the wrong way to define the basic packages of essential health care rights. Too much is at stake to leave these decisions to incomplete and partial criteria. How this debate about allocation of resources takes place may be illustrated by the way in which heart transplantation was introduced into the Dutch care system.

The major ethical issues in heart transplantation are: survival, quality of life, and financial costs. For our discussion we will focus on the last issue. Should the government decide on total coverage for heart transplantation, it then must find the money for it from other areas of the social welfare budget. Should the latter prove impossible, then the government must justify why, given limited means, one group of patients is to receive a specific treatment at the expense of other groups. Next to these dilemmas of macro-allocation, there are decisions to be taken at the level of meso- as well as micro-allocation. Local, regional, or national policies have to be set by developing criteria which will allow individual physician teams to make responsible decisions about individual patients. These individual patients are expected to muster the courage and strength to integrate the transplantation, physically and emotionally, but also to accept a possible refusal because others may have better chances for a successful outcome ([16], p. 39).

The moral dilemma of resource allocation often seems to be brought about by external factors, in particular by financial dearth.

This dilemma is nothing less than a new example of a fundamental tension which may be as old as medicine itself. L. Edelstein showed how ancient medicine was torn between a medical ethics from within and a popular ethics from without the profession [26]. In the context of medical anthropology I have advanced the hypothesis that medicine today may be about to recreate its ethic from within. Most conditions necessary for the elaboration of such an ethic have already been fulfilled. At the end of the 20th century we are about to acknowledge that medical "deontology" is a matter of clinical judgement at least as much as it is a question of philosophical thinking. Today's crisis of medical ethics may stimulate a maturation of ethics in medicine ([18], pp. 116–117; [9], pp. 1210–1212).

*The Search for Criteria to Limit Care*

Several Dutch authors from the fields of medicine, law, and ethics are making efforts to develop new criteria for limiting health care. Such criteria seem to promote the internal setting of limits above the external imposition of them. An effort is being made to find the norm within the human person consistent with the view of man to which he or she adheres.

A neurologist, for instance, who affirms that "by acting upon the body one touches not only matter but the soul as well" will readily conclude that "therefore, there are obvious limits to what one may do to one's own body" ([10, p. 849). The fundamental orientation of this statement is toward our own responsibility. Different philosophies of life and diverse religious convictions may lead to different conclusions in the matter, but common to all is the principle that man should decide for himself [11]. A similar approach is taken in health law. Self-determination includes at least the possibility of determining limits to the care one wishes to receive. The only restriction to this right to make autonomous decisions is to avoid *harm to oneself* or to others.

This restriction, however, has of late suffered serious blows through repeated and highly publicized euthanasia cases. The Department of Justice itself is rather cautious. Between 1982 and 1984 it decided that out of 36 cases only six were admissible for prosecution. Only a few of these led to a condemnation of minimal severity, e.g., conditional imprisonment for one week. These physicians who had been brought to court for having given deadly injections at their patients' request were, for all practical purposes, acquitted by the courts, in one case by the High Court of the Netherlands [12] in its verdict of November 27, 1984. The Royal Dutch Medical Association has issued guidelines and set conditions for physicians who might consider euthanasia. The medical association's main concern is to establish legal consensus, possibly through a change of law, allowing physicians openly to face a patient's request for euthanasia. Meanwhile a State Commission issued a Report on Euthanasia (August 1985). A majority of this commission holds that under certain conditions both active euthanasia and rational suicide must be allowed, and that the time has come for the legislator to speak up in the matter. The basic conditions mentioned by the State Commission are the following:

(1) it is essential that euthanasia be performed only at the patient's request;

(2) the situation must be one of extreme pain and suffering;

(3) euthanasia should be performed with great care, mostly after consultation with colleagues.

The arguments used in the courts were: absence of material injustice, carefully considered medical exception, psychic need, and conflict of duties. Interestingly enough the High Court of the Netherlands recommends that courts examine "whether according to right medical judgement, measured by general norms of medical ethics" there really was a case of utter distress or a conflict of duties. Whatever the arguments and their validity, it would seem that the principle of patient autonomy stands stronger than ever.

Finally, in ethics, health is seen as a value but not necessarily as a value which would override all other human values for this particular person who autonomously decides about his or her claim to health care. Moreover, since self-determination is narrowly linked with solidarity, the very exercise of autonomy may lead to the judgement that "I am not the only nor the most important individual in society," and therefore that "there may be situations where it is my duty to offer to another his or her chance" ([16], p. 43).

Since our interest is in the comparison of health policies between nations, it seems noteworthy to mention how one and the same notion, e.g., solidarity, functions differently in different countries when people use these concepts to work toward a conclusion. Thus, A. Capron, in a context of resource allocation, calls on solidarity to justify general access to health care ([3], p. 107). This is strikingly different from Sporken's position just mentioned where solidarity is a reason for self-exclusion from care. Capron talks about health policy set by the authorities, whereas Sporken talks more about individual self-determination in a context where not all the burdens are shouldered by the authorities. In other words, Capron seems to argue within the framework of the ideal welfare state, where the symbolic meaning of health care appears as expressing and nurturing bonds of empathy and compassion. Health care is a symbol for individuals and society alike; to let someone go without care and, as a result, to die would offend against solidarity.

## III. JUSTICE AND HEALTH POLICY

A. W. Musschenga has focused upon the problem of health care distribution at the meso-level by using the principles of utilitarianism and of distributive justice [14].

A consistent application of cost-benefit analysis to health care policy is not possible, he argues, except against the background of the goals of health care. Recall that these goals are life prolongation and improvement in the quantity of life. For the cost-benefit analysis to work well, it must be decided which one of these goals is to prevail. Nevertheless, both goals will justify the following priorities:

(1) diseases which can be successfully treated;
(2) diseases which occur frequently;
(3) diseases which appear early in life.

Similarly, Dutch economists have pointed out that as a result of the utilitarian approach little attention is being paid to the chronically ill, the mentally handicapped, the elderly, or the physically handicapped. They consider these omissions to be unethical and, therefore, plead for justice as the right principle for distribution ([7], p. 1089).

The ethicist then goes on to define justice with D. Miller: "The just state of affairs is that in which each individual has exactly those benefits and burdens which are due to him by virtue of his personal characteristics and circumstances" ([13], p. 20). There are three criteria for the application of the principle of justice: the individual's rights, his merits, and his needs. The last being the most relevant criterion for health care distribution, we now have a formal principle and a practical standard which we may apply to health policy. The highest priority resulting from the application of this standard goes to disease which, if left untreated, would lead to the worst possible state of health. Again as with the utilitarian principle, this egalitarian approach is not without its problem. What is that disease whose victims are the worst off? Furthermore, which is the worst disease, the one that threatens life or the one that threatens its quality?

Musschenga's conclusion, as expected, is that both principles are needed. More specifically, though, he prefers a trade-off between reduction of relative inequality (that is, the needs of the worst off) and a reduction of aggregate inequality (that is, the needs of all who suffer). His reason is that he achieves better results than when he applies either utilitarian or egalitarian principles alone.

For the way in which this combination of health care goals applies to policy decisions, Musschenga suggests tentatively that:

(1) with regard to the relationship between the prolongation of life and its improvement.

(a) the life-threatening diseases of youngsters be given priority over life-threatening diseases of the elderly;

(b) quality-threatening diseases of the elderly receive priority over life-threatening diseases of the elderly.

This consideration would lead, therefore, to the following list of health care priorities:

(a) life-threatening diseases among the young,
(b) quality-threatening diseases of all ages,
(c) life-threatening diseases of old age.

(2) with regard to each of these categories one may set further priorities on behalf of such criteria as seriousness, frequency, and chances of cure ([14], p. 10).

### *A Provisional Advice to the Government*

The National Council for Public Health functions as an advisory body to the Dutch government in matters of public health. The council relies on permanent and temporary committees in preparing its advice. In the matter of "limits to health care" the council formed a Working Group on Ethics as part of a wider commission.

In January 1985 the Working Group on Ethics initiated a series of working sessions on two questions: (1) What criteria may be used for the acceptability of medical acts? (2) Given the need to set financial limits to health care coverage, what ethical criteria may be used to select among new technologies those which are desirable or deserve high priority?

At an early stage the Working Group expressed its conviction that both questions have value in themselves, even outside the context of any budget restriction. Limit setting always raises ethical questions; too much health care as well as too little may represent the unethical practice of medicine. In order to enable health care workers to develop practical guidelines for decision making, the Working Group recommends these general moral principles: human dignity and autonomy in individual decisions; solidarity and equity in policy decisions. The first two principles speak to the human requirements in medicine allowing determination of the minimum which must be defended when medicine suffers budget cuts. The last two principles, i.e., solidarity and equity, govern decisions about health care cost coverage, and relate directly to our discussion of resource allocation.

Solidarity was understood by the Group as a new word for justice, yet with a strong emphasis on "feeling responsible for others," particularly for the poor. In the past solidarity referred to the individual's duty to care for others in need or distress; as such, the notion is comparable to charity. Today, however, solidarity points to the responsibility of the group, to a collective duty to care. It follows from this shift that policy-makers (who organize the common good) have a right to impose the sharing of this duty as a burden for all. Charity, so to speak, has become mandatory and is now called solidarity.

As for equality, the Working Group believed that in general the covered health services must be equally accessible to all in the strict egalitarian sense of "none should be helped if not all can be helped." The Group warns against abuse as a consequence of commercialized medicine which splits society into have's and have-not's.

Another important aspect of the Group's advice concerns priorities. Quite understandably the Group explored the applications of utilitarian and of egalitarian rules to the setting of priorities. Since most known applications of this kind are at the micro-level, the Group chose to present three categories in which both micro- and macro-dimensions apply, such as AIDS, IVF, and neonatology. These categories, it was hoped, would allow a framework for setting priorities by referring to (1) individual benefit, (2) societal benefit, (3) a common benefit for both.

The final report of the Working Group on Ethics was issued in April 1986. Its basic recommendations was that health care, if effective and ethically acceptable, in principle be made available to all, and that exceptions to this principle be explicitly defined and justified [25].

*Public Policy Development*

From a North American viewpoint, A. Capron stated recently that "for those making decisions about resource allocation, ethical pitfalls can best be avoided if policy formulation is differentiated from clinical decisions about specific patients' access to care" ([2], p. 103). In general Dutch policy-makers would go along with this viewpoint. Instead of talking about individual patients, they prefer to discuss groups of patients (or diseases) and groups of treatments in order to reach macro-decisions on scarce resource allocation.

Since the mid-1970s the Netherlands has felt the consequences of a long-lasting economic recession. The question of priorities in resource distribution have become paramount. The increasing costs have been accompanied by no measurable returns in terms of public health. This is partly because more health care needs are now caused by environmental conditions and life style. Dutch policy-makers share the WHO concern about a coming health care cost crisis unless environment and life styles change. Meanwhile, a shift in the nature of health care provision must be developed in order to meet the new needs.

The procedure for policy development is planned in such a way that social and economic factors are not the only decisive elements for future planning. Instead, the emphasis is on planning to meet future health care problems. Current and future planning is, in Holland as elsewhere, worked out through policy analysis, studies on cost-effectiveness, and predictions of the future. In this manner it is possible to select policy options and plan actual choices. Currently the government's Bureau for Policy Development has prepared a report on "Health in the Netherlands by 2000" ([21], p. 1524).

This forthcoming study analyzes (1) economic prospectives, (2) demographic developments, (3) cultural and social trends with their impact on environmental conditions, and (4) life style. We turn briefly to each of these items.

The overall economic predictions are grim. The structure of markets, for instance, and of conservative policy-making foresee no improvement on a short-term basis. While investment in new technology remains low, the collective burden, including health care, increased from 52% in 1975 to over 60% of the net national income in 1983. No growth in the support of health care is to be expected; meanwhile, the problems of poor health increase in part because of unemployment (16%).

Demographically the Dutch population keeps growing. This is not due to high birth rates but to the relatively young age of the population. Nevertheless, by the year 2010 the elderly will account for 15% of the population; to cover their health costs insurance premiums must be raised by 20%.

Cultural and social trends are leading to more unemployment. Fatalism and dissatisfaction cause the unemployed to become socially isolated, and thus the risk to their health increases. But other groups as well run these risks: large cities in particular will develop areas of deprivation where the mortality rate is double what it is elsewhere in the country.

Life style changes are badly needed, but there is little evidence that such changes are occurring in the Dutch population. Dietary habits, alcohol consumption, smoking, and the lack of exercise have not significantly changed, even with increased efforts to inform and educate the general public.

Predictions about health care in the 21st century suggest a growth in emphasis on prevention and an increase in home care, with a resulting decentralizing effect. Yet, at the same time concentrations of high technology tertiary care are planned. Decentralization would lead to the development of small centers for primary care, covering populations from 5,000 to 20,000. These must be flexible and sensitive to local needs, say of drug control, or with special units for chronic diseases when there are many elderly, or specific services for child care in areas with young families. The overall objective of the small scale primary care center is to bring providers and consumers of health care closer together. As for the simultaneous development of concentrations of high technology care, the future will show that their specific and costly services can be limited only if the primary care centers cooperate with these specialized centers. The quality of care, it is hoped, will improve.

Currently, the Bureau for Policy Development is trying to evaluate the effect of proposed changes in health care policy by developing models of centralized and decentralized care.

## IV. CONCLUSION

The description of current developments in Dutch health care policy allows for some conclusions:

*First*, this country early recognizes that more health care expenditure does not necessarily result in improved health. At the same time the debate reveals that lack of funds should not be the only reason for cost containment. While it is perhaps a sufficient reason in times of dearth, it certainly is not a necessary reason at all times. In other words, imposed cost containment could become obsolete once we have all become sufficiently cost conscious.

*Second*, the government's initiative to consult with advisory bodies on the issue of health care limits simply provides an example of how policy is made in this country; but it may also provide an example for other countries. In the mid-1980s the issue of resource allocation is universal. Each individual country, however, should create its own way of stating the question as well as arriving at answers which will manage scarcity and provide for a just allocation. This is where the preferences and specific approaches of each nation play an important role; comparing differences among countries then becomes both instructive and useful.

*Third*, a major concern of Dutch ethicists is the dilemma of health care needs. Should these be defined on the basis of expectations that may be nothing more than an outgrowth of the welfare state, or should they be defined in terms of a "decent maximum" of health care. This dilemma is to be solved not only by policy decisions; it may require implementation through regulation, education, self-discipline, and, last but not least, marketing strategies. The intricacy of this network constitutes a challenge to ethical analysis.

*Fourth*, with regard to the four major concerns of health care, described in Dr. Engelhardt's introductory paper, the current health care policy in the Netherlands shows the following characteristics: goal three, i.e., costs should be contained, is the leading principle; goal two, i.e., equality in health care, is a leading philosophical principle in a country that cherishes egalitarianism; goal four, i.e., consumers' and providers' choice, is acknowledged as desirable, and it is possible that this goal will take on more importance in the coming years; goal one, i.e., the best care for all, is only a conditional goal for policy-makers, who must always subject this goal to realities of cost-containment and to egalitarian concerns.

*Institute for Bioethics,*
*Maastricht, The Netherlands*

APPENDIX[1]

| | | |
|---|---|---|
| Population in (1983) (1 Jan.): | | 14,339,551 |
| Proportion aged 65 and over: | | 1,688,114 (11.8%) |
| Live births: | | 11.8 per 1,000 inh. |
| Mortality rate: | | 8.2 per 1,000 inh. |
| Percentages of total deaths: | | |
| Circulatory diseases: | | 44.9% |
| of which: ischaemic heart disease: | 21.1% | |
| acute myocardial infarct: | 17.0% | |
| Cerebrovascular conditions: | 10.1% | |
| Malignant neoplasms: | | (27.3%) |
| lungs and pleura: | | 6.9% |
| breast: | | 2.5% |
| cervix uteri: | | 0.3% |
| Respiratory diseases: | | 6.9% |
| Chronic non-specific respiratory disorders: | | — |
| External injuries and poisoning: | | 5.2% |
| Road accidents: | | 1.5% |
| Diseases of the digestive organs: | | 3.6% |
| Average period spent in hospitals: | | 13.0 |
| Hospitalization coefficient per 100 inhabitants: | | 11.3 |

Health care institutions (1983):

| | *Number* | *Beds* | *Beds per thousand inhab.* |
|---|---|---|---|
| Hospitals | 225 | 69,583 | 4.8 |
| Psychiatric hospitals | 76 | 24,761 | 1.7 |
| Institutions for the mentally defective | 129 | 29,722 | 2.1 |
| Nursing homes | 327 | 48,036 | 3.3 |

*Number of doctors, etc.* (1984):

| | Absolute numbers | Numbers per 100,000 inhabitants |
|---|---|---|
| | 29,951 | 208 |
| Doctors of which: | | |
| general practitioners: | 5,736 | 40 |
| specialists: | 10,336 | 72 |
| institutional doctors, etc.: | 1,557 | 11 |
| Dentists: | 6,586 | 46 |
| Pharmacists: | 1,728 | 12 |
| Obstetricians: | 1,022 | 7 |

*Total health care costs* (1983):

| | Amount (millions of guilders) | % |
|---|---|---|
| In-patient health care: | 19,429 | 59 |
| Out-patient health care: | 12,053 | 36 |
| Ambulant services and other health care: | 1,634 | 5 |
| Total cost: | 33,116 | 100 |

*Financing of health care* (1983):

| | Percentages |
|---|---|
| Sickness Fund Law: | 43 |
| AWBZ (National Insurance): | 27 |
| Private contributions: | |
| Subsidies (government, etc): | 5 |
| Other sources: | |

NOTE

[1] I am grateful to Drs. G. de Wert for updating the statistical data in the Appendix.

## BIBLIOGRAPHY


1. Andel, H. van den: 1984, 'Ten geleide', in Stichting gezondheidszorg 1982, *De grenzen van de gezondheidszorg*, Van de Ridder BV, Nijkerk, pp. 5–6.
2. Capron, A. M.: 1984, 'An Ethical Obligation to Ensure Access to New Medical Technologies?', *Journal of Health Care Technology* **I**, 103–120.
3. Centraal Bureau voor de Statistiek: 1984, *Vademecum gezondheidsstatistiek Nederland 1984*, C. B. S. -publikaties, 's-Gravenhage.
4. Galston, W.: 1980, *Justice and the Human Good*, University Press, Cambridge.
5. Groot, L. M. J. and Visinescu, C. B.: 1982, *Medical Technology, Legislation, Policy and Cost in the Netherlands*, Report for the European Economic Community, Brussels (manuscript).
6. Groot, L. M. J.: 1984, 'De grens tussen financierbare en niet meer financierbare zorg', in Stichting Gezondheidszorg 1982, *De grenzen van de zorg*, Van de Ridder BV, Nijkerk, pp. 69–80.
7. Jong, G. A. de and Rutten, F. F. H.: 1983, 'Justice and Health for All', *Social Science and Medicine* **17**, 1085–1095.
8. Juffermans, P: 1984, 'Economic Crisis and Health Policy in the Netherlands', *International Journal of Health Services* **14**, 147–158.
9. Lalonde, M.: 1974, *A New Perspective on the Health of Canadians*, Government of Canada, Ottawa.
10. Lange, S. A. de: 1984, 'Over de grenzen van de curatieve gezondheidszorg', *Medisch Contact* **27**, 847–851.
11. Leenen, H. J. J.: 1984, 'Grenzen van het recht op gezondheidszorg 1982', *De grenzen van de zorg*, Van de Ridder BV, Nijkerk, pp. 59–67.
12. Leenen H. J. J.: 1985, 'Euthanasie voor de Hoge Raad', *Nederlands Tijdschrift voor Geneeskunde,* **129** 414–417.
13. Miller, D.: 1976, *Social Justice*, Oxford University Press, Oxford.
14. Musschenga, A. W.: 1984, *Utiliteit, rechtvaardigheid en gezondheidszorgbeleid* (manuscript).
15. Schuyt, C. J. M.: 1981, 'De zin van de verzorgingsstaat: verdelende rechtvaardigheid in het licht van de toenemende schaarste', *Sociaal Maandblad Arbeid* **36**, 612–621.
16. Sporken, C. P.: 1984, 'De grenzen van het ethisch toelaatbare in de gezondheidszorg', in Stichting gezondheidszorg 1982, *De grenzen van de gezondheidszorg*, Van de Ridder BV, Nijkerk, pp. 29–46.
17. Vries, M. J. de: 1984, 'De grenzen van curatieve zorg binnen het kader van helingsprocessen', in Stichting gezondheidszorg 1982, *De grenzen van de zorg*, Van de Ridder BV, Nijkerk, pp. 47–57.
18. Wachter, M. A. M. de: 1985, 'Euthanasie et coût social', in V. Boulanger and G. Durand (eds.), *L'euthanasie, problème de société*, Fides, Montreal, pp. 107–117.
19. Wachter, M. A. M. de and Roy, D. J.: 1985, 'Médicine, éthique, anthropologie', in J. Dufresne *et al.* (eds.), *Traité d'anthropologie médicale*, Presses de l'Université du Québec, pp. 1189–1217.
20. Werff, A. van der: 1976, *Organizing Health Care Systems, a Developmental Approach*, Dissertation, University of Utrecht.

21. Werff, A. van der: 1984, 'Rolverdeling arts en overheid bij beleidsontwikkeling', *Medisch Contact* **47**, 1522–1524, and 1554–1556.
22. W. H. O.: 1982, *Regional Strategy for Attaining Health for All by the Year 2000*, Regional Office for Europe, Copenhagen.
23. Wolf, J. P. M. van der and Zenderen L. A. M. van: 1984, 'Adviesorganen', in E. W. Roscam Abbing, *Bouw en Werking van de gezondheidszorg in Nederland*, Bohn, Scheltema & Holkema, Utrecht-Antwerpen, pp. 451–483.
24. Andel, H. van: 1984, 'Ten geleide', in Stichting gezondheidszorg 1982, *De grenzen van de zorg*, Van de Ridder B. V., Nijkerk, pp. 5–6.
25. Kuiper, J. P.: 1984 'Wat kan nog gezondheidszorg worden genoemd?', in Stichting gezondheidszorg 1982, *De grenzen van de zorg*, Van de Ridder B. V., Nijkerk, pp. 7–28.
26. Edelstein, L.: 1967, 'The Professional Ethics of the Greek Physician', in O. Temkin and C. L. Temkin, *Ancient Medicine*, The Johns Hopkins Press, Baltimore, pp. 319–348.

MURRAY FESHBACH

# HEALTH IN THE U.S.S.R.: ORGANIZATION, TRENDS, AND ETHICS

When the Soviet regime took power in late 1917, health conditions then prevailing were abominable. Disruptions caused by wartime and revolutionary events led to interruptions in service, as limited as they were even under prewar conditions, to interruptions in delivery of medicines to medical facilities and especially so to individual medical practitioners. The situation was so difficult that when Lenin confronted the depth of the problem in his address to the December 1919 meeting of the Soviets (the administrative councils) he noted that the interruptions in public services, governmental operations as well as medical issues, were so bad that because of these problems "either the lice will defeat socialism, or socialism will defeat the lice"; at the time upwards of 20 million persons may have been ill with typhus.[1] The 25,000 doctors providing medical service to the 160 million population were insufficient in number overall, their distribution was concentrated in cities, and the quality of training depended heavily on the place of study in or out of Russia. When these factors were combined with the conditions of the time, it is no surprise – as terrible as the numbers are – that the League of Nations estimated that over three million persons in the Soviet Union died from cholera, typhus, typhoid, and dysentery during the period 1917 to 1923. Happily, this disease and death pattern no longer occurs.

With desperate conditions, desperate measures were taken in an attempt to provide medical services to the population, to train medical personnel with at least a modicum of medical knowledge, to admit students who met less rigorous requirements than even minimally required later by Soviet medical authorities, to provide minimal services in rural areas through resident paramedics (feldshers), and to utilize large quantities of medical relief provisions and services offered by foreign nations. While initial medical intervention significantly shifted the pattern of causes of death, and deaths from diseases such as those cited from estimates made by the League no longer occurred, the population suffered dramatically from both endogenous and exogenous causes. Included among the demographic catastrophes which impinged on the population of the country were the civil war, emigration of the

*Hans-Martin Sass and Robert U. Massey (eds.), Health Care Systems,* pp. 117–132.

Whites, foreign intervention, famine (of the early 1920s), collectivization, famine (of the early 1930s), labor camps and the purges, war with Finland, invasion of the country in the Second World War, forced movement of large numbers of Soviet workers to German factories and voluntary emigration of others during the retreat from the U.S.S.R., and another famine (in the second half of the 1940s). Thus, the Soviet population has undergone enormous losses, suffered poor health conditions, and witnessed destruction and disruption of its economy and society. Since 1950, however, we can consider that the country's population development – as reflected in fertility and mortality measures – has been a result of individual family decisions and national priorities without interference from exogenous factors

Before proceeding to describe and analyze the current trends, especially morbidity and mortality, but not totally excluding fertility, it is necessary to describe the Soviet health organization as presently constituted from a formal point of view. (Informal aspects, insofar as they are contributory to our understanding of medical ethics in the country, will be discussed below.)

The usual Western belief is that the Soviet Union has a single, centralized public health system. This belief, however, is far from correct. It is true, of course, that the Ministry of Public Health is the central, primary health deliverer. From an overall point of view, health services in principle are available to the entire population. In reality, however, there is enormous variety of types and quality of health services provided the population. From the administrative-territorial perspective, there is a ministry at the national level, at each of the 15 union republics, within the 130 or so oblasts, krays and autonomous republics at the next instance, and some 5,000 additional territorially-related units and the lower urban and rural levels.

An alternative way of looking at this issue is to dissect the Ministry of Health's institutional structure from the perspective of health delivery to individual populations. Thus, at the very top, the Kremlin hospital is run by the Fourth Directorate of the Ministry of Health (under Dr. Chazov), hospital admissions are restricted to the members of the highest leadership. With foreign medications, technology and services reportedly readily available to this service group, the level of health delivery is much different from that for the remainder of the Soviet population. Below the level of the Kremlin hospital, there are urban hospitals and medical facilities available only to restricted populations attached to various institutions; these services are administered by medical authorities not directly subject to the U.S.S.R. Ministry of Health's control. Thus, medical institutional facilities are under the auspices of the ministries of railway transportation, defense, internal affairs, and civil air transport, as well as the ministries of foreign trade, finance, higher and specialized secondary education, foreign affairs, aviation industry, the KGB (Committee for State Security), the Academy of Sciences of the U.S.S.R. and GUM (the department store on Red Square) ([18], p. 15).

For the population as a whole, there are two additional urban systems of the U.S.S.R. Ministry of Health as well as the overall rural portion of the Ministry. The urban hospital facilities are divided into teaching and non-teaching hospitals, with the latter reportedly much less well-equipped, with a lower staff quality, and generally overcrowded. In addition, located in both urban and rural localities, there are medical units within industrial and other economic sectors (other than those listed above) of varying sizes and quality. Open only to employees of the particular organizations, the facilities also vary in quality by region and by relative adherence to standards set by the U.S.S.R. Ministry of Health. Thus, it is difficult to generalize about their health delivery and practice.

It is not difficult, however, to generalize about the rural health system. It is very poor, frequently located far from population points in central regional hospitals and also frequently much less well-equipped than the urban polyclinics and hospitals. One of the consequences of this differential in service quality is a growth in the share of the population gaining hospital admission to urban medical units staffed by doctors with specific specialities, and supplied with actual, let alone better, equipment. As roads improve in rural areas, access to urban facilities will be enhanced and the share of such admissions from among the rural population undoubtedly will increase. Lack of fully trained medical personnel and of modern technology in the local (rural) institutions will continue to exacerbate the situation.

Resources allocated to the health sector have increased in absolute amounts and in *per capita* terms. The number of doctors and beds has increased apace. Nonetheless, these input figures do not accurately reflect the results of health services of the Soviet Union. The 16 billion rubles allocated from the state budget of the U.S.S.R. in 1982 to the health sector is about two-and-one-half times greater than the amount assigned in 1965 (16,035 and 6,669 million rubles, respectively) ([20], p. 524). Population of the country grew simultaneously from 232.2

million to 271.2 million between the end of 1965 and 1982 ([20], p. 5), for an increase of 17 percent. Regardless of the increase in *per capita* budgetary allocations, however, the share of the state budget assigned to health has decreased sharply: in 1965, the share was 6.6 percent, in 1982, 4.7 percent ([20], p. 521) – a decrease of almost 30 percent in its relative share. As we see later, measures of health status such as the death rates changed in the reverse pattern of budgetary allocation by increasing dramatically in the last two decades.

The number of doctors is currently reported in Soviet statistical sources as being just over one million at the end of 1982 ([20], p. 497). This number is exaggerated by about 10 percent, since the Soviets include dentists and stomatologists (dental surgeons) in their definition of "doctors of all specialties." At a rate of about 380 per 100,000 population (unadjusted), the Soviets can boast that they have a much higher ratio than other countries. The same is true of the number of beds per 100,000 population – the standard unit for such measurements. However, quantity does not make for quality and is not indicative of the output side of health services. And these measures of output (read "efficiency") leave very much to be desired.

This output perspective is manifested in very negative morbidity and mortality trends which are remarkably lengthy in duration and of profoundly large dimension.

After major success in reducing the incidence of disease to much lower levels than at the time of accession to power, and throughout the period until 1950, the Soviet leadership could rightly boast about their achievements in this area. However, since the mid-1960s, and particularly in the more recent period since the late 1970s, something seems to have gone awry in the health status of the Soviet population. From the overall point of view of mortality as measured by the crude death rate (i.e., the number of deaths per 1,000 population regardless of age or sex), there has been a remarkable increase in the last two decades. Thus, as officially reported in Soviet statistical yearbooks, the crude death rate reached its lowest level in 1964. The reported rate for that year was 6.9 deaths per 1,000 population. Since then, however, the rate has inexorably increased to a point almost 57 percent higher, or 10.8, in 1984. This rate followed a period of a slight decline after the 1980 rate had increased to 10.3, then a slight decline in the next two years by 0.1 per 1,000 population per year to 10.1. But then again an increase in 1983 to 10.4 and an even further increase in 1984 to the postwar historic high of 10.8. What is driving this rate? Is it solely the changing age structure of the population as fertility declined overall? Or are there other causes? In this case, it is undoubtedly the shift in trends leading to increases in infant mortality, to increases in premature mortality of young males, and in problems associated with health services, with health delivery, and with changes in ethical behavior of Soviet medical personnel.

Thus, after reductions recorded in many infectious diseases, a shift upward seems to have occurred over this period of time. One major example relates to diphtheria. In 1979, the reported number of cases of diphtheria in the country was 200; by 1983, the number was officially reported to be 1,410. Perhaps the earlier figure was too low, as inaccurate diagnoses may well have failed to identify this unusual illness in many instances, but some may still be undiagnosed, adding to the 1983 level as well (to 1,500? to 1,600?). Moreover, this illness can be expected to increase further inasmuch as recent Soviet sources indicate that some 30 to 50 percent of the children in their first year of life are not inoculated against this disease in the republics comprising about two-thirds of the total population of the country ([25], p. 7).

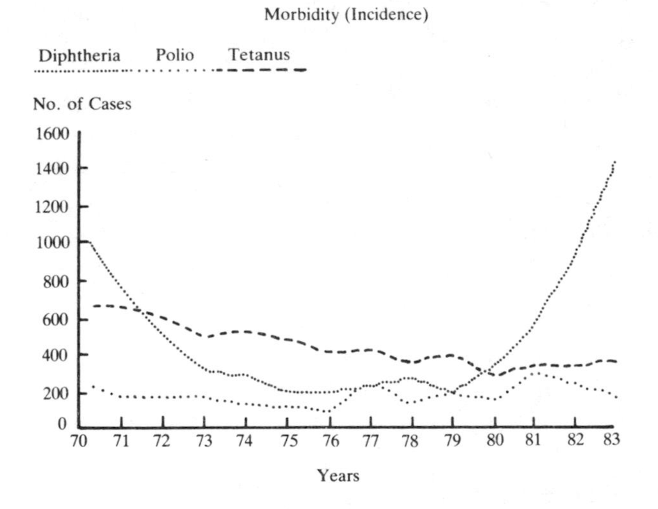


Source: Murray Feshbach, *A Compendium of Soviet Health Statistics*, Washington, DC, 1985, Table 27. Polio is expected to increase because of new evidence of non-inoculation of some 30–58 percent of young children in five republics with a total of about two-thirds of the entire population.

Similar rates of non-inoculation of young children are noted for polio immunization. The effectiveness of other immunizations also is very low, namely influenza, measles, mumps, typhoid, and others. The number of cases of mumps reported to the Centers for Disease Control in the United States is now at a level of about 2,000 cases in 1983 and again in 1984 among a population of about 235 million persons; in 1968, the first year in which mumps became an officially notifiable disease, there were 152,209 cases reported. The Soviet Union, with about 15 percent higher population size (of slightly over 275 million), recorded some 600,000 cases in 1966 (the last reported year). On the basis of incidence rates (per 100,000 population), I have estimated that the level of mumps cases based on patient visit data had increased to over 900,000 by 1979, and more recently is no less than 1,300,000 ([5], p. 12). (Patient visit data have been criticized by Soviet as well as other country sources as being far short of actual illness incidence, perhaps by one-half overall.)

Increases in hepatitis also are significant in this period of time. Again, the Centers for Disease Control of the United States showed a figure for all hepatitis cases of about 65,000 cases last year (1984), comprising some 30,000 each for hepatitis A and B, and the remaining 5,000 among the non-A, non-B and unspecified categories. In the Soviet Union, the figure may be no less than 1,400,000 in recent years, and growing. Typhoid is now recorded at over 18,000 in 1983 in the U.S.S.R. whereas in the United States the number ranges from 100 to 500. These figures and others which could be adduced tend to show that the number and direction of disease rates are of a very unexpected dimension and vector.

One of the possible reasons for increases in illness rates is the reduction in the share of the state budget devoted to health sector expenditures, as noted earlier. This low share can be demonstrated by an analysis of the very low expenditures per patient bed-day, which may well underlie the poorer medical care and inability to cope with a growing problem. Thus, a 1980 source provides the base cost of health delivery for patients in the surgical ward of a cancer hospital in three parts:

(1) Direct expenditures for special treatment (including wages, amortization, and wear and tear of fixed assets, as well as current expenditures for instruments) – 1.98 rubles per bed-day in a cancer hospital;

(2) Other direct expenditures for treatment and provision of services

for the ill – food (1.35) and medications (1.27), or 2.62 rubles per day; and

(3) Indirect costs – 5.89. The sum of the parts for the surgical division of the hospital, therefore, is 10.49 rubles per bed-day ([16], p. 95).

This sum (converted to dollars) of $17.00 per day is unbelievably low in its components, and especially so for food costs, let alone for medications. (One ruble is currently – Dec. 1987 – officially equal to about $1.70).

Even with administered prices for goods and services, it is hard to credit such figures. If Communist Party and Kremlin special hospitals spend much more to obtain services relatively equivalent to those generally available in the West for food and medication or patient bed-day, then perhaps the entire Soviet health expenditure figure should be so multiplied by a factor of 7, 10, or even higher so as to bring it closer to "outside country" service levels. Frequent reference is made to the cost of antibiotics, complex medications such as rifampin (used for the treatment of active tuberculosis patients) which should not be continued if not successful soon after initial administration, or at least showing a significant impact within a very few months; instead surgery should be resorted to rather than exceeding the budgetary limits for medication costs. Perhaps this is also one of the underlying causes for the reports of rampant bribery of medical personnel to administer the prescribed medications which they might otherwise not have to record as a cost item. Capitalization of medical facilities also varies widely among the 15 union republics. In 1970, for example, the fixed capital in the health sector ranged from a low of 266,000 rubles per 10,000 population in Uzbekistan to triple that rate, or 792,000 rubles per 10,000 population in Latvia ([2], p. 243). Presumably even within Uzbekistan there is a significant disparity between urban and rural rates of capitalization, and therefore the very large proportion of rural population – particularly composed of Uzbeks rather than the Russians (who would be among the urban population) – would be even less well equipped and provided for than the republic by republic disparity shown above. Even in cities we find frequent reference to conversion (albeit poorly done) of old buildings to medical facilities and their poor design for the purpose of a medical unit.

Thus, we can think of this resource allocation principle as being a reflection of the ethics of the state in choosing between medicine and, say, defense. The type of medical service required of doctors is a similar state choice. In this case, the initial abolition in 1917 of the Hippocratic Oath under the new regime, thus doing away with bourgeois approaches to doctor-patient relationships, is a state-directed choice of much importance. It was not until the latter part of the 1960s that the first conference on medical deontology and ethics was held. At that time, the discussions centered on the duty and behavior of the Soviet physician mostly related to doctor-patient relationships. Only in March 1971 was a Soviet doctor's oath legislated. This oath affirmed the priority of the state in obligating the physician to submit to the state's interest in the first instance, and secondarily to the patient's. Thus, the doctor is a state employee, as is the nurse, and relationships with the individual patient are at a lower level of priority. Regardless of platitudes made at frequent intervals, the state is overwhelming focused on the health of the individual as contributor to the economic functions of the country, rather than on the individual *per se*. Obviously, there are many individual doctors and other medical personnel who are sincerely interested in the individual's health, wish to do no harm, are considerate and endeavor to do their utmost within the constraints of short supplies of modern equipment, medications, or even bandages. Nonetheless, they may well be the exceptions.

Moreover, given the enormous drop in the increments to the working age population of 20 to 59 year olds in the last 15 years of the century (dropping from 30 to 6 million in the periods 1970 to 1985 and 1986 to 2000, respectively), the attitude of the state toward the use of medical personnel as "medical police" is understandable if not regrettable. Inasmuch as each person is worth more at the margin (given lesser numbers), their return to work as an economic producer is the first priority and not the individual's health *per se*. Even military leaders are publicly expressing concern over the lack of increase in average life expectancy of young males at birth. Thus, in February 1983, the First Deputy Chief of Defense of the Main Political Administration of the Ministry of Defense Admiral Sorokin, made explicit his concern over the lack of increase in this longevity measure and therefore of the health of young males ([22], p. 9).

In 1977, the Second All-Union Conference on Medical Deontology was held in the U.S.S.R. In July of the same year, a major legislative act on improvement of health delivery, services, and needs was adopted. Nonetheless, in the Accountability Report of the then General Secretary, Leonid Brezhnev, in February of 1981, he noted the inappropriate behavior – that is, lack of good physician-patient relationships, corruption, bribery, stealing of food from patients, nepotism, lack of adherence to sanitation norms, etc. – and this continues to the present. Frequent stories appear in the Soviet press to complain about these issues in one place or another, by one individual hospital or doctor or another, by the Ministry of Health of one place or another. Thus, in sum, improper behavior – whether it be by an institution or an individual – is still widespread. In addition, the medical establishment seems to have become alarmed at the apparent rejection of standard Soviet medical practice as evidenced by the frequent public media attacks on alleged quacks, faith healers, shamans, or their equivalents. In part, this is also associated with a return to religion as an alternate solution to the god of the state.

Potential patients are aware of the lack of medical supplies and the insufficient attention given to individual patients by the primary physician who has to "go through her appointments without looking up," because she "has to examine 36 people in four hours. On the average, she has seven minutes per patient" ([13], p. 11). There are multiple examples of "callousness, hardheartedness, and even criminal negligence committed by personnel at medical institutions which have led to grave consequences and have cast a dark shadow over representatives of this humane and noble profession." This 1984 summation of the dark side of Soviet medical practice goes on to note that:

> Public health organs note with alarm that the careless attitude of some physicians toward their obligations has been significantly reducing the effectiveness of many measures designed to improve the level of public medical services. The number of complaints about activities incompatible with medical ethics has not been decreasing in recent years ([14], p. 1).

If only some, then why the alarm? How could only some supposedly small number of physicians "significantly" have an impact on the effectiveness of legislation to improve an apparently bad situation before the adoption of such legislation? The numbers must be of a large magnitude if in the RSFSR alone (the largest republic of the country with about half the total population) in 1983 "more than 400 physicians . . . were deprived of their licenses for serious negligence, 72 physicians were found guilty of actions inappropriate to their duty position, and many employees of medical institutions have been subjected to disciplinary action" ([14], p. 1).

This attack on medical incompetency and unethical behavior, moreover, may be understated if a front page report several months earlier in the national medical newspaper is credible. According to this article in *Meditsinskaya gazeta*, young physicians in their first three years of practice after graduation are forgiven ANY mistakes they make, even if the procedures undertaken are not particularly complex or the clinical problems "fully within the competence of an ordinary physician to make a correct diagnosis" ([11], p. 1). The article continues:

> Apparently there exists some unwritten code or stipulation that 'protects' the young specialist during the first three years of work. [This attack on the ability of the young doctor to practice] frequently . . . turns out that [he or she] still needs to fully master resuscitation . . ., is sometimes at a loss on elementary questions of diagnosing acute cardiovascular diseases or acute diseases of the abdominal cavity. . . . Today, post-internship certification is, to all extents and purposes, a rubber-stamp procedure. . . . Why should we have to baby-sit a mature 25-year-old for three years, make allowances for his 'youth' and put up with his negligence ([11], p. 1).

This article is a major condemnation of the training process, of the attitudes of young doctors, and of the system which allows this to occur.

A survey of 303 pediatricians who were enrolled in a post-graduate course in Kazakhstan during 1982–83 was most revealing about the abominable lack of knowledge by these important health providers. Almost three quarters (71 percent) of the pediatricians were between 30 and 50 years of age; they held jobs ranging from the chief pediatrician of rural regions to heads of pediatric services of a medical institution to local pediatricians in urban as well as rural areas ([3], pp. 9–12). They had had from five to twenty years of practice in the field of medicine. Only one half (47.4 percent) knew as many as six of the "16 named modern medications of wide application." One-quarter (23.4 percent) knew four. Therefore, less than half knew even this small percentage of these new medications. Even worse, 18.9 percent knew not a single one ([3], p. 10). One-third could not name a single medical book which they had read in the previous two years. One hopes that this survey is not representative of the entire profession of pediatricians or of all doctors in the Soviet Union, but . . .

This lack of knowledge on the part of the young as well as older Soviet doctors may also underlie the reports on the behavior of former Soviet citizens who recently arrived in the United States toward health providers in their new place of residence. Thus, a survey of the conflict in the minds of the individual emigrants toward medical services in their new location (in and near San Francisco) is very revealing of their expectations, their behavior toward following prescriptive advice or instructions of the doctor, and their knowledge ([23], pp. 900–904). It appears that there is complete distrust toward the young physician, although he may be well trained and knowledgeable of new diagnostic and therapeutic technology. In the emigre patient's mind he cannot by definition know how to provide clever if not proper medical services. The physician must be a specialist who has grey hair and is flexible enough to be bribed to do the "right" thing. With age comes wisdom and experience. The patients do not expect to be told the underlying diagnostic reason for the operation which has been performed on them; they do not expect any continuity of service by the operating physician; pleasant, smiling attendants – be it doctor or nurse or helper – are not expected; the drawing of blood by one nurse, the combing of hair by another, and the making of the bed by a third as regular routine are instead interpreted as "assault" or "ridiculing" of the patient, and not as serious attitudes toward their duties. Any lack of authoritarian behavior by the doctor, for example, means that the individual does not know what to do and discussion of possible alternatives only confirms this lack of knowledge and/or assurance. The referrals to older specialists also can be interpreted as a substitute for knowledge or for the lack of technology. The use of a workup for a full diagnosis implies a depth of illness that lack of significant symptoms does not call for, and, therefore, they are generally opposed to this standard procedure. ". . . the only possible benefit is to the physician who must be using them as a guinea pig" ([23], p. 902).

Compliance with prescription usage also is different based upon their experience in their former country. Thus, if a medication is readily available at the hospital or local pharmacy, then it cannot be efficacious. If so easily available it is not worth anything, as only that which is difficult to obtain on the second economy is good. Moreover, as this California survey demonstrated, the emigrants (former members of the urban Soviet population) will cease taking medication at the first sign of improvement in their condition. They "firmly believe that all drugs are poison in some way." That may explain or be explained by the use of galenicals to such an extent in the Soviet pharmacopeia. Perhaps one-third of their medicines are derived from plants raised on specialized state farms belonging to the Ministry of Medical Industry of the U.S.S.R. In years in which the harvest is bad in general (as in the last six years) or in particular regions where these farms are located, the proverbial shortages of medicines are rationalized by the suppliers as based on poor weather. This does not help the individual patient in need of digitalis or other plant-derived medications. Nonetheless, some of these herbal drugs are effective, even if non-technology based.

Ex-Soviet patients expect to stay in hospitals for a lengthy, minimum period and not be hurried out of the medical facility. They are there for bed rest, the most common treatment in the Soviet Union according to the information derived from the survey. This fact may also explain why hospital admissions have grown to one-quarter of the Soviet population in 1983 (about 67 million admissions out of 270 million population, in round terms, assuming no repeat admissions).

From a more philosophical point of view, it is instructive to cite Soviet sources about the differences in their opinion of medical norms in an East-West perspective. Thus, a medical philosopher from the Kazan Medical Institute explicitly contrasts the two as follows: "The ethics of a Soviet Doctor is qualitatively different from the ethics of a bourgeois doctor inasmuch as the goals and tasks of the physician's activities in the two social systems are diametrically opposed" ([4], p. 3). Another commentator writing in the same journal links the Soviet approach more directly to the communist philosophy: "Modern medicine . . . [in its] . . . search for general theory and definition . . . inevitably turns to dialectical materialism, the only scientific philosophical system that interprets correctly the fundamental problems of the essence of man . . . " ([1], p. 241). Further, "In Soviet science, which is based on Marxist-Leninist methodology, the thesis that there is an organic relationship between dialectical materialistic philosophy and medicine emerges as an important theoretical principle that guides scientists in their research" ([8], p. 27). From this, as Loren Graham has pointed out, ethics as well as values also are considered to be subjects of scientific study as is biology, and that this further places Soviet Marxists in what he notes is a "naturalistic fallacy," a position unlike Western views of biomedical ethics. He proceeds to explain, in this light, why Soviets have not treated or have not revealed their interest in ethical issues and dilemmas related to human genetics, such as XYY chromosome debates; not formed ethical review boards at medical facilities; not discussed issues related to use of aborted fetuses for research, and so forth. Using the classification devised by Beauchamp of resolution or closure of an issue, he notes how "diametrically opposed" – to use the Soviet term cited earlier – are their choices, particularly choosing the "sound argument closure" most frequently, and negotiated closure rarely ([8], p. 30). Since a state employee in the U.S.S.R. does not, openly at least, question the standard procedures, attitudes, and protocols predetermined at higher levels, all negotiation is virtually ruled out, because the soundness of the Soviet higher authorities' choices is virtually uncontestable.

One of the standard procedures reflecting system choices for medical ethics in the U.S.S.R. is manifested in the expected behavior of physicians regarding confidentiality. While, in general, one could recognize their behavior regarding the release of confidential information about a patient as being restricted to those who have need to know, this does not extend to revealing to the patient any possible poor and/or fatal prognosis. Thus, patients with cancer in general, especially those who can be expected to die, are not told of their illness, and not informed about the treatment beforehand, and are offered euphemisms which sound less threatening. Thus, a faculty head of the Ryazan Medical Institute writes in the national medical journal of the Ministry of Health of the U.S.S.R. the following:

> The question of keeping the diagnosis of the illness secret from the patient has large practical significance. If the illness is not dangerous or the prognosis is positive, then such a problem does not arise; in this case, there is no need to hide the character of the illness from the patient. It is a completely different matter in regard to an illness which the present level of medical development cannot cure and has an unhappy prognosis ([12], p. 48).

Michael Ryan provides illustrations of the euphemisms used for cancer patients. The patients are told that they have tumors, or neoplasms, or blastomas, or malignant ulcers, not the word "cancer," because (it is asserted) those terms "do not have a sharply negative psychological impact." An interview with the head of the leading cancer institute cited by Ryan demonstrates that the purpose of avoidance is to allay fear and to provide reassurance, to tell the patient the truth "does not mobilize the will – it paralyzes it" ([17], p. 480).

Frequently, it appears that medical institutions will refuse admission to terminally ill patients, which not only avoids the ethical dilemma inherent in the medical secrecy issue, but also avoids "a certain secret statistic." This secret relates to the success indicator of each institution in which the trend in the in-hospital mortality rate is a basic measure. Writing about this secret statistic in *Izvestiya* in early 1982, the author berates the chief physician of the hospital to which a gravely ill person was refused admission. Without any hope for recovery, the author of the article was told, the hospital would not admit the individual; "preference is given to patients who might recover their health" ([21], p. 3).

If admitted, surgery may be performed without informed consent by the responsible person. For example, two years later *Izvestiya* reported that the parent of a 14-year-old child was not informed about an early morning surgery performed without permission. The response indicates that this should have been done for all persons under 16, but was not illegal if the patient's medical condition required surgery immediately even without notifying the parent or guardian ([15], p. 3). It is true that some doctors and institutions are sued for various types of alleged malpractice. According to the figures for 1979, for the U.S.S.R. as a whole, however, 89.0 percent of the allegations were dismissed at the preliminary investigatory stage, 3.5 percent of the accused were vindicated in the court proceedings, and a total of 7.3 percent were guilty as accused ([19], p. 149).[2] More openness about these data may indicate that complaints are so prevalent that they cannot remain undiscussed is a systematic way.

Another change "in the wind" can be seen in the more open, albeit guarded, discussion of euthanasia. Thus, in June of 1984 the Academy of Medical Sciences' journal published an article with the apparent purpose of rejecting this procedure in the U.S.S.R. ([9], pp. 72–77). Noting that euthanasia is strictly prohibited by the Hippocratic oath (which, as noted earlier, Soviet physicians do not utilize) and was "pioneered by Nazi physicians," it nonetheless also notes that it has received much more attention in the West in the last two decades ([9], p. 73). In addition to attacking such decisions because there is the possibility that the diagnosis could be wrong, the authors also note that they believe it is a reflection of "bourgeois individualism" ([9], p. 73). Nonetheless, in the December 1984 issue of *Voprosy filosofii* (*Questions of Philosophy*) another article appeared attacking Western practice while simultaneously reviewing Western literature on the subject. Is this an indirect way of introducing the possibility for discussion and rationalization of future changes not previously discussed also be opened? Given Soviet medical problems in a wide range of areas, perhaps this potential for new approaches may not be totally precluded.

*Center for Population Research,*
*Georgetown University,*
*Washington, D.C.*

## NOTES

[1] Unless otherwise noted, the citations for this and other statements are found in [5] and [6] and unpublished paper on Recent Research on Soviet Health.
[2] Another report on the number and types of complaints is given for Tadzhikistan in the Republic's journal ([24], p. 86).

## BIBLIOGRAPHY

1. Baltanov, R. G.: 1984, 'Pressing Problems of the Relationship Between Marxist Philosophy and Modern Medicine', *Kazanskiy meditsinskiy zhurnal* No. 4, July-August 1984.
2. Burenkov, S. P. *et al.*: 1982, *Zdravookhraneniye v period razvitogo sotsializma. Planirovaniye i upravleniye*, Meditsina, Moscow.
3. Davidovskiy, L.Ya. and Il'yasova, R. N.: 1984, 'Aktual'nyye voprosy povysheniya professional'noy kvalifikatsii pediatrov', *Zdravookhraneniye Kazakhstana*, No. 10, October 1984.
4. Fedyayev, A. P.: 1983, 'Kommunisticheskaya moral' kak osnova vrachebnoy etiki', *Kazanskiy meditsinskiy zhurnal*, No. 1, January-February 1983.
5. Feshbach, M.: 1985, *A Compendium of Soviet Health Statistics*, CIR Staff Paper No. 5. US Bureau of the Census, Washington, DC, January 1985, p. 81.
6. Feshbach, M.: 1982, *The Soviet Union: Population Trends and Dilemmas*, Population Reference Bureau, Vol. 37, No. 2, Washington, DC, August 1982.
7. Feshbach, M.: 1984, testimony for the Joint Economic Committee and House Foreign Relations Committee, US Congress, *Political Economy of the Soviet Union*, May 1984.
8. Graham, L. R.: 1982, 'When Ideology and Controversy Collide: The Case of Soviet Science', *The Hastings Center Report*, **12** (2) 26–32.
9. Ivanyushkin, A.Ya. and Dubova, Ya.A.: 1984, 'Evtanaziya: problemy, suzhdeniya, poisk al'ternativy', *Vestnik Akademii meditsinskikh nauk*, No 6, June 1984.
10. Kaverin, N.: 1984, 'Your Opinion, Doctor? WITHOUT ALLOWANCES FOR YOUTH,' *Meditsinskaya gazeta*, 16 March 1984.
11. Kupov, I.Ya.: 1981, 'Nekotoryye aspekty problemy vrachebnoy tayny', *Sovetskoye zdravookhraneniye* No. 6, June 1981.
12. Ministerstro zdravookhraneniya RSFSR, Leningradskiy nauchno-issledovatel'skiy institut epidemiologii i mikrobiologii imeni Pastera, *Zakonomernosti epidemicheskogo protsessa*, Trudy Tom 61, Leningrad, 1983, p. 12
13. Palkin, A. and Silina G.: 1978, 'The Sector Physician', *Literaturnaya gazeta*, 27 September 1978.
14. 'A Physician's Responsibility', *Sovetskaya Rossiya*, 30 June 1984.
15. 'Pravo patsienta', *Izvestiya*, 25 July 1984.
16. Rutgayzer, V. M. *et al.*: 1980, 'Ob izmerenii effektivnosti zatrat na meditsinskoye obsluzhivaniye', *Seriya ekonomicheskaya* **5** May 1980.
17. Ryan, Michael.: 1979, 'Ethics and the Patient with Cancer', *British Medical Journal*, 2(6188), (25 August), p. 480
18. Ryan, M.: 1978, *The Organization of Soviet Medical Care,* Basil Blackwell, Oxford.

19. Tsaregorodtsev, G. I. and Ivanyshkin, A.Ya.: 1983, 'Meditsina i etika', *Voprosy filosofii* 9 (September), 149.
20. TsSU SSSR: 1983 *Narodnoye khozyaystvo SSSR v 1982 godu, statisticheskiy yezhegodnik*, Finansy i statistika, Moscow.
21. Tutorskaya, S.: 1982, 'A Want of Compassion', *Izvestiya*, 24 January 1982.
22. *Voprosy filosofii*, No. 2, February 1983.
23. Wheat, Mary, M.D., Harold Brownstein, M.D. and Vadim Kvitash, M.D., Ph.D.: 1983, 'Aspects of Medical Care of Soviet Jewish Emigres', *The Western Journal of Medicine*, 139, No. 6, December 1983.
24. *Zdravookhraneniye Tadzhikistana*, No. 2, March-April 1982.
25. *Zhurnal mikrobiologii, epidimiologii i immunologii*, No. 7, June 1984.

ALAN MAYNARD

# THE PUBLIC AND PRIVATE REGULATION OF HEALTH CARE MARKETS

Policy-makers in health care systems throughout the world are facing similar problems and their attempts to grapple with these problems are leading to a sharper definition of the universal policy issues and the alternative ways of dealing with them. The principal issues of health care in the 1980s seem to be cost-containment, best subsumed into the general problem of efficiency and distributional equity.

Once it is recognized that the historical role of health care in the reduction of disease levels and death rates has been limited and less important than improvements in nutrition, water supply, and sewerage [40, 41] and that the majority of health care therapies in use today have not been evaluated in a scientific fashion and hence may lack efficacy despite their high cost [10], the true magnitude of the problems involved in evaluating competing health care policies can be seen. It is difficult to identify and eradicate existing inefficient practices, and it is difficult for policy makers to oppose the introduction of new therapies. The tradition of the medical profession and policy makers has been to assume that innovations are efficacious and to commit large amounts of resources on the basis of faith rather than on the basis of evidence of relative advantage in outcomes and costs.

The design of health care systems, both "socialized" and "private", has tended to reduce the price barrier to consumption to the patient, care is provided at zero price or at a highly subsidized price, and has ensured that the cost constraints on producers, in particular doctors, are minimal: the costs are borne by insurers, sickness funds, or the taxpayer, i.e., the third-party pays. Thus health care systems encourage patients to maximize consumption regardless of costs, and producers (doctors) to maximize, within the constraints of their ignorance about efficacy, benefits regardless of costs. Both demanders and suppliers have no incentives to compare and trade-off the costs and benefits of competing health care therapies at the margin.

The behavior of demanders and suppliers has inevitably led to a rapid escalation in the costs of health care and the adoption of cost containment policies. These policies, usually quite naive and concerned with

*Hans-Martin Sass and Robert U. Massey (eds.), Health Care Systems,* pp. 133–166.

the limitation of increases in expenditure and not with the effects of such policies on the price, quality, and quantity of health care delivered, have been adopted in a variety of guises in a majority of countries. In West Germany the Concerted Action program has sought to limit budget expansion by mutual agreement between producer, financer, and beneficiary interests. In Belgium cost inflation has led to the discussion of radical changes in public provision and the development of private provision. Similar developments have taken place throughout the European Community ([2, 3]; see also [45, 27]) and in North America where Diagnosis-Related Groups (DRG) price controls are, like all such mechanisms, directed at controlling costs and inputs rather than ensuring efficiency, i.e., trading off costs and benefits (improved health status or increase in quality adjusted life years [QALYs] at the margin.

Some of these measures have been discussed with remarkably little attention being paid to the need for more micro-evaluation (cost-benefit and cost-effectiveness analysis) and for the development of incentive systems which induce decision makers, both patients and doctors, to take action to minimize costs and maximize benefits. The lack of such incentives has contributed to the rapid escalation in costs and expenditures in health care, and to the situation where many of the therapies in use remain of unknown or dubious validity [10, 8].

Any discussion of cost-containment and efficiency is usually complicated by the propensity of policy-makers to introduce, often implicitly, distributional issues. These take a variety of forms. Most health care systems are characterized by explicit policies to secure advantages to particular geographical areas and socio-economic groups.

Geographical inequalities in the provision of health care facilities have led to policies to equalize either financial capacity or, more directly, the provision of facilities, especially beds. In 1970, the first resource allocation formula was introduced in England, and since then a variety of formulae have been evolved for the constituent parts of the United Kingdom [37]. In 1971, the French adopted the "health map" which is a mechanism used to direct hospital bed resources to under provided areas. In 1972, the Germans began to provide federal funds for hospital beds, provided new construction, and reduced geographical inequalities in provision. More recently the Italians, the Dutch, and the New Zealanders have begun to discuss RAWP-type, formulae, and similar attempts to control the geographical allocation of resources have been implemented in North America.

Whilst the degree of government financial involvement in the provision of health care varies across countries, all developed countries provide some subsidies to give special support to poor groups. Although the magnitude of involvement varies, no country has failed to recognize that some groups, usually the poor, the aged, and the chronic sick (here used as a collective term to cover the mentally ill, the mentally and physically handicapped, and those with long-term illnesses such as cancer and TB), require special treatment which is bound to involve special funding. This concern manifests itself as Medicare and Medicaid in the United States, federal and state finance of health care for the "needy" in Australia, the comprehensive "heavy risks" insurance program in the Netherlands, and many similar programs elsewhere. The experience of these countries is that a substantial section of the community is unlikely to be able to buy health care insurance coverage in the market, and that as a consequence the state must offer substantial subsidies of one form and another. With such subsidies comes government concern about "value for money" (efficiency) and access (distribution). This concern seems to generate inevitably further state regulation of the health care market.

This propensity for the need for state regulating policies to develop is seen clearly in the pro-competitive environment in the USA. The Reagan administration's pro-competitive policies are reducing the capacity of providers to cross subsidize care groups and provide hospitalization for the poor. Thus out of the competitive market is arising the acute need to provide cover for 35 million poor Americans who have little or no insurance. Competition creates new needs for state regulation.

It is against this international background of common problems and the similarity of policy responses in nations with very different health care systems, that this paper is written. The current debate about health care policy, wherever it takes place, is set against a background of severe public expenditure constraints, little knowledge about the efficiency with which care is produced, and ill-defined (and usually unevaluated) distributional goals. Because of these background characteristics, in particular because of the numerous, generally perverse, incentives which do not induce economical behavior in decision-makers, the state has sought to increase its regulation of this market place, or divest itself of some of its responsibilities by allocating them to the private sector.

Whether health care is provided and financed publicly or privately, these problems are ubiquitous. If the response of the state is to regulate, it will seek to use existing and develop new policy instruments to affect the prices, quantities, and qualities of health care services. This task is fraught with difficulties because of our ignorance of the nature of the input/output link and because of the economic power of many of the groups demanding and supplying health care.

If the state fails in its efforts to regulate the public sector's provision and finance of health care, it can divest itself of these functions, i.e., they can be privatized. However, whilst such policies may reduce public expenditure, the overall effect on total (social) expenditure on health care is unlikely to lead to economy, as the fundamental problems involved in the delivery of health care remain. The private financers and providers will ultimately face the same pressures that are generated in a public system, as can be seen in the United States. They will respond to these pressures by self-regulation, i.e., the private insurers will seek to control the prices, quantities, and quality of health care.

So whether the state or private agencies finance and provide health care, the fundamental problems of this market are similar, and regulatory activity, defined as action by financers and providers to manipulate prices, quantities, and quality, is unavoidable and ubiquitous [37].

This thesis will be developed below. After a discussion of ideology, which demonstrates how much of the contemporary debate about health care is dominated by values, the nature of the health care market will be elaborated. This is followed by a discussion of the nature of the liberal/market and collectivist/socialist options in health care and the development of the thesis that regulations in the health care market are ubiquitous and unavoidable.

## I. THE IDEOLOGY AND THE OBJECTIVES OF THE REFORMERS

A consistent ideological position characterizes those who, over the last twenty years, have sought to reform, or even to abolish, the National Health Service. The section will summarize these arguments and then compare and contrast the competing ideologies in an attempt to identify the objectives of the reformers and the defenders of the NHS.

The first carefully articulated critique of the NHS by a member of the reformers group was published in the early 1960s and written by Dennis Lees [30]. Lees argued that the market was a superior mechanism (to the ballot box) for registering consumer choices and that the most efficient means of providing care was in a competitive market which created incentives for decision-makers to take appropriate action to minimize costs rather than in a state monopoly where there were limited incentives. He argued that patients could be subsidized if necessary and that the state "would retain a special responsibility for the mentally ill and the chronic sick" ([30], p. 77). Lees argued that because of the insensitivity of the vote mechanism, health care had been underfinanced in the NHS and he advocated the common Institute of Economic Affairs preference for decentralized rather than centralized decision making.

The "Jones" Report [28] also asserted that the NHS was underfinanced, although it offered no means of determining what this term meant in any exact sense. The writers of the Report argued that additional finance could not be acquired from taxation and that the bulk of everyday health care for the normally healthy should be financed out of private insurance. This insurance would be compulsory for all to some (government determined) national minimum but consumers would be free to purchase cover over and above this level. The premia of these policies would be related to costs and the Report emphasized the need for state regulation of the insurance industry to prevent cost escalation ([28], p. 156). Those unable to pay the compulsory premiums would be subsidized via the tax system: the poor, the aged, and the chronic sick would be credited with compulsory insurance cover, and their family income subsidies would decline *pari passu* with income increases until they paid the full cost of the premium. These subsidies would be financed out of government taxes.

The more recent reformist literature on the financing of the NHS tends to be more polemical and generally less carefully articulated. Two books in this literature [23, 33] are written for American audiences as contributions to the debate about national health insurance in the USA, and seek to demonstrate the weaknesses of the NHS. This they do in a highly imperfect manner; the Royal Commission on the NHS [15] provides a much more analytical approach to the Service's defects than is available in the limited, and sometimes inaccurate, work of Goodman in particular. (This type of analysis is also used in Aaron and Schwartz's [1] comparison of the US and UK health care systems.)

Goodman and Lindsay contribute to the Litmus Papers [47]. The style of these papers is polemical and very superficial. No serious issues are discussed and over twenty authors offer often emotive and always very short "pyrotechnics" on disparate issues. The themes which come out of these papers are familiar: rejection of the "monopoly bureaucratic power" of the NHS, a preference for "choice", decentralization, and private insurance, and an assertion that the NHS is a more inefficient and inequitable health care system than some of its private and insurance financed rivals.

Many of these preferences are consistent with what, following Donabedian [16], we might call the market ideology. The implications of this ideology, and its collectivist alternative, are discussed in the context of health care by Culyer, Maynard, and Williams [11]. Following the framework used there it is possible to distinguish between the two ideologies in relation to their beliefs about personal responsibility, social concern, freedom, and equality:

*Personal Responsibility*

*Market view.* Personal responsibility for achievement is seen as very important, and this is weakened if people are offered unearned rewards. (e.g., "free" health care). Furthermore, such unearned rewards weaken the motive force that assures economic well-being (e.g., incentives to work to save and to take risks). The weakening of the motive force which generates economic growth also undermines moral well-being, because of the intimate connection between moral well-being and the personal effort to achieve.

*Collectivist view.* While accepting the desirability of some personal incentives to achieve, economic failure is not equated with moral depravity or social worthlessness.

*Social Concern*

Social Darwinism generates an apparently cruel indifference to the fate of those who fail in the economic system. A less extreme position is that charity expressed and carried out under private auspices is the proper vehicle to meet social concern. Charity, however, needs to be exercised under carefully prescribed conditions, for example, such that the potential recipient must first mobilize all his own resources and when helped must not be in as favorable a position as those who are self-supporting (the principle of "lesser eligibility"). Without such provisions, incentives, economic growth, and moral well-being will be reduced.

Private charitable action is rejected, and is seen as potentially dangerous morally (because it is often demeaning to the recipient and corrupting to the donor) and usually inequitable. It seems preferable to construct social mechanisms that create and sustain self-sufficiency, and are accessible according to precise rules concerning entitlement which are applied equitably and explicitly sanctioned by society at large.

*Freedom*

Freedom is to be sought as a supreme good in itself. Compulsion attenuates both personal responsibility and individualistic and voluntary expressions of social concern. Centralized health planning and a large governmental role in health care financing are seen as an unwarranted reduction of the freedom of clients (patients) as well as of health professionals (providers), and private medicine is therefore viewed as a bulwark against totalitarianism.

Freedom is seen as the presence of real opportunities of choice, and although economic constraints are less openly coercive than political constraints, they are nonetheless real and often the effective limits on choice. Freedom is not indivisible, but may be sacrificed in one respect in order to obtain greater freedom in some other. Government is not seen as an external threat to individuals in the society, but the means by which individuals achieve greater freedom of choice (that is, greater real freedom).

*Equality*

Equality before the law is the key concept, with clear precedence being given to freedom over equality wherever the two conflict.

Since the only moral justification for using personal achievement as the basis for distributing rewards is that everyone has equal opportunities for such achievement, then the main emphasis is on equality for opportunity, and, where this cannot be assured, the moral worth of achievement is thereby undermined. Equality is seen as an extension to the many of the freedoms actually enjoyed only by the few.

The critics of the NHS have a strong preference for freedom as elaborated in the market ideology, and from this prime objective it is possible to identify their preferred means to attain this goal:

*Selectivity*

Health care is part of the reward system and the well-being of society depends on this reward system. As a result access to health care is to be determined largely by ability and willingness to pay. Those without the ability to pay would be guaranteed some minimal level of access to health care. Thus this approach regards inequality as efficient: it is the engine of economic growth and the guarantee of freedom. Also it is implies private (insurance) finance of health care.

*Private Ownership*

Freedom requires decentralization and private (usually non-profit making) ownership of the means of production (e.g., hospitals) with only minimal government control of finance and resource allocation.

*Rewards*

Providers of care (e.g., doctors) to be directly rewarded according to market forces, usually with a fee-per-item of service system of remuneration for doctors and *per diem* fees paid to hospitals.

Those rejecting this ideology (e.g., the collectivist supporters of the NHS) regard equality of access to health care and of health status as their prime objectives. The means by which this goal can be attained are:

*Universality*

Health care is not a part of the reward system, and should be allocated on the basis of need or the patients' capacity to benefit from care regardless of their willingness or ability to pay. Universality implies public finance of health care.

*Public ownership*

Equality requires centralization and public ownership of the means of production (e.g., hospitals) with extensive government control of finance and resource allocation.

*Rewards*

Providers of care will be rewarded by the outcome of bureaucratic bargaining procedures between monopoly providers (e.g., doctors) and monopsonistic financers (the NHS).

The goals of "freedom" and "equality" are imprecise. The means by which these goals will be attained are generally asserted to be efficient. Usually health care systems have goals whose nature is imprecise and even where some degree of precision is attained in goal-setting, the discussion of the efficiency of the means is a matter of supposition rather than of fact. The contending ideologies contain empirical statements whose verity could and should be tested. Unfortunately, at this level of argument such scientific rigor is noticeable by its absence and, as a consequence, we have, through time, the recurring advocacy of privatization or socialization. Whilst such advocacy may be the stuff of politics, it guarantees, by its dominance and use of scarce intellectual and policy making talent, that policy formulation is superficial and that scarce resources in policy analysis are wasted.

The naive pyrotechnics of the contending ideologists must be set aside. All parties, whatever their ideological predilections, need to recognize the precise nature of the health care market and the ramifications for policy, public and private, of the imperfections of this market.

## II. THE NATURE OF THE HEALTH CARE MARKET

### *Monopolistic non-competitive markets*

The advocates of the market solution recognize that a major obstacle to the smooth running of a competitive market is monopoly power, or the ability of the sellers of health care services to influence (to their advantage) the prices of their services. As Adam Smith observed over two hundred years ago:

> People of the same trade seldom meet together, even for merriment and diversion, but conversation ends in a conspiracy against the public, or in some contrivance to raise prices ([48], I, p. 117).

Thus the health market is not competitively organized: it is highly imperfect and characterized by powerful monopolies which inhibit the free play of the competitive "hidden" hand of the market.

The market for health care is characterized also by uncertainty: the patient does not know when he will be ill, and the patient may be relatively inefficient in determining the appropriate diagnosis and treatment. Generally the comparative knowledge of doctors may be greater and society, in order to protect its members from "quacks" and uncertainty, has created social institutions and professional regulation to ensure that doctors are "expert" in their trade and do offer efficacious health care [4].

The counter argument to this position has been set out by Friedman [21] who argues that professional regulation is used to further the interests of the profession. Citing US experience, he argues that the profession by controlling entry into medical schools, has restricted the supply of doctors, and their inflated incomes over the life cycle. Furthermore, it is argued that professional control has permitted discrimination of a racial, sexual, and color nature, and that the profession has inflated the "quality" of doctors (measured in terms of duration of training) regardless of its costs and in the absence of scientific evidence that quality of physicians (measured in terms of their impact on the health status of patients) is adequate.

Whether the Arrow or Friedman position is adopted, it is clear that the state, in all western countries, has given the medical profession considerable monopoly powers to influence the quantity and quality of its services and its own remuneration. The evidence about the effects of this seems to indicate that the profession has used its power both in public interest and in its own self-interest [20, 31].

Thus professional organizations use their power to influence their remuneration and hence all such organizations will seek to protect health care expenditure, as it is their income, and to advocate its (and their!) increase.

The uncertainty associated with health care, which led to state enforcement of professional power, creates additional problems. The individual patient is typically quite poorly informed about diagnostic techniques and treatment. Although it is generally an individual patient decision to make the first contact with the health care system, after that much of the demand decision-making may be delegated by the patient to the doctor because of the asymmetry in information between patient and provider. Thus the patient uses the doctor as his agent to make health care demand decisions for him and it can be argued that the supplier of health care (the doctor) is the demander rather than the patient. (For a summary of the literature see [50].)

If this agency relationship is used neutrally by the doctor, that is, in the interests of the patient only, there may be limited implications for the efficient use of resources. The doctor may, however, use this relationship non-neutrally, e.g., to increase his income by raising the demands for his services, by increasing the size of his "empire", and by pressing for other ends which are not conducive to efficiency. This theory of physician induced demand, that doctors can create demand for their services, has been tested imperfectly. Some (e.g., [22]) argue that a 10 percent increase in the supply of surgeons, all else held constant, leads to a three per cent increase in surgical activity. Other evidence indicates that not only can physicians generate demand for their own services, they may also be able to raise their fees as their number increases [50].

The effects of professional power and the agency relationship cannot be ignored in any discussion of the supply of health care in particular and health care policy in general. Professional power has given doctors control over their training and a strong influence over the supply of doctors and their earnings. This power has led to increases of inputs (training time) in training but little evaluation of this training and little investment in managerial and economic skills, or in scientific evaluation, despite the fact that doctors are managers of scarce resources. Furthermore, monopoly power may be used to create demand for doctor services. Thus the existing market for doctors is highly imperfect (i.e., non-competitive) and this fact cannot be ignored in policy discussions. Not only is the market for doctors' services characterized by monopoly, there are also severe restrictions on competition elsewhere in the health care market. The pharmaceutical companies are highly regulated by the state and have been given monopoly powers by it, e.g., patient legislation. The hospital sector is either state owned, at local or central level, or run by private non-profit-making bodies. The incentives to seek out cost-effective practices in both the state sector and the private non-profit making sector may be similar and weak; in both types of organization the managers get no direct reward in any cost savings so why should they seek them out? Not only are incentives weak, there is also scope for "producer agreements", often protected by state legislation, which may limit competition in the market for hospital beds. Thus the supply side of the health care market is characterized by a lack of competition and powerful, well established, state supported monopoly power. It is not surprising that in this environment we know little about the efficiency with which resources are used: the active pursuit of efficiency is a threat to the income of providers! Efficiency, in economic terms, means that resources are used in a way which ensures the least cost production of these goods and services which are most highly valued by society. In the context of the health care market the doctor may regard a procedure as "efficient" ( = clinically effective) if it is the most effective therapy available to improve the health status of the individual (increase in QALYs) suffering from a particular complaint; often evidence to reach this conclusion will be absent due to the lack of evaluation of practice. A procedure may be cost-effective if it is the cheapest way of achieving a given therapeutic (level of QALYs) end, with the value of this end (outcome) being unquestioned. A procedure is economically efficient if cost is minimized per unit of socially desired QALY or if the output of socially desired QALYs is maximized from a given budget. In the health care market, the supply side forces which induce the achievement of this outcome, the efficient use of resources, are limited. Clinical freedom and professional power have led to a failure to create mechanisms which ensure that practitioners evaluate clinical outcomes and their cost implications.

Such behavior is not only inefficient but it is also unethical: resources used inefficiently are not available to treat those who are in the queue and who could potentially gain more, in terms of health status, if they were treated. Inefficiency deprives potential patients of care from which they could benefit.

*Doctor demands, not patient demands*

The effects of these "supply side" failures in the market mechanism have been compounded by demand side institutions. The effects on the demand side of the NHS and private insurance-type institutions is similar. The NHS is tax financed and removes the price barrier to consumption. With the demand for health care infinite, and supply finite and fixed by goverment fiat, the NHS inevitably manifests excess demand, i.e., waiting lists. Scarce resources (supply) are allocated amongst competing demanders by doctors according to imprecise notions of "need" [51], e.g., scarce kidney-machines are allocated in the NHS to young married people rather than to single elderly people.

The insurance system allocates resources according to willingness and ability to pay. Insurance coverage may remove the price barrier to consumption just like the NHS. Excess demand is, however, rationed in this system by prices (co-insurance). Prices reduce the demands of the poor more than the rich, and of the less ill more than the severely ill [9, 36, 43], and such effects have ambiguous consequences for health status [6]. Prices remove waiting-lists, those unable to finance care can be left in pain and/or to die without benefit of treatment in due course! Those who can pay the price of the co-insurance and the insurance premiums can enter the market, and, subject to the agency relationship, will get access to care.

In both the NHS and the insurance system the doctor has a crucial role as an allocator of resources. In the NHS he decides who will be treated or not treated, and how. In the insurance system, subject to willingness and ability to pay (which may induce some limited competition by doctors for customers), the doctor decides who will be treated or not treated, and how. The doctor is the "guardian" who determines treatment patterns. Because of the ethic, which dominates his training, to provide the best care for the patient in his charge, the doctor seeks to maximize the benefits of health care regardless of cost. In the NHS the state meets the cost of care. In the insurance system, the insurance fund or company meets the cost of care. In neither regime does the doctor have to evaluate costs and benefits at the margin and make efficient decisions. The demand and supply sides of the health care market contain muted and limited devices to persuade doctors and patients to be efficient users of society's scarce health care resources: the bill is picked up either by the long-suffering tax-payer or by the insurance company; the third-party (not the supplier or demander of care) pays.

### *The many imperfections of the health care market*

Those people who have adopted the market ideology tend to believe that markets will be more efficient than governments in allocating resources because the competitive process rewards those who wish to minimize costs and provides incentives for decision-makers to behave efficiently. This paradigm may, however, be defective in the health care sector. Significant market imperfections, professional power, the agency relationship, and the nature of the insurance mechanism blunt the competitive system considerably, making it likely that the outcomes of the market will be uneconomic and not necessarily superior to those of alternative institutions such as the NHS. The superiority, if it exists, of the market has to be identified and quantified, not merely asserted as a matter of faith rather than scientific fact.

## III. THE CHARACTERISTICS OF THE ALTERNATIVE SCENARIOS

The market and collectivist ideologies identify objectives and associated means which could be used to design a health care system. However, at the level of generalized ideology, such designs are imprecise and beg greater definition and careful analysis.

### *The market alternative*

The market ideology continues to be held and articulated by prominent members of the Thatcher Conservative Government. For instance, Sir Geoffrey Howe is convinced that high public expenditure is financed by high tax-rates which "stifle incentives and inhibit economic growth" ([26], p. 10). He favors the development of the private sector and the possible wider use of charges in the NHS to make patients aware of the costs of care. All these arguments are consistent with the views expressed by the "Jones" Report [28] of which Howe was a member. They have been reflected by Leon Brittan, Chief Secretary of the Treasury, when in a speech to the Institute for Fiscal Studies [5] he advocated extensive privatization of, *inter alia*, the NHS. Such assertions continue to be made by members of the Thatcher administration.

*The creation of competitive conditions*

The efficacy of the private health care market depends on reform and the development of new institutions and incentives which would make the market work in a competitive rather than a monopolistic fashion. The achievement of competition could not be acquired without some cost.

As Milton Friedman argues, "there is no such thing as a free lunch"; what would you have to pay in order to achieve the Friedmanite Nirvana of competition and consumer sovereignity? The first necessary reform as set out by Friedman, ([21], Chapter 9), and elaborated by Green [24] would be the abolition of professional licensure as its exists in the UK, the USA, and Europe today. Friedman argues that professional power has been used to raise professional incomes above their market rates, and to inhibit the exploitation of substitution possibilities; although we know nurses could do many of the tasks of doctors, and dental assistants could do many of the tasks of dentists equally well and cheaper, such substitution of cheap inputs for expensive inputs is not permitted. Indeed the British Government, advised by the dental profession, closed the only training school for dental assistants in the early 1980s.

If the barrier to specialization and cost minimization were removed by the abolition of professional licensure, health care might be provided more efficiently but doctors' incomes could be reduced considerably. This outcome would arise from competition and substitution reducing doctors' fees and reducing their opportunities for employment. Thus the implication of Friedman's radical policy proposal is potentially severe income losses for powerful professional groups.

Let us assume that Geoffrey Howe and Leon Brittan are sincere in their wish for privatization and competition, and that they are prepared to accept, as a necessary cost of achieving this goal, a reduction in the affluence of powerful health care professionals. If monopoly power is mitigated, competitive forces being permitted to generate uncertainty about tenure and salaries for professionals, can the market work efficiently?

Only a theoretical answer can be given to this question since the evidence is lacking. In theory it could, however, work over time. The first reform after the abolition of professional power would be the adoption of the use of the price mechanism as a means of allocating scarce health care. This allocation method, which is regarded as functional by liberals as it creates incentives which, it is alleged, are the engine of efficient resource use, would require private financers of health care to use co-payments to limit demand.

*Co-payments cost-sharing*

At present the UK provident companies are not fully developed insurance carriers in that their monitoring and rationing of demand is limited and unsophisticated. In the present limited UK market for private health care, BUPA and its rivals generally meet all expenditures, and the financial burden on patients, apart from contributions, is minimal. The absence of a price barrier to consumption may induce increased utilization, what the Americans term "moral hazard", where the cost of marginal units of health care may exceed their value. If the market for private health care expanded, and insurers carried a wider burden of risks, escalating expenditure would lead to the use of co-payments as a device to ration the demands of patients.

The effects of such co-payments, co-insurance (paying a percentage of the bill) or deductibles (paying a given lump sum) on the demand for health care by patients seems to be quite clear and consistent with the predictions of economic theory: a rise in price will reduce the quantity demand and the magnitude of this effect, the price elasticity of demand, will depend on the extent of the price rise and the nature of the service which is being priced (e.g., the demand for essential emergency care is less likely to be reduced by a price rise than cosmetic surgery) [9, 36]. These results were, however, generated by experiments and statistical analysis of events which were imperfect in scientific design and execution. In an attempt to improve the available data about elasticities and, in particular, to determine the nature of the utilization-health status link, the United States Government financed a study by the Rand Corporation, consisting of multi-center trials with varying co-insurance rates and in varying geographical locations. This work shows that expenditure per person does respond to variations in cost-sharing: it is

50 percent higher with no cost-sharing compared to 95 percent co-insurance up to a maximum of $1,000, and that as cost-sharing declines patients seek more care. Furthermore, this study shows, contrary to earlier *ad hoc* studies, that the poor (the bottom one-third of the income distribution) and the rich (the top one-third of the income distribution) respond to cost-sharing in similar ways. However, the response of the poor would have been greater than that of the rich if the cost-sharing arrangements had not been income related. Hospital utilization does not appear to be affected by coinsurance.

Thus, the Rand study shows that some health care cost inflation can be contained by cost-sharing, that for situations where cost-sharing is related to income the utilization effects on rich and poor are similar, and that hospital cost inflation cannot be contained by copayments. These utilization effects tell us nothing, however, about whether changes in use affect, adversely or beneficially, the health status of the patient. A familiar argument in the UK and the USA is that many patients present "unnecessarily". This argument has led to the joke that a person who presents to a doctor with nothing wrong, should go and see a doctor, i.e., the definition of an "unnecessary" presentation by the doctor may not coincide with the patient's definition of his problems. The use of cost-sharing to reduce "unnecessary" demands may in fact reduce the demand for care of people who have identifiable illnesses. To what extent does cost-sharing (copayments) reduce the demands for care of people who are ill and whose delayed treatment may lead to greater ill-health and higher health care "repair" costs in the long run? The Rand study offers some insights into these questions [6] but it is clear from these data that a simple, definitive answer is not possible.

The effect of co-payment is that utilization patterns are affected. The patient faces a price and has an incentive to economize in the use of health care services. He can either decide to reduce his consumption or seek out cheaper alternative forms of health care provision. This latter impulse could generate demands for the substitute (to doctors and dentists for instance) practitioners and develop the fruits of the Friedmanite destruction of professional monopolistic power.

*Alternative insurance plans*

Another effect will be that the patient will be faced in a competitive market by a choice of alternative insurance plans. These plans will offer varying degrees of coverage at various prices. The patient will shop around to identify and select the policy of his choice and could be offered regular (e.g., annual) reselection of coverage from competing insurance options. The insurers will have an incentive to market policies for which there is a demand at a price which the patient is prepared to pay.

In order to compete for the custom of the cost-conscious consumer, the insurer will have an incentive to minimize costs. One way of achieving this goal is to evaluate health care provision and rigorously monitor the practice and fees of the providers. Thus, the insurers will create information systems, evaluate provision, establish practice norms for practitioners, monitor behavior, and develop incentive systems to control deviant behavior, and ensure that practitioners are efficient users of scarce resources. Such behavior can be seen in some United States hospitals already where administration costs are high, and used in order to police with great care the operation of the hospital's budget. It can be seen also in the United Kingdom, where BUPA has now recognized that some providers (e.g., Hospital Corporation of America) are high cost. In an effort to reduce costs, BUPA is embarking on the difficult process of evolving cost and quality control mechanisms which are commonplace in North America.

Yet such intervention is likely to raise administrative costs. The Organization for Economic Cooperation and Development [44] has shown that the NHS is cheap to run: its administrative costs were 2.6 percent of total health care expenditure in the mid 1970s compared to 10.8 percent in France, 10.6 in Belgium, 6.5 in the Netherlands and 5.0 in Germany. These costs contain two elements: revenue collection and resource management. The costs of revenue collection for the NHS are small as the tax system is cheaper than an insurance system. It is difficult to separate out resource management costs from these figures but it is likely that they would be higher in an insurance system with duplication. Whether the resources available to finance management are used effectively and are adequate is unlikely.

The reintroduction of the price mechanism gives the patient an incentive to seek the lowest cost method of providing efficacious care (i.e., substitution possibilities are exploited) and the lowest cost provider of insurance cover. Price searching (the seeking of the lowest price) by consumers gives the providers and the insurers an incentive to compete and to minimize costs. The predicted effect of the competitive process is that prices will reflect costs at the margin and the surplus profits of inefficient and monopolistic providers and insurers will be dissipated.

All these conclusions are theoretical statements based on the famous economic assumption of *ceteris paribus*: other things being equal. The things which are being held equal are no monopoly power (i.e., a reduction in income for existing powerful professional groups, and no cartel-type behavior by insurers), the exploitation of substitution possibilities (i.e., reduced employment opportunities for some providers) and the availability of alternative and competing providers and insurers for consumers (i.e., choice is available and sustained).

These conditions require a radical change in the existing medical care market. Friedman is clearly aware of this as are, no doubt, our politicians. The political costs of disturbing the *status quo* should not be underestimated: powerful producer and financer groups will not take kindly to their members' income and jobs being destroyed by the competitive process.

Furthermore, it is debatable whether this competitive Nirvana is attainable. The quotation from Adam Smith above indicates that capitalists are the enemies of capitalism. Capitalists can gain, in income and in a quieter life, by rigging the market, covertly or explicitly, and acquiring monopolistic powers.If capitalists manage to acquire such powers they destroy the competitive process because they obtain the ability to fix prices and to determine outputs and the quality of care. In any market the forces of competition and monopoly are in continual conflict. The monopolist rigs prices to his advantage and acquires profits from his action. These profits provide an incentive for a competitor to enter the market place and compete for these profits. This process does take time and the monopolist can create many barriers to prevent new firms entering the market. Two important conclusions can be drawn from this analysis. First, it is theoretical and its validity has to be demonstrated with evidence from operating markets. The lack of correspondence between the predicted outcomes and the evidence of the market place indicates that health care markets do not conform to the competitive ideal. No doubt this evidence could and should be elaborated by carefully designed and implemented controlled trials. In the United States the main thrust of reform is to "make the market work" (e.g., [17]) and conform more closely to the competitive ideal. However, many Americans are recognizing the difficulties involved in this

this approach and it is uncertain whether the political will necessary for its adoption in any significant way will be present.

The second conclusion is that the competitive market, even if it could be attained, would have to be sustained by considerable government intervention. The natural inclination of a competitive market is for it to become monopolistic. Thus Enthoven *advocates legislation* to ensure that consumers are offered an effective choice of insurers annually. Similarly, Friedman's advocacy of the removal of licensure would almost certainly be challenged by the self-interest of professionals. Would the competitive market survive with the removal of licensure alone or would it need to be evaluated and monitored continually by anti-trust bodies? The liberal-marketeers admit a role for the state as the "policeman of the market place" who protects the competitive process. It is likely that this role would be substantial and executed with some difficulty in the health care market.

The nature of outcomes of the competitive process and the need for government supervision of the market are conjectural: the evidence from competitive markets is absent and many liberals doubt whether politicians could take on and survive a conflict of vested monopoly interests. However, let us assume that politicians were men of principle, that they could convince their electorates that these principles were advantageous, and a competitive market was somehow created and sustained, what other problems arise with the liberal alternative in health care?

*Competition and distribution*

The competitive market is unlikely to create a distribution of income and wealth that conforms with the distributional goals of any Western society, e.g., the competitive experiment in the United States leaves up to 35 million Americans with little hospital cover and virtually no primary care. The liberal response to this problem is not to abolish the market but to let it operate freely and redistribute income directly by the tax system. Unfortunately, redistributing financial capacity to buy goods and services from the rich to the poor will not necessarily lead to greater equality in health care consumption or even perhaps health status. Such redistribution of cash, if it is to be effective, must be spent on health care and this health care must be the more efficient means of reducing the inequalities in health which society wishes to mitigate.

The fear that people might spend cash on the "wrong things," such as alcohol and tobacco which may worsen health status, has led liberals such as Lees [30] and Seldon [46] to advocate redistribution by vouchers. Vouchers would tie redistributed resources to health care: they represent a judgement by liberals that in the short run, at least, paternalism is necessary although the liberals hope that, in time, the consumers will learn to be responsible, value health care appropriately, and be given cash rather than tied (voucher) resources.

The use of the voucher is not only a deviation from the "pure" liberal-market position, it is only tied to health care and this may not be the most efficient way of improving health status. As has been argued in economic theory [25, 42] and in the historical-epidemiological literature [40, 41] there is good reason to believe that many inputs, of which health care is only one, affect health status, and that higher income, better nutrition, and public health measures which induce environmental and behavioral change may be more productive than medical care in improving the quality and length of life. Cash redistribution would, perhaps, be more efficient in these circumstances.

Another complication in the liberal-market model is that it regards incentives as the engine of economic growth and moral well-being. Any redistribution by fiscal measures will reduce the incentives to save, to work, and to take risks. The more significant the redistribution that takes place the more the liberals enter into conflicts with their own ideology and have to face problems similar to their collectivist colleagues. Clearly the preference for redistribution among liberals varies considerably (as can be seen from the inspection of output of the Institute of Economic Affairs) from the relatively limited (Friedman) to the more substantial (Peacock and Wiseman).

The liberal would not favor redistribution through institutions, rather he prefers redistribution to individuals. Thus, rather than adjust the existing budget allocation formulae, the typical liberal, for instance Friedman, would prefer direct redistribution to individuals by a negative income taxation. Within a competitive market system with private production of health care this may have many attractions, but in a mixed public-private system the administrative costs may be substantial and the problem of devising a sensible and sensitive method of redistribution might be considerable.

The preference of the liberal for the private production of health care and decentralized decision making is likely to make the competitive market difficult to regulate with regard to distributive goals. The logic of the liberal position is for cash redistribution and private production of health care.

Whether such arrangements would enable society to achieve its distributive goals is a moot point. As liberals have been the first to note (e.g., [49]) the motivation of bureaucrats and politicians is such that the achievement of the goals of social policies are often frustrated. The radical reforms and limited redistribution envisaged by Friedman are likely to injure many powerful pressure groups and their response may be to oppose measures which improve efficiency and permit the effective pursuit of distributional goals. The risk is that we may get some form of uncompetitive market and no effective redistribution because of these pressures. Such an outcome, like all the liberal arguments, is theoretical and awaits empirical verification.

*Incentives, producer behavior, and expenditure inflation*

An important aspect of the market mechanism is the effect of incentive structures and the behavior of producers. Coinsurance (prices) causes consumer behavior to alter. What is the effect of remuneration systems on the behavior producers? In many health care markets, and in the minds of many liberals, incentives should be determined by the 'competitive process'. This process, influenced, if not determined, by powerful professional influences, usually results in doctors being paid on a fee-per-item of service basis (piece rates) and hospitals being financed by *per diem* payments.

It is generally accepted amongst economists that both theory and evidence lead to the conclusion that a system of payment on a fee-per-item of service basis has both costs and benefits. If we assume that the doctor is an individual concerned with maximizing his income and if he is paid on an item-for-service basis, it is predictable that he would maximize the number of items-per-service delivered to the patient. Thus the payment system induces the doctor to maximize quantity, perhaps with little regard to quality. The testing of these predictions has not been precise but arguments have been put forward to the effect that surgical acts are more numerous, more prescriptions are written, and many other indicators show a comparatively higher level of activity than in health care systems without such payment rates.

Such an outcome raises questions about "clinical necessity:" is the higher level of surgical activity in the Eastern United States indicative of greater illness in the United States or is it due to the fact that there are more surgeons than in England and these surgeons are paid fees-per-item of service? (Bunker, 1970). Because of the lack of evaluation no sensible conclusion can be drawn about the quality of care in both diagnosis and treatment in regimes where such payment schemes exist. There is a belief on the part of government in Britain that piece rates affect behavior; after all, such payments were introduced into the remueration of general practitioners in order to induce them to provide preventive services such as vaccination, immunization, and ante-natal care.

Fees-per-item of service may induce doctors to work longer hours and to produce more medical "acts" (for which they are paid). Such inducements must be introduced with caution when we know so little about the efficiency of the services provided. Also such activity leads to greater expenditure. Clearly doctors do not have an infinite desire for income; they value leisure, too. Yet fees-per-item of service may make leisure more expensive, in terms of fees foregone, and the effect on overall health expenditure may be inflationary.

The payment of hospitals on a *per diem* (or patient daily rate) basis guarantees income for the hospital provided they can keep their beds full.

The predicted effect of this from that dismal scientist, the economist, is that the hospital would have an incentive to keep its beds full and that bed stocks would be high and lengths of stay longer. Why should the hospital administrator seek to use his bed stock efficiently if, by removing patients from hospital quickly, he is left with empty beds and no *per diem* income? Once again we have a perverse set of incentives which may induce hospital managers to use scarce resources inefficiently.

In health care markets using fee-per-item of service and *per diem* payment systems, the financers (the insurance companies) of health care are often price-takers, i.e., they accept the usual customary and reasonable rates (UCR) set by the providers. The inflationary pressures on expenditure created by the perverse incentives inherent in these payment schemes have led to the financers of care to question the UCR principle and, instead of taking prices as given by providers, seek to regulate fees by using their buying power. The literature on these developments is vast but the success of the insurance companies in regulating this market place is limited.

More recently, hospitals in the USA and elsewhere have accepted prospective payment systems (e,g., diagnosis-related group payments or

DRGs). These may have perverse effects because they induce cost minimization in the hospital system and this may lead to premature discharges and the shifting of costs onto other health care providers. Minimizing hospital costs, the result of DRGs, may not generate efficient use of resources.

The failure of the insurers to curb expenditure inflation, the incentives which induce providers to increase their utilization and bills, and the fact that co-insurance is used generally only in a limited way, means the existing market arrangements are characterized by cost inflation. By and large, health care systems, other than the NHS, cost more and their expenditure is growing more rapidly. Perhaps the main cause of this is that providers and consumers have little incentive to economize, and the financers have been slow to implement effective means of evaluating the prices, quantity, and qualities of the health services whose costs they meet.

In summary, the market alternative is concerned with altering the incentive structure in the health care market so that producers, consumers, and financers are rewarded for behavior which is efficient and the "hidden hand of the market" produces an optimal allocation of resources. These alternative incentive structures are not without their costs. Copayment means that patients pay part of the cost of care. This has known effects on utilization and little known effects on health status. Some schemes of copayments, particularly if they are not related to income, lead to utilization reductions which may be larger amongst the poor, a group which typically has an inferior health status. Before co-payments are used, it is necessary that policy goals be identified clearly. Otherwise, it may be found that the outcomes produced by such incentives are unwanted: if a market health care is rationed with reference to willingness and ability to pay, patients in clinical need of care may not be able to get access to it because of the price barrier.

Such distributional outcomes could, according to the liberals, be rectified by redistribution directly to the individual. Whether this redistribution would be in cash or vouchers and whether it would be very substantial are matters of heated debate. This debate is based on the images in competing crystal balls rather than on evidence.

The marketeers have failed to resolve the obvious problems created by the perverse effects of incentives on providers. The incentive for doctors to carry out too many medical acts and perhaps have too little regard for the quality of their work is an inevitable outcome of payment by fees-per-item of service. The payment of hospitals on a *per diem* basis or with DRGs may also lead to inefficient use of resources.

The market mechanism in health care is fine in theory. In fact, it works in a highly imperfect manner. To get it to work more efficiently requires radical reforms (e.g., the removal of professional power), the use of coinsurance with its uncertain attributes, and the reform of perverse market incentives which encourage providers to be inefficient. To the extent that liberals have remedies for these problems they are usually radical, untested, and will require, if introduced, to be evaluated carefully and sustained by much private and perhaps public regulation of the health care market place. Such characteristics should make potential innovators very cautious about implementing the privatization proposals; good evidence is required that private incentives (regulation) of the market produce better outcomes than public regulation of the market.

*The collective alternative*

The preceding criticisms of the liberal Nirvana do not imply that the alternative to the market, collectivist or socialist organization, is not without defects, sometimes of a serious nature. These defects are similar to those in the private market place and have survived often because of the failure of the public regulators to monitor and evaluate their regulatory activities.

*Allocation according to need*

The collectivist seeks to allocate scarce health care resources on the basis of need rather than on the basis of willingness and ability to pay. But what is need? Need may be a demand or a supply concept, as Williams [51] has pointed out. The patient may need care according to his own evaluation. An agent of the patient, for instance the doctor, may demand care on his behalf. These needs may be very different. A supply concept of need would concentrate on the productivity of health care as seen by the professional for the individual patient. It is perhaps this latter concept that is in the mind of some collectivists. Competing patients are to be appraised in the light of the potential productivity of health care allocated to them. The scarce resource, health care, is to be allocated on the basis of the maximum possible gains in outcome, improved health status (QALYs).

Even if this view were accepted, it contains certain difficulties. How do we measure outcomes? There are few good health status measures in use today. The intellectual problems associated with these measures seem to have been overcome but they have not been implemented extensively in operational conditions (see Williams [52] which offers a preliminary use of such measures). The lack of outcome measures together with the lack of evaluation of health care means that little is known about the efficiency of competing therapies for a given illness, let alone the relative values of therapies for different illnesses, e.g., does a heart bypass operation give greater benefits in health status than an operation for varicose veins or renal dialysis?

As a consequence, the allocation of care is governed by principles that are implicit and imprecise. Thus the feeling of solidarity inherent in the collectivist ideology means that care may still be given to the terminally ill without adequate reference to cost. This care might be radical surgery, which has little or no effect on life's length and possibly negative effects on its quality, or nursing care in a hospice. No rules have been estabished to allocate resources effectively in this or most other cases. Often the allocation is based on past practice, the supply of surgeons or hospice beds, and random chance. It does not and cannot conform with the collectivist ideal because of ignorance about the benefits of competing therapies. Until such information is available it will not be possible to detect whether resources are allocated in a manner consistent with the collectivist ideal: according to capacity to benefit from care regardless of the individual's resources.

*Collective allocation, public ownership, and planning*

If the technocrat's dream was fulfilled and it was possible to identify outcomes and those patients who would benefit most from the use of the scarce collectively provided resources that are available, the attainment of this collectivist Nirvana would require careful regulation of the means of production. Despite the fact that the ideology of those who created the NHS was collectivist, there was little attempt to devise regulatory devices and monitor their efficiency in relation to the goals of greater equality of health status. The transfer of the ownership of hospitals to the state and the institution of tax finance in 1948 changed some of the fabric of the "health care shop" but it did not immediately lead to a radical change in the production or distribution of the wares of that shop.

The removal of the price-barrier to the consumption of health care in

1948 did not mean that health care was provided comprehensively at zero cost to consumers. Significant non-price cost-barriers remained, in particular time-costs and costs associated with work income-losses related to the use of the NHS. These costs are likely to affect income groups in different ways. The worker at Rowntrees Chocolate factory who consults his GP may lose income and have to use slow public transport to get to his doctor, the pharmacy, and the hospital. The academic at a University will lose no income and have access to a care which reduces the time costs of consumption. Location of new hospitals and clinics in 'green field' suburban sites tend to accentuate these cost differentials and explain, in part, different health care consumption patterns.

To overcome such cost-barriers requires careful research to identify the cost-characteristics of the alternatives and effectively control capital information in the NHS. Such controls are noticeable by their absence. Only recently (1981–'86) has the Department of Health refined investment appraisal techniques (Capricode) and suggested that capital decisions take account of the full social costs (including time-costs) and benefits of the alternatives.

This lack of control of capital expenditure is paralleled by the relative lack of control of revenue expenditure. Until the early 1970s the NHS revenue budget was allocated on the basis of the following criteria: what was needed last year *plus* an allowance for inflation *plus* a small growth allowance *plus* a residual item related to public scandals (usually in the mental illness sector) which were covered by the media! During the 1970s explicit revenue allocation formulae were devised for each of the constituent parts of the United Kingdom [14, 38].

These sought to equalize the financial capacity to provide health care by allocating monies in relation to need, adjusted by population weighted by standardized mortality rates. Whilst the NHS had positive growth rates in total resource levels, territorial redistribution of resources was possible without much trouble. With the tightening of cash-limits, however, this first attempt systematically to equalize resources is running into increasing opposition when significant inequalities in geographical financial capacity remain.

Attempts to meet need, observed however crudely, by shifting resources out of acute care and into care of the elderly, the mentally ill, the physically and mentally handicapped, and the chronic sick has proved also to be difficult to implement [13]. Despite regular reiteration of this "priorities" policy, the system has been relatively unresponsive to these statements and the moral persuasion of the Department of Health. The department's influence is not improved by the relative autonomy of the Regions and the casual way in which Regions have monitored the Districts until recent years.

The work of the Department of Health since 1982, in terms of a decision to attempt to establish "performance indicators" and call each Region to account each year, is novel, naive, but well intentioned. The ministers wish to determine whether they are getting "value for money" and whether priorities criteria are being followed, but this is impossible to know with much certainty because of the NHS's ignorance and unwillingness to monitor and evaluate what it is doing. The risk is that an admirable venture may fail because of optimism about the ease of the task and hope for quick payoffs which will prove impossible to attain.

Thus the paradox of the collective alternative, in the form of the NHS, is that public ownership of the means of production and public finance has not led to the achievement of collective goals. The Black Report [15] and Le Grand's work [32] has demonstrated seemingly that significant inequalities continue to exist nearly 35 years after the creation of the NHS. Planning, monitoring, and evaluation have hardly been tried in the NHS! The socialization of the health care system has not brought the collective Nirvana much closer, and the NHS, like all other health care systems, has not resolved the problems associated with the achievement of efficiency or equality.

*Collective allocation and incentives*

The failure, or relative absence, of "top-down" planning from the center has been supplemented by an unwillingness to develop "bottom-up" incentives by accepting that decision makers respond to inducements and that care has to be used in designing and evaluating appropriate incentive systems.

The effects of the NHS on producer and patient incentives are very similar to those produced by private insurance systems. The patient consumes health care and the price at the point of consumption is usually zero (except for the few items such as prescriptions, dental care, spectacles, and some appliances for which there are charges); he has no incentive to economize. The main decision maker on the supply side, the doctor, has no incentive to economize either, especially as he has no idea of the cost of his decisions anyway. Thus, whether it is believed that the patient or his agent, the doctor, is the demander, neither has an incentive to economize. Thus all decision-makers have little incentive to be careful in their use of society's scarce resources.

Indeed, not only are the incentives inadequate, they are, as in the insurance system, sometimes perverse. Some elements of doctor remuneration may induce less than efficient responses. For instance, *per capita* payments to general practitioners may give doctors an incentive to economize on their use of their time because, whilst they cannot increase their income, they can increase their leisure with such a payment system. The fact that GPs give average consultations of five minutes duration may be the outcome, in part at least, of the way they are paid. Similarly, hospital doctors are paid a fixed salary. This means their income is not affected by their work rate and if they prefer "on-the-job leisure" to do research or private practice, they may press for more appointments, particularly of juniors who do not compete for private practice and can carry out their NHS functions for them.

These outcomes are predicted and the causal evidence available does not seem to refute them. It is clear that any payment-system, salary, piece-rate, or capitation, may have effects on behavior. These effects require evaluation continually in order that the payment systems can be "tuned" to generate behavior consistent with the objectives of policy: we need to pay the piper appropriately if we are to get the tune that is wanted!

*Possible reforms*

The reform of the collective allocation system on health care is not about changing institutional structures (e.g., reorganizing the NHS), the usual pre-occupation of politicians. Such policies change the facade outside the NHS 'shop' but do not affect behavior within the shop. At present the NHS is under-managed (i.e., inadequately regulated) and the meagre management resources are badly trained and ill-informed.

In the NHS there is no information about speciality costs; there is little evaluation of the process of providing care or the outputs, if any, of the caring process; and the idea of quality evaluation is regarded as novel and radical. For a trader to use resources efficiently in a market such information is essential; imagine an oil company that did not know the cost and quality of the oil and petrol it sold! Such a situation is unlikely in the oil market but commonplace in the health care industries of most countries including Britain.

The basic minimum for any collective system to be efficient and to meet its distributional targets is that it collects, manipulates, and evaluates information about costs, quantities, and qualities. Such information is required at a specialty level so that it is possible to identify which actors are least cost, quantity, and quality maximizers, and which doctors use resources inefficiently. The costs of collecting and using this information will be considerable but efficiency can be bought only at the price of using scarce resources.

The production of such information will enable all health care actors, doctors, nurses, physiotherapists, and others, to establish explicit professional standards of practice. The establishment of standards or norms of practice would enable the professions to use peer review and medical audit systems to ensure that professionals adhered in their everyday practice to the standards of their trade.

The problem associated with standard setting is that it tends to be conservative. It thus needs to be reviewed continually, and even if deviants are identified, they may not always be prepared to alter their practice. This latter effect has to be substantiated by research but if it is a major problem then "sticks and carrots" would have to be devised to induce decision makers to behave efficiently. The most obvious, but contentious, method of doing this would be to make decision-makers budget holders.

For instance, the use of Health Maintenance Organization structures within the NHS has been advocated [18, 39]. The US literature on HMOs indicates that these institutions may be up to 40 percent cheaper than their rivals [34, 35] although this relative advantage may be eroded as other buyers of health care services force down costs.

The NHS-HMO could consist of the patient being given an annual choice between competing general practice (primary case) HMO-firms. The signing-up of a patient would generate a state financed income (contribution) payment of say £200 for the provider. The GP-HMOs income would be determined by its ability to attract patients and its costs would be determined by its ability to control outlays in general practice and the costs of care in hospitals and elsewhere. The GP-HMO would buy in services from the cheapest hospital provider, public or private.

It is argued that the savings generated from HMOs are the result of changed incentives. The providers have an incentive to minimize costs as surpluses of income (contribution) over expenditure which can be used at the providers' discretion to augment their income or develop their HMO. This monetary incentive induces careful management of resources and rigorous clinical review of practice. Furthermore, such organizations, in an effort to minimize costs, may invest in screening and health promotion activities if they are cost effective.

The problems which HMOs generate are much debated in the literature. Does the drive to lower costs reduce the quality of care and lead to careful selection of healthy patients in the annual enrollment? The evidence suggests that this is not so. Furthermore, cost cutting of this nature might affect annual re-enrollment rates and HMO income detrimentally so care has to be taken by resource managers in their trading off of costs, activity, and outcomes.

The HMO has been advocated strongly by Enthoven [17] in his health care plan for the U.S.A. More recently he and others [39] have advocated experiments in the use of the HMO design within a state financed NHS. Such a proposal envisages a structure where health care might be publicly financed but provided *via* private and public HMO organizations and public and private hospitals. The exact public-private mix in the provision of health care would be determined by the relative efficiency of public and private providers: the latter could only expand if they generated QALYs at a lower cost than NHS hospitals. Whether such a radical change in the structure of the NHS would ensure that its objectives are achieved more efficiently only much needed experimentation can tell.

To summarize this brief discussion, it seems clear that the collectivistic alternative, as it manifests itself in Britain and elsewhere, is far from perfect in the efficiency with which it uses resources and of its achievement of distributional goals. The regulatory devices that have been used have been few in number and introduced only recently after the rediscovery of defects which the Service was created to remedy.

Existing collective health care systems have been reluctant to develop means to analyze efficiency, and the distributional goals have not been monitored effectively. The means to remedy these defects are difficult to identify and there is an apparent unwillingness to experiment. Until experiments are implemented and evaluated in Britain, the NHS will continue to use resources inefficiently and will fail to achieve the distributional goals which have been set. Principle and practice in the collectivist system, like principle and practice in the market, bear all too little relation to one another.

## IV. CONCLUSION

To assert that the market is more efficient than the National Health Service is naive and unhelpful. Both the market and the NHS have severe limitations and the achievement of the objectives of liberal freedom and collectivist equality is obstructed by similar forces: monopoly power, lack of evaluation of costs and benefits, and perverse incentives for providers and demanders. It is inappropriate to judge the NHS in the light of market objectives or the market in the light of collective objectives, but it is necessary to be aware of the characteristics of each of the alternatives, both in terms of the goals and the means by which these goals can be attained.

It is doubtful whether the market could achieve the objectives of its liberal architects. The power of the monopoly interests and the income losses these interests would suffer if competition existed make it likely that any market for health care will be dominated by monopolies. Such concentrations of power inhibit freedom as much as a large public sector and it is likely that, to protect freedom, the state would have to regulate the monopolies and actively try to sustain competition. Whether the state could do this, when the frustration of monopolies will cost votes, is a contentious subject.

It is unlikely that a market, as set out in the liberal paradigm, could achieve collectivist ends. Although economists in the past [29] have believed that all that is necessary is to set out goals, adjust prices, and then set the market to work, this economic theory of socialism has been shown not to work in practice.

The socialist model is fraught with difficulties which arise from the same problem that exists in markets. The monopoly power of producer groups is a major obstacle to change. The removal of health care from the private market does not remove the problems of inefficiency and inequality. Active intervention (planning) and the creation of new incentive structures consistent with the goals of policy are necessary. However, in the NHS these challenges have been ignored until recently and even now the efforts in these areas are feeble.

The definition of efficient policy, whether the market is "better" than the NHS, begs the definition of objectives. If social objectives are being redefined away from collective action, the changed public/private mix will raise many new regulatory problems for the state and private institutions. If objectives are static, then the existing public/private mix awaits effective regulatory action to improve performance of both its constituent parts. Probably the problems of the health care market are insoluble whether publicly or privately organized. Whatever organization exists and whatever the policy goals, regulation, by which is meant intervention by public and private interests to determine the price, quantity, and quality of health care, is unavoidable.

*Centre for Health Economics*
*University of York*
*Great Britain*

BIBLIOGRAPHY

1. Aaron, H. and Schwartz, W. B. G.: 1984, *The Painful Prescription: Rationing Hospital Care*, Brookings, Washington, D.C.
2. Abel-Smith, B.: 1984, *Cost Contamination Health Care*, Bedford Square Press, London.
3. Abel-Smith, B. and Maynard, A.: 1979, *The Organisation, Financing and Cost of Health Care in the European Community*, Commission of the European Communities, Social Policy Series, No. 36, Brussels.
4. Arrow, K.: 1963, 'Uncertainty and the Welfare Economics of Medical Care', *American Economic Review*, 1iii, December.
5. Brittan, L.: 1982, 'Health Service Pressure Point', *Guardian* **11** (May), 20.
6. Brook, R. and *alia*: 1983, 'Does free care improve adults' health? Results from a randomized controlled trial', *New England Journal of Medicine* 309 (23) 1426–33.
7. Bunker, J. P.: 1970, 'A Comparison of Operations and Surgeons in the United States and in England and Wales', *New England Journal of Medicine* (January 15), 136–143.
8. Bunker, J. P., Barnes, B. A., and Mosteller, F.: 1977, *The Cost, Benefits and Risks of Surgery*, Oxford University Press, New York.
9. Cairnes, J. and Snell, M.: 1978, 'Prices and the Demand for Health Care', in A. J. Culyer and K. G. Wright (eds.), *Economic Aspects of Health Services*, Martin Robertson, England.
10. Cochrane, A. L.: 1972, *Effectiveness and Efficiency: Random Reflections on Health Services*, Nuffield Provincial Hospitals Trust, London.
11. Culyer, A. J., Maynard, A. and Williams, A.: 1981, 'Alternative Systems of Health Care Provision: An Essay on Motes and Beams', in M. Olson (ed.), *A New Approach to the Economics of Medical Care*, American Enterprise Institute, Washington, D.C.
12. Department of Health (New Zealand): 1980, *Funding for Health: An Allocation Formula*, Special Report No. 58, Wellington.
13. Department of Health and Social Security: 1976, *Priorities for Health and Personal Social Services in England: A Consultative Document*, HMSO, London.
14. Department of Health and Social Security: 1976, *Sharing Resources for Health in England: Report of the Resource Allocation Working Party*, HMSO, London.
15. Department of Health and Social Security: 1980, *Inequalities in Health: A Report of a Research Working Group* (The Black Report), HMSO, London.

16. Donabedian, A.: 1971, 'Social Responsibility for Personal Health Services: An Examination of Basic Values', *Inquiry*, viii, 2, 3–19.
17. Enthoven, A. C.: 1980, *Health Plan*, Addison-Wesley Publishing Company, Reading, Massachusetts.
18. Enthoven, A. C.: 1985, *Reflections on the Management of the National Health Service*, Occasional Paper No. 5, Nuffield Provincial Hospitals Trust, London.
19. *Financial Times*: 1982, 'Some Hospitals to be Ruled Out for BUPA Cover', by R. Snoddy, May 12, p. 7.
20. Frech, H. E.: 1974, 'Occupational Licensure and Health Care Productivity: The Issues and the Literature', in J. Rafferty (ed.), *Health Manpower and Productivity*, Lexington, Massachusetts.
21. Friedman, M.: 1962, *Capitalism and Freedom*, University of Chicago Press.
22. Fuchs, V.: 1978, 'The Supply of Surgeons and the Demand for Operatives', *Journal of Human Resources*, xiii, Supplement, 35–36.
23. Goodman, J.: 1980, *National Health Care in Great Britain: Lessons for the USA*, The Fisher Institute, Dallas, Texas.
24. Green, D.: 1985, *Which Doctor?*, Institute of Economic Affairs, London.
25. Grossman, M.: 1972, *The Demand for Health: A Theoretical and Empirical Investigation*, National Bureau of Economic Research, New York.
26. Howe, G.: 1981, 'Health and the Economy', Speech to the Royal Society of Health (mimeo).
27. Jamieson Report: 1980, *Report of the Commission of Inquiry into the Efficiency and Administration of Hospitals*, three volumes (report).
28. Jones, I. M.: 1970, *Health Services Financing*, British Medical Association, London.
29. Lange, O. and Taylor, F. M.: 1983, *An Economic Theory of Socialism*, University of Minnesota Press.
30. Lees, D.: 1965, 'Health through Choice', in *Freedom or Free for All?*, Institute of Economic Affairs, London.
31. Leffler, K. B.: 1978, 'Physician Licensure: Competition and Monopoly in American Medicine', *Journal of Law and Economics*, xxi (1), 165–186.
32. LeGrand, J.: 1982, *Strategy for Equality*, Allen and Unwin, London.
33. Lindsay, C. M.: 1980, *National Health Issues: The British Experience*, Hoffman La Roche, New York.
34. Luft, H. S.: 1981, *Health Maintenance Organizations: Dimensions of Performance*, Wiley, New York.
35. Manning, W. G., Leibowitz, A., Goldberg, G. A., Rogers, W. H., and Newhouse, J. P.: 1984, 'A Controlled Trial of the Effect of Prepaid Group Practice on the Use of Services', *New England Journal of Medicine* **310**(23), 1505–1510.
36. Maynard, A.: 1979, 'Pricing, Insurance and the NHS', *Journal of Social Policy*.
37. Maynard, A. and Ludbrook, A.: 1980, 'Budget Allocation in the National Health Service', Journal of Social Policy, July.
38. Maynard, A. and Ludbrook, A.: 1980, 'What's Wrong with the NHS?', *Lloyds Bank Review*, October.
39. Maynard, A.: 1985, 'Policy Choices in Health', in R. Berthoud (ed.) *Challenges to Social Policy*, Gower Press, Brookfield, Vermont.
40. McKeown, T.: 1977, *The Modern Rise of Population*, Edward Arnold, London.
41. McKeown, T.: 1979, *The Role of Medicine*, 2nd edition, Blackwell, Oxford.

42. Muurinen, J. M.: 1982, "Demand for Health: A Generalised Grossman Model', *Journal of Health Economics* **1**(1).
43. Newhouse, J. E. *et al.*: 1981, *Some Interim Results from a Controlled Trial of Cost Sharing in Health Insurance*, Health Insurance Experiment Series, R-2847-M.MS, Rand Corporation, Santa Monica, California.
44. Organisation for Economic Co-operation and Development: 1977, *Public Expenditure on Health*, Paris.
45. Schweitzer, S. O. (ed.): 1978, *Policies for the Containment of Health Care Costs and Expenditures*, U.S. Department of Health, Education and Welfare, U.S. Government Printing Office, Washington, D.C.
46. Seldon, A.: 1977, *Charge!*, Temple Smith, London.
47. Seldon, A. (ed.): 1980, *The Litmus Papers: A National Health Dis-Service*, Center for Policy Studies, London.
48. Smith, A.: 1976, *An Inquiry into the Nature and Causes of the Wealth of Nations*, Everyman Editors, London.
49. Tullock, G.: 1976, *The Vote Motive*, Institute of Economic Affairs, London.
50. United States Department of Health and Human Services: 1981, *Physician Induced Demand for Surgical Operation*, Health Care Financing Administration, Washington, D.C.
51. Williams, A.: 1978, 'Need: An Economic Exegesis'. in A. J. Culyer and K. G. Wright (eds.), *Economic Aspects of Health Services*, Martin Robertson, London.
52. Williams, A.: 1985, 'The Economics of Coronary Artery By-Pass Grafting', *British Medical Journal* **291**, August 3rd, 326–329.

CORINNA DELKESKAMP-HAYES

# JUSTICE AS FAIRNESS OR FAIRNESS AS PRUDENCE?

## (A Commentary on Alan Maynard's Essay)

It is instructive to consider Alan Maynard's [41] discussion of the British and North American health systems in the light of Murray Feshbach's [19] description of Soviet health care policies. Soviet state medicine historically has served as a model for British defenders of nationalized health administrations;[1] however, the present deterioration of medical care in the Soviet Union argues the need for reconsidering those theoretical foundations which Maynard claims underlie the British National Health Service.

This need seems more pressing in view of certain difficulties raised by Maynard's account. They concern the general problem of how to devise a satisfactory conceptual justification for health policies in Western democratic societies. The following essay will therefore (I) briefly describe these difficulties, (II) suggest some guidance for the needed justification, and (III) draw few selected practical conclusions relating to Maynard's concern with efficiency. My thesis in these last two parts will be that only a profound Kantian modification of Rawls' "Justice as Fairness" will satisfy all crucial requirements, but that such a modification would need another essay in itself if it were to be executed properly.

### I. THE DIFFICULTIES OF MAYNARD'S ACCOUNT

Let us (1) look at Maynard's reasoning taken by itself and then (2) in light of Feshbach's essay. As a result, (3) a number of requirements for a more adequate theoretical foundation for health care policies will emerge.

#### 1.

Maynard juxtaposes two ways of financing health care: state and market medicine. He also undertakes the somewhat formidable task of characterizing the "ideologies" underlying both positions, namely, collectivism and liberalism (or "libertarianism" in an American context). He claims

*Hans-Martin Sass and Robert U. Massey (eds.), Health Care Systems,* pp. 167–198.

that neither position is tenable, because their practical implementations fail to realize their prime objectives.

Thus collectivists set a high value on humanitarian goals, including, for example, equal access to health care. But the corresponding system of state medicine (e.g., the NHS) fails to achieve this equality, both on the regional and on the social levels. On the other side, libertarian marketeers set a high value on individual freedom of choice and achievement, but the corresponding more privatized sectors of health care in the United States are equally deficient in realizing this ideal.

The difficulty with Maynard's account does not appear in the first part of his argument: the shortcomings of collectivism are well established on the basis of the British (and some selective American) experience; it appears, however, in his critique of the market alternative. His evidence for that alternative failing to guarantee the kind of individual freedom claimed by the "ideologues" of private financing rests on a single quotation taken out of context from Adam Smith's *Wealth of Nations* [53] which he uses to support a position contrary to that endorsed by Smith. Whereas Maynard thinks it is the free market system which is responsible for the generation and support of undesirable monopolistic tendencies, Smith castigates the mercantilist system of state intervention for encouraging such tendencies which he thinks would cause no lasting harm, if they were only left to the forces of the market.

Thus, whereas Maynard invokes the danger of monopolies in order to advocate government intervention, Smith does so in order to oppose such intervention.[2]

Likewise, it is argued by many modern scholars[3] that cartels or monopolies arise and exist only with government support or protection; that such a concentration of power cannot develop in a normal market.[4]

Thus Maynard has not established his thesis that the market, if left to itself, will impede individual freedom. This is aggravated by a conceptual muddle arising from Maynard's identifying regulation and intervention, and from his lumping together government regulation and private activity. The final statement of Maynard's paper reads:

> . . . whatever organization exists and whatever the policy goals, regulation, by which is meant intervention by public and private interests to determine the price, quantity, and quality of health care, is unavoidable.[5]

It seems obvious that regulation is indeed intervention, but this does not hold the other way around. The "intervention" which Maynard takes to characterize the market system as well is actually nothing but participation in the market, and thus of course unavoidable. But this unavoidability and hence (as a matter of principle) harmless nature of such economic intervention covers activities by single consumers as well as by large scale producers or capitalists. Whether I be who I am or Mr. Rockefeller, my participation in the market helps to determine price, quantity, and quality of health care, of toothpaste, of what have you. There is no difference in kind.

Hence it is quite misleading of Maynard to claim that individual freedom is hampered by monopoly or even by insurers' self-regulation (both of which are instances of harmless participation in the market) in the same way as it is hampered by government regulation. The first is expressive of economic freedom, the second restricts that freedom. Maynard's thesis that in both systems "regulation" is required in order to achieve efficiency, equates what is fundamentally different and therefore reduces to a truism.[6]

This conceptual difficulty can be traced to an incompleteness in Maynard's description of two opposing "ideologies." He restricts this description to the "pyrotechnics" of merely utilitarian claims commonly advanced on each side, but fails to notice that the "market" view is – at least in philosophical discussions – often based on a consideration of rights. The issue of "efficiency," which occupies the center of Maynard's essay, is thus not relevant to the libertarian perspective. This perspective arises, instead, from a reflection on the limitations of state or government authority in pursuing any, howsoever persuasive and generally accepted, humanitarian goals (e.g., equality, redistribution of income, and a needs-based notion of government activity). Such reflection is entirely absent in Maynard's argument. He seems to suppose that issues of medical efficiency, satisfaction of needs, and security are automatically the concern of the state. His very juxtaposition of "opposing ideologies" is thus biased toward the collectivist alternative.

## 2.

This limitation of Maynard's description becomes especially significant if we consider Feshbach's essay. The deplorable picture Feshbach presents of recent trends in Soviet health care administration can be interpreted on two different levels:

First, the population of the USSR is insufficiently provided for. But this state of affairs may differ only in degree from that characteristic of the United States health care system (as becomes obvious, for example, from Buchanan's essay [9]). After all, western countries also limit their health care budgets in view of economic and other considerations.

Second, the underprovision of health care in the USSR is caused and aggravated by the totalitarian government itself. While it claims to be merely pursuing generally accepted humanitarian goals, it trespasses on the individual citizen's freedom and self-determination.

The question thus arises whether Maynard (and the collectivist theorists he seems to sympathize with) has established within his own position any theoretical safeguards against such totalitarian abuse of humanitarian values. Or, to put it another way, do these theorists provide any conceptual grounds on which to criticize a totalitarian state, and totalitarian state medicine? If not, how can they be sure that the policy suggestions implied in their theoretical presuppositions are viable in view of their own obviously liberal commitments?

3.

Questions such as these suggest that what is needed is not so much an abstaining from theoretical quibbles (as Maynard would have it), but that these quibbles should be carefully reconsidered. Such a reconsideration would have to meet the requirements implied in Maynard's own account as well as those which arise from its shortcomings. That is, it would have to provide a conceptual foundation for the evaluation of health care systems which could account both (1) for our commitment to humanitarian goals (including efficiency, at least insofar as the economic feasibility of such objectives is concerned) and (2) for our respect for individual freedom. At the same time, (3) the desired opposition to the totalitarian option would have to be rendered theoretically compelling. The second part of this essay will be devoted to sketching such a foundation.

## II. SKETCH OF A MODIFIED RAWLSIAN FOUNDATION FOR THE EVALUATION OF HEALTH POLICY

The conceptual grounds needed to secure respect for individual liberty and self-determination are hard to render compatible with conceptual grounds endorsing human solidarity and extensive social obligations. One has to look for a side-by-side account of rights determining duties and values determining goals, or of a theory containing both deontological and teleological elements.

John Rawls' political theory promises the desired double orientation. The goal of this theory is to define "justice" as "fairness," with the first notion suggesting a deontological, the second a teleological concern. More particularly, the three main aspects of his theory correspond to the three requirements noted above for the needed conceptual foundation:

(1) His notion of "justice as fairness" ([47], pp. 14f) accounts for our humanitarian acceptance of social obligation towards the disadvantaged. His maximin rule ([47], pp. 152 ff) states that differences in economic and social position should be made to work out in favor of the least well off.[7] (At the same time, Rawls' particular definition of "fairness" allows for a view towards economic and political feasibility. In rejecting strictly egalitarian measures he makes sure that the capital producing elements can be conceived to retain the upper hand over redistributive considerations.[8])

(2) His theory is deontological. He accepts an unconditional duty to respect certain human rights and liberties.

(3) In view of these rights, totalitarian governments which endeavor to determine the conditions for citizens' happiness and self-determination can be criticized as transgressing the limits of state authority.[9]

Unfortunately, Rawls says very little about the issue of health care.[10] One can, of course, attempt to extrapolate his view on these matters from the general tendency of his reasoning. Yet, wherever such extrapolations have so far been attempted, they lead to more egalitarian or even utilitarian conclusions[11] than is compatible with Rawls' opposition to these positions. Nor would such conclusions be helpful in view of the requirements just laid down for the desired foundations.

My argument in this essay will be that this incompatibility is due to problems inherent in Rawls' theory itself, and that these problems can be avoided by attending more carefully to the Kantian roots of his thinking.[12] As a result, I shall hope to show that the notion of fairness can be modified in such a way as to keep the desired extrapolation within the bounds of the requirements.

I shall first sketch Rawls' basic ideas; second, indicate the difficulties hampering their theoretical realization; third, suggest a modification of his theory which avoids these difficulties.

## 1.

Rawls' political philosophy may profitably be seen as designed to determine the conditions under which governmental power can be rationally justified. These conditions concern certain changes in the organization of Western democratic and capitalist societies. The desired justification can then be framed on the model of a Kantian rationality.[13]

At the same time, Rawls wishes to render that justification intuitively comprehensible by introducing a "procedural analysis" of "justice as fairness". This amounts to the claim that "just" is what is agreed upon as a result of a discussion held under specifically fair conditions.

Thus, Rawls' notion of "fairness" operates on two levels ([47, p. 15): It first distinguishes a supposed "original position," i.e., a fictitious situation in which mutually disinterested but respecting rational beings can be thought to meet as equals in order to reach agreements concerning just institutions ([47], pp. 12, 17). Second, it distinguishes those institutions which they agree upon ([47], pp. 302 f). We have then a *fairness antecedent* to and characterizing the original position, and a *fairness consequent* to and characterized by the original position.

(a) The quest for antecedent fairness is linked with the deontological character of Rawls' theory. Rawls, following Kant, assumes that human persons have inviolable rights (*Metaphysik*, A45; [47], p. 32) which carry with them the unconditional duty for everyone to respect these rights.[14] These rights are grounded in the inherent dignity or respectability of rational beings whose rationality in turn is linked to their moral personhood ([47], p. 19). They concern each individual's freedom of unhindered self-determination in finding out and pursuing his own terms of happiness (*Gemeinspruch*, A235f; [47], p. 129).

(b) As the people in the original position are obliged to respect one another's rights, they are obliged to design their social arrangements, and thus the conditions of consequent fairness, according to that respect. That is, within the state, each must be granted the basic liberties (*Gemeinspruch*, A265).[15]

These comprise not only the realm of moral but also of practical self-determination, and thus the free disposal of those worldly goods which provide the condition for the realization of one's goals ([47], p. 255, line 24; cf. *Metaphysik*, A163). In particular, this implies for Rawls, as for Kant, that human beings should never be used exclusively as means but should also be considered as ends in themselves (*Grundlegung*, A64f; [47], p. 180). That is, they should only be treated in ways to which they could rationally consent (*Grundlegung*, A69; *Metaphysik*, A166; [47]; pp. 13, 180).[16] Hence, states and laws are morally acceptable or just only (but not always) if they satisfy the condition of such consentability.[17]

In addition, such "consequent fairness" is characterized by the Difference Principle. Unlike Kant, who settles for some sort of liberal *laissez faire* view of capitalist society and government, Rawls argues that social and economic inequalities require some compensatory action by the state: Each gain in the prospects of those who are better off must, by appropriate redistributive measures, be matched by corresponding improvement in the prospects of those at the bottom of the social scale ([47], pp. 60, 75, 152).

With this additional qualification of "consequent fairness" Rawls has gone beyond what Kant would concede. For Kant, the state is unconditionally obligated exclusively in view of the "formal side" of human persons, i.e., their rationality and resultant self-determination. As a consequence of this exclusiveness, Kant can (so to speak) afford letting this obligation also encompass the citizens' worldly property, as the principal means of realizing their liberty. Hence, redistribution would be a breach of that duty.[18] For Rawls, the state is seen to be obligated in view of the "material side" of human persons as well, i.e., in view of their actual well being. Rawls therefore must balance the claims of both sides against one another.

Thus, while he agrees with Kant both in opposing egalitarianism and in favoring the capital producing elements of society, Rawls disagrees with him on the issue of social sharing. At the same time it is here that the difficulties of Rawls' theory arise.[19]

## 2.

As a result of the changes Rawls has introduced within the Kantian framework, his theory is now meant to be deontological in respect of both the "formal" and the "material" side of persons. To mold these two respects into one consistent account, he must somehow link the (Kantian) concept of antecedent fairness with his own (non-Kantian) concept of consequent fairness.

The obvious middle term is the notion of liberty as comprising the worldly conditions of its realization. It is, after all, the purpose of providing the less advantaged with a bit more of such "conditions" which the Difference Principle is intended to serve.

But how, if the Kantian side-constraint of respecting (wealthier) citizens' property is waived, can one hope to keep the (state's, contractors', society's) unconditional duty towards the worldly-condition-liberty from implying the duty to guarantee an equal amount of those conditions to each citizen, and thus an "equal worth" of everybody's freedom ([47], p. 204)? In other words, how can one keep the equality of respect which one owes each rational being from carrying along with it the concession of an equality of worldly goods? Or, how can the reasoning underlying Rawls' Difference Principle be rendered compatible with his avowed anti-egalitarianism? After all, the whole project of rendering "justice" as "fairness" rests on the contention that, rather than abolishing inequality, it should and could be dealt with "fairly".

Rawls offers three main arguments designed to prove this compatibility, and they can all be shown to be unsuccessful.

(a) There is no need to abolish all social and economic inequality because some such advantages arise from individual virtuous effort. They are morally deserved.[20]

Not all of them are deserved, however. Some differences in social position arise from individually undeserved differences in starting position, and these distort the results of fair competition.

It is in order to alleviate this distortive element that the Difference Principle is introduced ([47], pp. 301, 511 f): Some of the superior gains achieved on the basis of antecedent inequality must be used for improving the prospects of those more unfavorably situated.[21]

Rawls specifies those undeserved advantages as arising not only from inherited social position but also from education, family background, natural assets, and psychological factors ([47], pp. 15, 74, 104, 312). All these are presented as conditions for the possibility of an individual's successful exertion. Yet this very conditionalist model renders the notion of "moral desert," which underlies Rawls' rebuttal of egalitarianism, altogether obsolete.[22] The ascription of such desert would presuppose the ascription of individual autonomy (as in [47], p. 519) or accountability. Quite regardless of any merely scientific model for human behavior, it would require a moral acknowledgement of moral freedom.

Hence, the conceptual means by which Rawls endeavors to defend the Difference Principle are incompatible with the theoretical framework that was designed both for his libertarian[23] and his anti-egalitarian intentions.[24]

(b) Society is a cooperative venture for mutual advantage ([47], p. 84,

529).[25] In such a venture the unconditional demand for treating human beings not as means only but as ends in themselves is not violated when, as a result of such cooperation, some wind up better than others. Hence egalitarianism is not warrented by the required mutual respect among persons.

Such respect is violated, Rawls argues, wherever the gains from this cooperation remain constant for some individuals while they rise for others. In this case he finds the former "exploited" by the latter and derives the claim that compensatory gains should be adjudicated to those who profit least from the common endeavor, i.e., the Difference Principle.

With this argument the concept of instrumentalization originally taken from Kant has assumed a wider range of application. On a Kantian understanding, no person could be said to be exploited unless he reaped no benefit at all from the common endeavor.[26] As long as he is better off by participating than by not participating, and regardless of the different gains among the team members, he has a reason to consent. Thus the requirement that one must respect persons' rational self-determination is satisfied in such cases. To be sure, everybody wishes to increase the advantages accruing to him from his cooperation with others, and all (except those gaining more) usually wish to have these gains more evenly distributed. Yet there exists no argument by which states (or contractors) can be thought unconditionally obligated in view of any universal, let alone merely group-specific, wishes of their social members. (Rawls' obvious desire to suggest such an obligation with regard to the "material" aspects of human persons is not sufficient.)[27]

Hence, while the instrumentalization argument does effectively exclude any egalitarian conclusion from the specifically Rawlsian implications of the worldly-freedom-premise, it is just not strong enough to justify his rendering "justice" as "fairness."

(c) In order to render that "material" side of personhood a proper subject for unconditional obligation as well, the notion of rationality is extended to include a "deliberative" aspect. The contractors' societal agreements are now additionally justified in view of their securing certain generally accepted advantages ([47], pp. 15, 176–9, 277, 348, 440, 476).

Here Rawls steps outside the Kantian framework.[28] The fictitious contractors' deontological concern for individual rights is supplemented by a teleological view towards universally desirable consequences expected to result from a much more particularly contrived societal setup.

In view of this new brand of rationality the egalitarian option is again easily ruled out: Since all members of a society would certainly wind up as losers in point of goods available ([47], pp. 50, 62, 546), no rational contractors could feel tempted to settle for such a solution.

At the same time, Rawls' Difference Principle acquires considerable persuasive force: Since the contractors legislate concerning a society in which they will themselves partake, and since they remain ignorant of the place they will occupy in that society, it is rational, so Rawls argues, that they, facing the risk of winding up at the lower end of the social scale, will wish to make sure that the accruing disadvantages remain as small as is compatible with the overall existence of inequality itself. Thus it is claimed inescapably rational that they should secure a maximum advantage attending any minimally favorable social position. (Rawls' Difference Principle is also described as the "maximin rule.")

Rawls is ardent in recommending the overall acceptability of such a decision (see for example [47], p. 513). Those promised advantages are, however, bought at a price on the theoretical level. The merely formal notion of rationality, which for Kant had served as validating ground for the fictitious social contract, and which was simply too meagre to justify any conclusions concerning the more particular contents of justice, let alone its definition as "fairness," was supplemented by a material notion of rationality, encompassing a regard for all socially relevant humanitarian goals. This latter notion, while permitting one to draw all the desired social policy conclusions, is at the same time, Rawls hopes, so narrowly circumscribed ([47], pp. 153 ff, 172, 585) that no reasonable being could fail to consent. The unconditionally compelling nature of the resulting government entitlements, which was guaranteed in the Kantian model, is thereby believed to have been preserved.

On closer scrutiny these hopes and beliefs prove unfounded. We must recall that Rawls' Fairness Principle is not intended as a vague humanitarian appeal leaving everyone free to determine his own practical involvement. As a principle implying taxation policies along with a redefinition of property rights ([47], pp. 275–8) and endorsing sanctions for the discouragement of noncompliance ([47], pp. 575 f), its social implementation is to be effected by force. More fundamentally, therefore, the issue revolves around the extent and limits of government authority as well as around the fictitious contractors' range of devisory competence[29] and Rawls' political theory conspicuously shuns the reflection on such limits.[30]

Originally, Rawls had followed the Kantian argument that every citizen has a right to let himself be constrained, for example by a state and its executive organs, only to the extent of securing the conditions for everybody else's unhindered enjoyment of personal freedom. That is, persons could be constrained only according to those principles to which they, insofar as they themselves will demand respect for their own freedom, must necessarily consent. Presently, however, both the notion of "rational being" underlying the Kantian view of personhood and the validating function of consent have changed. Every citizen now has a duty to let himself be constrained in view of the realization of goals which he can only be enjoined to endorse by persuasive appeals to his worldly prudence.

These goals, in addition, are by no means as universally acceptable as Rawls would have it ([47], pp. 153 ff, 172, 585). An ample literature bears witness to the extent to which his supposedly inescapable dictates of deliberate rationality can be contested.[31] Thus, the rationality according to which the use of force is to be justified is no longer narrowly limited in view of the conditions for respecting persons as such but has been expanded so as to encompass debatable issues.[32]

To be sure, one can now expect that a utilitarian concern for universal acceptability will forestall any outrightly exotic value-decisions and resulting legal sanctions. But there are, on the other hand, no theoretical safeguards against disregarding the contingent value preferences of easily suppressed minority groups. Nor are there any rock bottom conceptual grounds against a state abusing the contract model for determining "proper kinds" of such preferences, and thus for legitimizing totalitarian "educational" strictures.[33]

As a result, Rawls' "adding on"[34] a deliberative extension to the Kantian rationality which was to serve as ground for the philosophical justification of governmental authority according to a theory of justice has defeated its purpose.[35] His lack of reflection on any principal limits to contractors' entitlements and to state authority opens the way for transgression.[36] The appeal to some fictitious contractors' persuasive recommendations justifies too much.[37] Our Western commitment to individual and political liberty, which Rawls had endorsed by employing the libertarian social contract approach, is rendered obsolete by the body of his theory.

3.

In the previous three subsections I have discussed three attempts by Rawls to link the worldly-freedom-premise both with the libertarian framework of his theory and with its additional humanitarian (and efficiency related) commitments. All three attempts have failed.

I shall now (a) derive this failure from the unfortunate use Rawls has made of the notion of "social cooperation," (b) exhibit the importance of this notion for his theory, and (c) suggest a modified use which accounts for that importance.

(a) The reason for Rawls' failure lies in his trying to accommodate the libertarian and the humanitarian, or the deontological and the teleological element of his thought within one single part of his theory, namely the fictitious contract. This endeavour, which runs counter to the Kantian elements of his theory, is motivated by Rawls' uncritical understanding of society as a cooperative venture.

The term "cooperation" carries with it connotations which render intuitively acceptable the moral conclusions Rawls has drawn and which he recommends for enforcement on the political level. The most important of these connotations concerns the *voluntary character* of cooperative endeavors. This character is, however, lacking in relations of citizenship, where every one is made to join by forces inherent in the social situation as a whole. To speak of society as a "cooperative venture" is thus merely to use a euphemism for existing power structures.

At the outset of my presentation of Rawls' basic ideas it was suggested that he should have wished to render such power philosophically justifiable by employing a *contractarian approach* to political theory. This task can now be more thoroughly matched with Rawls' actual intentions.

Rawls claims that if society could be arranged according to principles which every rational being would have to accept, then the submission of citizens to political power would become innocuous in view of that rational acceptance. Such consent, so Rawls could hope, would in addition introduce, in the reconstructive manner typical of contractarian thinking, that very element of voluntariness into the societal endeavor, which could justify the cooperative model as well as its moral-political implications. Yet it has been shown that the notion of rationally compelling consent could not be made to cover these implications (in particular, of course, the Difference Principle as a maxim for policy decisions). Hence, Rawls' hoped-for transformation of the cooperative euphemism into philosophical reality has not worked out.

(b) Nevertheless, Rawls' insistence on "society as a cooperative venture" and on the resulting rendering of justice as "fairness" is responsible for much of the persuasive force of his theory. After all, for a society to be worth living in the many individual cooperative ventures cannot be left to themselves, for two reasons.

First, these ventures are in most cases geared towards profit. The view towards profit does not necessarily, or not even frequently, amount to any Rawlsian deliberative rationality. Such rationality considers long-term stable interest, with all the relevant circumstances taken into account ([47], pp. 401, 408). By contrast, at least in economic ventures, the goal is to secure immediate, short term advantages. Moreover, if such advantages are reaped at certain social costs, it is not these ventures exclusively, and usually not even very directly, which suffer from the consequences. They will naturally tend to disregard any view towards the social system as a whole.

Second, these ventures frequently comprise unequal partners. Those on a socially more elevated level, when negotiating freely with those of lesser status, are at the same time in a position of superior power. This position permits them, in pursuing their own advantage, to dictate conditions which secure their permanent superiority and keep their partners' expectations as low as possible. Thus, employers usually occupy a superior position with respect to their employees (except, of course, in cases where the employees possess special qualifications).

Thus, both in view of securing some deliberately rational arrangements among the particular cooperative ventures, and in view of securing some elementary fairness within these ventures, Rawls had wished to reserve some regulative authority to those responsible for the whole of a society's stability. We find here an appeal to both human solidarity and enlightened self interest which agrees with our considered value judgements. By contrast, any alternative ways of eliciting the needed willing cooperation of the less advantaged groups (such as by sheer suppression) are universally held to be unacceptable.

(c) If Rawls' theory is to accommodate these value judgements, perhaps it could accomplish this task not so much by considering the "cooperative nature" of society as something as yet unreal, such that philosophy would have to prescribe the practical and theoretical conditions for making it real, but rather as an ever ongoing task, such that philosophy could prescribe the theoretical conditions under which the political system can, in a more enlightened manner, be challenged to pursue its practical implementation.

His theory, understood in a thus modified sense, would have to prescribe these conditions in two respects:

(1) It would have to distinguish, within the political realm, between a primary area of *strict obligations* and corresponding entitlements, and a secondary area of merely *prudential considerations*. Such considerations, while focussing on issues which are left undecided by a government's more narrowly circumscribed unconditional duties, would yet remain subject to these duties as a side constraint.[38]

Thus, the regulative authority Rawls had wished to reserve for governments (or for the contractors setting up governments) would have to be restricted to more indirect measures. Instead of simply redistributing property by invoking immediately effective coercive measures (e.g., taxation or forced insurance schemes), such governments would have to be thought of as instituting procedural rules which change the ordering of power positions, and thus more obliquely influence the distribution of levels or status.

Such rules might govern the ways in which businesses are set up and conducted. They might also (going beyond the obligatory tasks of permitting and protecting) encourage, for example, the establishment of unions and legitimize certain means of their applying pressure. A government committed on duely humanitarian prudential grounds to Rawls' Fairness Principle could be thought to arrange the system of those rules in such a way that, even though the social and economic partners following these rules are motivated by a narrow regard for profit and for the power to increase it, yet the outcome of their negotiations, or of the balancing of their respective powers, will conform to the requirements of that principle.[39]

Such measures would remain within the libertarian (or Kantian) constraints of state authority (and of contractors' entitlement) insofar as they, first, affect the individual citizen not as such but only in the particular social and economic activities he has voluntarily elected to pursue (or affect the individual not as a compulsion from which he cannot escape, but only conditionally, concerning activities which he might also forego when considering their boundary conditions)[40]; second, even though these measures would go beyond the mere necessity of instituting any workable civil law rules for rendering everyone's exercise of liberty compatible with everyone else's they would still concern not so much that liberty as such but rather the advantages expected from its use.

(2) In addition, even if within such side constraints the possibility of political interference with social activities could be considered licit, still a sufficient reason for such interference according to howsoever prudential maxims would have to be established on independent grounds.

One way of securing such a constraint-oriented government authority with respect to a more extended realm of issues is provided by the Rawlsian theory itself. The contractors, regarding exclusively the realm of unconditional obligation (no longer valid given the Kantian reasons to the contrary, see note 26), had to be thought of as opting for a democratic setting. Hence the considered value judgements of a voting majority can be permitted to influence the more narrow level of prudential policy considerations. After all, such judgements, howsoever they may be accepted in a given community at a given time, do not thereby have to be accommodated within the very setup which is supposed to remain in force beyond any given time. Instead of including them in the fictitious contractors' deliberations, Rawls should have considered them as touching issues of merely common interest, to be settled in response to public sentiment.

To be sure, Rawls' quest for guaranteeing one enlightened and impartial court of appeal responsible for the stability of society as a whole and thus committed to deliberately rational precepts, cannot thereby be quite fulfilled. To introduce a "prudential realm" of government activity which will depend on majority opinions is not to guarantee the "prudence" of the resulting policy decisions. Responsibility for the whole can only be had as a response of the most, and these will contain "considered" along with "unconsidered" value judgements.

But this disadvantage is less consequential in the context of a modified Rawlsian theory than it would have been in the original. Unlike Rawls' original, this modification sharply separates obligations from interests. Any governmental measures aiming at implementation among the latter was said to be subject to side-constraints regarding the former. Hence, while these constraints prevent even deliberatively rational decisions from becoming as immediately effective as Rawls might have wished, they also will keep the more irrational votes from doing very considerable harm.

Under the perspective introduced by this modification of the Rawlsian theory, his own description of a deliberatively rational social contract, since it does no longer qualify as a philosophical reconstruction in its own right, would have to be understood as a persuasive appeal, directed at the voting public itself in an attempt to favorably influence its opinions. His "theory of justice" would appear less as an analysis than as an educational exhortation.

This *prudential-realm addendum* to Rawls' political philosophy amounts to permitting a non-obligatory zone of government regulation which is reminiscent of Maynard's more sweeping "regulative hypothesis." The crucial difference, however, arises from the clear separation in this addendum, but not in Maynard's account, of issues that concern rights (the deontological side of Rawls' theory) from issues that concern generally desirable goals (its teleological side). Or, it arises from our having juxtaposed an area where the government is obliged and entitled to protect by immediately goal-directed intervention the relevant liberties of its citizens[41] and an area where it can prudentially follow the value preferences of a voting majority only in such mediated and indirect ways, through carefully arranged incentives and disincentives which do not conflict with its obligation.[42]

What I have suggested here is a totally new addition to Rawls' political theory: While the Kantian aspect of his "justice as fairness" is reserved for the antecedent fairness of the original position, Rawls' particular definition of consequent fairness, his Difference Principle, is now allotted to a merely prudential realm of more restricted government authority. It is thus better transcribed as "fairness as prudence."

To execute my suggestion rightly, one would have to rebuild Rawls' theory from scratch, a task beyond the intent of this article. All I can hope to have pointed out is the general direction where I think a solution is conceivable.

## IV. CLOSING REMARKS

Returning to the issue of health care policy for which I have sought to develop a foundational framework, two major conclusions may now be drawn.

1.

We are in a better position to understand why Rawls has tended to disregard the issue of public health policy and in what sense the reasons for this disregard are no longer valid within his modified theory.

(a) Having superimposed deliberative issues on the original contract situation, Rawls must have noticed that the ensuing redistributive measures, if they were made to cover people's health care needs as well, would prove too costly.

However, these issues have now been allocated to a merely contingent area of government activity. Hence the extent of the desired financial involvement can be conceived to be balanced against the available means as well as other desirable objectives. There is no longer any need to exclude the health sector.

(b) Rawls' Fairness Principle was motivated by a regard for a universal sense of human solidarity operating to assure social cooperation. It is conceivable that Rawls thought the model of such cooperation inapplicable to the health sector: after all, the healthy and the sick do not "cooperate" with one another in the same sense in which the rich and the poor do. Hence neither is the political significance of humanitarian feelings in the first case as easily granted as it is in the second, nor does enlightened self-interest (which underlies that attribution of significance) become operative in the former case at all.

Yet this reasoning is not convincing. The point is not to deny that there exists a cooperative analogy between differences in health and differences in wealth, because the former unlike the latter need not be considered a substantial issue, anyway. Rather, these former differences (at least in some instances) are merely aspects qualifying certain forms of wealth-oriented cooperation. What individuals "receive" from their social involvement can be considered in view of its unintended bad effects, such as diminished health, just as well as in view of its intended good effects, such as increased wealth. The moral question of fair sharing can be applied to both sides: To the extent that it holds that those who gain more wealth should be induced to compensate those who gain less, one might argue that it should also hold that those who incur less damage to their health should be made to compensate those who incur more. (In fact, in his later article, Rawls has argued in this way, again however without considering the issue of health, [48], p. 528) In both cases the Fairness Principle is satisfied: Those winding up with better chances for future success are enjoined to share with those whose chances have diminished.[43]

As a result, the considered value judgements which Rawls had argued would favor redistributive policies in one respect should have also been argued to favor them in the other. Rawls' exclusion of the health care issue is thus, even on the basis of his own (un-modified) reasoning, not acceptable.

2.

The extent and limit of different kinds of government intervention in the health care sector can now be determined in such a way that the issue of *efficiency*, which Maynard rather undifferentiatingly has supposed to be a government concern, appears distributed more equitably among the various "social partners."[44]

I shall now approach this matter of efficiency by distinguishing several interpretations of health status, which determine different areas of relevance within the foundational framework.

(a) Health status can affect people's chances of survival. It thus raises the issue of the means for subsistence. Granting that a social minimum is clearly compatible with that minimal paternal function which even a libertarian state such as Kant's (cf. *Metaphysik* A 326f) must be allowed to perform, some minimal and reasonable emergency health care can be argued to be rightfully due each needy citizen. Since such care is provided out of common resources, the population in general has an interest in matching its expectations of efficiency with its charitable contribution.

(b) Health status can be what individuals carry with them as a matter of natural endowment. Political prudence suggests that people should be given incentives to use their endowments so as to render them effective social assets. In view of their health, these incentives could encourage preventive measures as well as compliance with therapeutic regimens. Thus it would be prudential for governments to render the receiver-side of medicine's efficiency a responsibility of each individual citizen.

(c) Health status may be influenced by what people can buy, at least insofar as effective care is available. To this extent, matters of efficiency are by the very nature of his consumer-role an immediate concern of the patient.

But health care may also be considered one of the socially primary goods.[45] The list of such goods, in Rawls' theory, determines the minimal conditions for satisfactory social life. By adding such care to the list, it becomes prudent (even though not obligatory, as Rawls would have it) for governments to provide incentives for the maintenance and development of such care as well as to enable patients (e.g., by permitting patient associations) to seek efficiency effectively.

(d) Health status can relate to what people bargain for when they negotiate for preventive measures and insurance schemes that accompany cooperative endeavors. According to "Fairness as Prudence" the stabilizing factor which consists in individuals feeling secure from major health hazards argues for state endorsement of social structures that permit the Difference Principle to come into play within the narrow confines of such schemes. The efficiency of these insurance models can then be left to the cooperating partners themselves.

(e) Health status can be the effect of societal factors (like industrialization, pollution, or insufficient working conditions) that result in the worsening of the chances of some while improving the status of others. According to "Fairness as Prudence" (in the modified sense specified above) the Difference Principle is applicable here as well. Just as in the previous case, so here as well a government's involvement concerns the securing of conditions for the possibility of fair negotiation. Again, the efficiency or quality of such care are issues to be negotiated.[46]

This sketch is composed of very abstract and general statements. The principal point which I have intended to make is that one can construct an acceptable theoretical framework for the moral and prudential evaluation of these matters according to which much less is, or should wisely be considered, an immediate concern of governments[47], and that the notions of "right" and "justice" taken in a strict sense have a much narrower range of application than is generally believed.

*Freigericht*
*Federal Republic of Germany*

NOTES

[1] See for example Hyde [30], Kazer [37], Lindsey [39], and Sigerist [52].

[2] The gist of Smith's argument becomes clear from the statements following the one quoted by Maynard, in particular: "But though the law cannot hinder people of the same trade from sometimes assembling together, it ought to do nothing to facilitate such assemblies; much less to render them necessary" [53], p. 145; see also pp. 135 ff.

[3] See, for example, Friedman [22].

[4] The plausibility of this theory is underlined by the recent attempt of the Hunt brothers to corner the world silver market. They bought rights to silver, but as they bought, so rose

the price, so that an ample supply of silver always eluded their grasp. The Hunts eventually ran out of funds, and the price collapsed. I owe this example, as well as much of this economic critique, to Michael Hayes.

[5] See also: "The private financers and providers will ultimately face the same pressures that are generated in a public system, as can be seen in the United States. They will respond to these pressures by self-regulation, i.e., the private insurers will seek to control the prices, quantities and quality of health care. So whether the State or private agencies finance and provide health care, the fundamental problems of this market are similar and regulatory activity, defined as action by financers and providers to manipulate prices, quantities, and quality, is unavoidable and ubiquitous" [41].

[6] Maynard's purpose is obvious. He wishes to disqualify those opponents of the NHS system of British state medicine who are advocating a greater market orientation. In order to do this he points out that the market alternative labors under the "same" difficulties as the collectivist alternative.

[7] In contrast to Maynard's implied notion of solidarity, Rawls still opposes needs-based accounts of claims against the state in general ([47], p. 88), even though he does accept a concern for needs where the security of a social minimum is concerned ([47], p. 276).

[8] Rawls recommends markets as prima facie suitable instruments for distributing goods ([47], p. 274) and emphasizes the significance of incentives ([47], p. 279). He agrees with Maynard in demanding proper regulations of market activities ([47], p. 309). Rawls' discussion of efficiency, however, links this concept not only with justice in general but also with liberty as a side constraint ([47], pp. 69ff).

[9] To put it in terms of Rawls' "original position": No rational creature would consent to a social arrangement which (given the temporal frame of reference for his decision, i. e., the fact that the communist paradise remains at a yet distant future) will either render him a living sacrifice to future generations, or grant him only a dwindling chance of occupying one of the all too few privileged positions presently available. Rawls' opposition to utilitarian accounts of the relationship between generations (see [47], pp. 285ff, 289) implies a similar critique.

[10] See, for example, [47], pp. 62, 97.

[11] See, for example, Green [23, 24]; Daniels [10, 12, 13]; Stern [55], and perhaps (depending on how he understands "welfare") Moskop [43]. Egalitarian readings of Rawls' theory as a whole are also frequent, see for example Wolf [58], p. 195, and Barry [2].

[12] The legitimacy of suggesting such a model arises from Rawls' own frequent references to Kant ([47], p. 140, and section 40 on 251ff; see also his later essay [48]).

In using this model as an interpretive tool, some caution is, however, needed. One must not simply apply Kant's ethical categories to Rawls' political framework and then, realizing that they don't fit, deny the very relevance of the Kantian elements of Rawls' theory [14]. The realm of ethics is, for Kant, both distinguished from and also connected with the realm of politics, and one must keep this two-fold sense in mind.

A similar conflation can be seen in Wolf's claim that since the original contractors have (even if only few) interests, they cannot be counted as Kantian rational subjects ([58], p. 113). For this claim could be used as a criticism only if it would indeed be necessary (as Rawls sometimes falsely thinks) to assume that those contractors constitute the concept of justice. Yet it is quite sufficient to see the contract situation as merely a particular political illustration of that mutual moral respect among persons which in turn is a constitutive element of the concept of justice.

If conducted with the appropriate caution the resulting Kant-oriented interpretation of Rawls' task could also prove useful in rebutting various criticisms which, insofar as they aspire to an *ad hominem* force, stand in the way of using Rawls' theory for the desired critical purposes. This holds, first, for various Marxist or quasi-Marxist arguments denying the relevance, and, second, for other arguments denying the tenability of that theory. Thus, first, (and disregarding such clearly extraneous critiques such as by Sneed [54], Brakemeyer [5], and Doppelt [16]) the reasonings used by Fisk [21] and Daniels [11] presuppose that Rawls' theory is merely apologetic of modern liberal democracy, that its underlying view of human nature is false, that the theory must therefore have bad practical consequences [21], and that the issue of equal political liberty cannot be separated from that of economic inequality [11].

Concerning the charge of apology, even though Rawls himself has in a later article [48] restricted his intention to merely reconstructing the values embodied in liberal democracy, and even though Kant's own favoring a republican framework has been convincingly argued to apply to modern constitutional democracy ([49], p. 29), this partiality in both rests on a philosophical view of human personhood the validity of which would have to be more thoroughly challenged before the charge of a merely parochial apology of capitalist society could gain some weight (cf. [17], p. 52). (In fact, Rawls himself has considered his "Justice as Fairness" to hold for the socialist context as well; [47], pp. 137, 280).

Concerning the issue of a false theory of human nature, Fisk has 1) confused the Kantian noumenal disinterestedness characterizing Rawls' "original contractors" with merely phenomenal self-interestedness (cf. [47], pp. 129, 147 ff, 583f). The necessity of coercive governments is essentially grounded for Rawls, as it is for Kant, in the struggle between human beings' noumenal and phenomenal side. It can therefore not be expected, as Fisk supposes, to vanish under conditions of a classless society. Fisk's neglecting this twofold viewpoint also, 2), renders his critique of Rawls' moral psychology pointless. He may be right in arguing that the human sense of justice is harder to render effective in social life than Rawls believes. But this does not repudiate the theoretical validity of an account which starts out with such a sense. (Quite similarly, Miller's quoting psychological evidence against the contractors' capacity to assume the viewpoint of justice ([42], pp. 211, 215, 219f, 225) disregards the strictly normative character of Rawls' contract model.) Fisk's further objecting, 3) to the anthropological atomism giving rise to – as he terms it – an atomistic model of society and to the view that all social union aims at stabilization ([21], p. 63) disregards the fact that Kant's as well as Rawls' noumenal atomism is, first, mitigated by a rational respect for humanity as an end in itself: second, by its compatibility with a priority of groups on the phenomenal level ([47], pp. 522 f, 584). Hence, as we shall see later, it appears quite possible to account for a certain "noumenal tolerance" for destabilizing group conflict characterizing phenomenal social life.

In what concerns the theory's bad practical consequences ([21], p. 57), the issue of institutions operating at "enormous social costs" will be discussed later (see note 44). Yet, even if such costs were unavoidable (which they are not), Rawls' Kantian opposition to utilitarian arguments not only on the object-level of moral and political theory, but also on the meta-level of theory-constructing itself renders Fisk's criticism pointless.

Thus, second, Nagel's finding fault with the fact that the original contractors are deprived of any notion of their good and that this deprivation is not in turn justified by reference to any good ([44], pp. 6–9) becomes less persuasive if the Kantian regard for humanity as an end in itself and to a kingdom of ends is duly considered. And Lyons' denying the binding force of a merely hypothetical contract ([40], pp. 156ff) can be rebutted in view of the categorial imperative reasons which were found to underlie this contract.

[13] This rendering of Rawls' philosophical project does not conform to his own formulation, as for example in [48], pp. 516, 532. In this later essay he claims that the *Theory of Justice* was merely intended to articulate certain shared beliefs concerning freedom and equality in such a way as to place them in an appropriate relationship. But this claim falls short of his "Theory"'s preoccupation with justificatory issues (and this holds even though he mistakenly believes that "articulating shared beliefs" could somehow by itself amount to a justification ([48], p. 519).

It is, in fact, for the very purpose of reviving the more fundamental issue concerning the grounds of political power, and thus concerning the scope and limit of state or government authority, that this Kantian interpretation of Rawls is being undertaken. After all, these issues must be decided before the further question concerning any "shared beliefs" determining the particular uses of that authority can legitimately be asked.

[14] Correspondingly, both agree in their opposition both to utilitarian ([47], pp. 15, 30) or (in Kant's term) teleological (*Grundlegung*, A15) and to emotivist ([47], p. 129; *Grundlegung*, A27f) or merely prudential moral theories. Rawls' opposition to utilitarian foundations of justice (and thus of morality), just as Kant's opposition to teleological foundations of morality (and thus of justice), leads to a very limited view of the function of states or governments. Just as Kant insists that the state's task is not to secure "the happiness of the citizens" (*Gemeinspruch*, A252) so Rawls argues for a similar restriction ([47], pp. 212, 566) and opposes perfectionist political theories ([47], p. 15).

[15] In contrast to Kant, Rawls includes political liberties (the right to vote and to be elected) in his list ([47], pp. 60, 61). That is, while Kant suggests a republican model (*Frieden*, A24ff), Rawls recommends a democratic setting ([47], pp. 66, 195). Yet this difference is not a crucial one. First, Rawls has only prudential, not fundamental reasons for his preference ([47], p. 233); second, he concedes that under certain circumstances, even lesser political liberties are compatible with the precepts of Justice as Fairness ([47], pp. 74 note, 152, 247).

Kant, on the other hand, while declaiming democracy as despotism (loc. cit), and thus as incompatible with justice, does so on grounds which can be disregarded. (Luf appropriately criticizes his restriction of voting rights as transcendentally not warranted, [39], pp. 153ff). He considers democracy, together with oligarchy and monarchy, as three different power structures, whereas the term "republic" is used to refer to a specific sort of justificatory enterprise. But the justification here rests on the very contractual fiction which is used by Rawls within the democratic context (cf. Reiss [49], p. 29). As will become obvious later, Rawls can be seen as having merely updated the principles Kant applied to the Prussian monarchy such as to render them applicable to the present times.

[16] At least for Kant, this does not imply that humans may always in fact be treated in such a way. As the limits of government authority suggest, there are other constraints to "treating" persons beside the condition for rational consentability. This point will become significant later on.

[17] Thus on a more general level, for Kant actions are morally acceptable if the maxim determining the will to their performance is capable of serving as a general law (*Grundlegung*, A17) of a sort which would be devised by purely rational beings legislating themselves (*Grundlegung*, A57, A70; *Metaphysik*, A166f, A212. Cf [47], pp. 131–5).

This is also why both Kant and Rawls employ the fiction (*Gemeinspruch*, A249f, 162; [47], p. 12) of an original contract for their justification of social and political institutions. In Kant, the significance of the presupposed rationality is determined by the notion of noumenality; in Rawls, this notion receives an imaginative interpretation in his description of "people in the original position". Or Rawls' "procedural" criterium for the moral acceptability of social institutions appears to be a particular society related interpretation of the first and second version of the Kantian categorial imperative ([47], p. 256, [58], p. 106) What I have termed the "antecedent fairness" of that position is thus characterized by a Kantian regard for humanity as an end in itself.

[18] Taxation, in that it involves taking part of a citizen's property away, would have to fall under the general Kantian notion of an invasion of those citizens' freedom. The state is unconditionally obliged to respect and protect this freedom (*Metaphysik*, A163f). This obligation does not speak against taxes as such. It implies, however, that they must be justifiable in view of the state's primary obligation (*Metaphysik* A187 f, [39], p. 118, who criticizes Kant for being so restrictive in view of the state's function.)

[19] In attending to the inadequacy of Rawls' (otherwise very admirable) theory, I shall restrict myself to the most fundamental issues. That is, I shall leave aside the more particular difficulties that encumber the specifics concerning, for example, what constitutes a "corresponding" chain, or what justifies his compensating only the worst off and relying on some chain-mechanism by which the more moderately disadvantaged groups can be reached. These issues have been competently discussed by others.

[20] (Cf. Kant, *Gemeinspruch* A237 ff; *Metaphysik* A 168 f.) This Kantian regard for the free exercise of skillful effort is matched by Rawls' concern for holding favored positions open ([47], pp. 61, 84) and by his authorizing government efforts at redistributing income insofar as this is required for the securing of equal liberty and open opportunity for all ([47], pp. 277f.)

[21] Thus "looking for a conception of justice that nullifies the accidents of natural endowments" ([47], p. 15), Rawls has designed his principles of fairness in order to "regulate" these inequalities ([47], p. 96) because "undeserved qualities call for redress" ([47], p. 100). Unequal inheritance of wealth and unequal inheritance of intelligence are deemed equally unjust ([47], p. 278).

[22] This critique agrees with Hoeffe ([28], 22f). Kant would of course concede that one can look at people on a phenomenal level as well, where their actions appear as conditioned by physical or psychic circumstance and where it therefore makes no sense to give someone credit for what he did. On a noumenal level, however, people must be considered exclusively in view of their freedom as moral agents (*Grundlegung*, A110f).

[23] The same point is made by Nozick ([46], pp. 213f).

[24] To be sure, a state can be considered unconditionally obliged in view of individual persons' freedom and at the same time consider groups of individuals in the light of statistical evidence as conditioned by social or economic factors. The problem with Rawls' argument is that he wants both aspects to enter into the "unconditional duty" realm of government activity. That is, he makes it impossible to assume that different viewpoints underlie these mutually incompatible considerations. To put it more bluntly: Rawls appears to require the state to violate the freedom-respect by taking money from some in order to satisfy the condition-respect by giving it to others.

[25] Thus contrary to the Kantian assumption of the economic independence of property-owners, not only do those at the bottom (say employees) depend on those at the top (say

employers) but the latter are equally dependent on the willingness of the former to accept their kind of work and salary and to perform it correctly.

This is significant in view of Kant's own motive for the restriction of voting rights: The supposed economic independence seemed to argue a corresponding conceptual independence of property owners' judgement on legislative matters (*Metaphysik*, A166). Since for Rawls the former supposition no longer holds, whatever reconstructive capacity to consent to legislation can be attributed to the "haves" must also hold for the "have nots". Consequently, it is only in view of differences in factual assumptions concerning society's social permeability that Kant's and Rawls' evaluation of democracy differs. (A similar argument is used by Luf [39], p. 127 f, in order to show why Kant's political theory must be modified in order to become applicable to modern times.)

[26] Thus, even if some one would find himself losing while others gain, he would still have to be thought of as having a reason to rationally consent to his cooperation as long as his losses – all things considered – do not make him worse off than if he would quit joining altogether. Kant, of course, does not pose the question whether some persons might conceivably be better off in a state of nature than at the lower social end of industrialized society. He seems to presuppose that humans realize their rational dignity only within the legal context of a state, and that such realization is all which should basically matter.

[27] Many modern Kant scholars have attempted to use an updated Kantian framework for the justification of distributive measures akin to those envisaged by Rawls. Thus Kuehl ([38], p. 271), Brehmer ([7], p. 147), Luf [39] and Volkmann-Schluck ([57], pp. 119, 127), for example, argue that under the condition of modern industrialization and capital accumulation there have developed social and economic dependencies which Kant did not forsee, and that these dependencies hinder the sort of individual freedom to the protection of which the Kantian state is essentially committed. Yet these scholars then have difficulties in accounting for Kant's amazing unconcern not only with the existence of merely "passive" citizens (who are denied the voting rights) but also with institutionalized social and economic dependence relationships such as between master and apprentice or between the head of a household and the (often lifetime) servants (Kant here significantly uses the term "dinglicher Besitz an Personen"). Kant considered such dependencies licit because they involve a reciprocal obligation: While the dependent member of a social unit provides his working power, the master is obliged to secure his welfare and protection. It is this reciprocity, along with the supposition that through individual effort and application those who wish to become independent will be able to do so, which explains why the state's protective function in view of individual freedom does not come into play in these cases. Accordingly, no similarly reciprocal relationship between modern employers and employees can be argued on Kantian grounds to call for state intervention.

To be sure, some have tried to evade this conclusion by invoking some "aristotelian" social prejudices in Kant ([39], pp. 160ff) or by arguing that Kant was concerned only with the minimal conditions for just politics ([49], p. 34) which call for amplification. But just as it would not make sense to conclude from the formal character of Kant's moral theory that its merely minimal strictures call for a fuller elaboration in order to make up a decent ethical system, so, it seems to me, Kant's abstaining from holding the state responsible for helping the dependent become independent (as well as, to point to a very different corner of his political theory, Kant's stand on the issue of resistance) instead refers back to the ascetic nature and reckless rigorism of his moral theory, which is the price he pays for the unconditional validity of his ethical precepts.

[28] Deliberative rationality differs from Kantian rationality in that its orientation is teleological, not deontological. Rawls claims that the latter must be supplemented by some regard to ends ([47], p. 30), and that both kinds of rationality have different "functions". But how could these different functions be specified? It is hard to determine what Rawls means. In ([47], p. 584) he distinguishes between the original position determining the content of justice, and the later considerations showing that justice is also "good" for us. This distinction reminds one of that between Kant's philosophical foundation of the validity of his moral norms (for example in *Critik*, Part I) and his (sketchy) comments on how to teach ethics, i.e., how to render compliance with these norms palatable (in Part II, his "methodology"). In ([47], p. 586) however, Rawls says that "the theory of justice provides a rendering of these ideas" (about the dignity of persons, for example) "but we cannot start out from them." This distinction reminds one of the problem in Kant: how to get content into the required "respect". Rawls seems to suggest on this page that he solved this problem by letting the conditions governing respectful behavior in some preliminary and formal manner determine the original situation, and then asking what choices would rationally be agreed upon by people looking at the givens of psychology and sociology while pursuing their good. In the end, Rawls seems to claim, these rational choices provide a deliberative reconstruction for the merely formal notion of respect initially posited. At least his claim to have "reformulated and established" Kantian themes in view of his "general" (i.e., deliberative) conception of rational choice or to have "clarified" the older contract theory ([47], p. 584) points in this direction. (See also "we must assume . . ., " on p. 434). One may perhaps deem such a "reconstruction" harmless, or compatible with Kantian reasoning as long as the question concerning the grounds of the validity of "justice as fairness" is distinguished from the question concerning its content. As will be seen shortly, however, Rawls does not maintain this distinction.

[29] A similar point seems implied in Nozick's questioning Rawls' very assumption that contractors have something to distribute, such that a fairness principle is called for ([46], p. 219) or that people should be treated at all, such that the question of treating them to equality can arise ([46], p. 223), or that the contractors even have the right to divide whatever there might be to distribute ([46], p. 199 note).

It is this lack in Rawls which renders him vulnerable to Barber's criticism. For in all of his references to exploitative relationships it is merely presupposed that these should be a concern for governments, or states, or contractors determining the institutional setup of states. But then Barber gives no reasons why this should be so either. Hence his criticism as well as his additional charge of apolitical thinking ([1], pp. 311–14) amounts to the complaint that Rawls is not quite up to the extent to his own (Barber's) weakness.

[30] (Cf. [58], pp. 202f). By contrast, such reflecting is central to Kant's concern. It is very much alive, for instance, in the way Kant links the law to the relationship between person's actions insofar as they affect each persons' discretion (Willkür) (*Metaphysik* A25, A32f, A312). He thus restricts the function of the law to securing that people's actions do not interfere with other people's liberty to act. He expressly denies the law to relate to persons' actions insofar as they affect other persons' needs or their desires (loc. cit.; see also *Gemeinspruch*, A251 with a merely utilitarian exception, where the existence of a state is threatened, A252f.). (It is, of course, those very – even though deliberately rational – desires which Rawls has linked with his view of the law.)

This opposition to any welfare- or happiness-oriented (paternalistic) state is central to Kant's concern (cf. Volkmann-Schluck [57], pp. 104, 116f). The conditions of human happiness are an empirical matter, they can never be determined in advance or even on the grounds of rational investigation (*Grundlegung*, A46ff); the advancement of humans' capacities cannot be made the object of purposeful action (*Grundlegung*, A83); the more one abstains from actively pursuing such a goal, the more one's moral behavior will by itself come to coincide with this goal. (*Frieden*, Anhang, A85) Thus Kant also emphasizes the illegitimacy of governments meddling with the terms on which citizens pursue their happiness. In particular, he warns of any state attempting to implement moral goals by the use of legal sanctions (*Metaphysik*, A47).

[31] See for example Benhabib ([3], p. 53), Hoeffe ([29], pp. 15, 27, [28], p. 24), Brehmer ([7], p. 143), Nagel ([44], pp. 3, 10–12), Hare ([25], pp. 103–7), Dworkin ([18], p. 139), Lyons ([40], pp. 161, 163), and Barber ([1], pp. 279 ff). Nozick also joins in with this criticism (see note 24.)

[32] To be sure, the contractors' deliberative rationality is restricted by the thin theory of the good. They are concerned exclusively with the fair distribution of societal advantages supposedly desired by all humans, whatever else they may desire. Or these advantages are conceived to be generally accepted means for the realization of prudent self-determination. And to be sure, this theory of the good is oriented around a view of human nature which agrees not only with the philosophical tradition at large ([47], pp. 414, 431), but can also, and in particular, be argued to coincide with the conditions of rational human self development and with the ever fuller realization of humanity as an end in itself, as these are conceived by Kant (*Grundlegung*, A68f).

Yet such traditions, taken as a whole, always encompass too many heterogeneous elements, thus precluding those appeals from having any distinctive meaning. And as to the selection of these elements which would serve the Rawlsian purpose, its adequacy as representative for the whole could again only be established on merely persuasive grounds. [The same problem reappears in Rawls' later article ([48], pp. 519, 527.)] For a similar critique see ([58], p. 116).

[33] To be sure, Rawls himself is innocent of any clearly totalitarian leanings. In fact, the details of his account can even be seen as intentionally designed to rule out any abuse of the contractors', and thereby the state's authority. First, Rawls opposes perfectionist theories of the state and thus clearly excludes at least the other extreme of avowedly non-libertarian arrangements. Second, Rawls always insists on the principle of equal liberty (and on open chances) as the essential basis for a societal setup ([47], pp. 62f, 396f).

Yet as has been shown, the theoretical grounds for this insistence are insufficient. Quite irrespective of Rawls' personal intentions, the implications of his theory point in the other direction.

[34] It has become clear then that deliberative rationality is not in the final analysis intended by Rawls to merely supplement Kantian deontological rationality. Instead, the former is "superimposed" on the latter. But this is just a nice way of saying that Rawls confuses the two. This confusion between a rights-based account and a goal-envisaging account explains Rawls' failure to even consider the limits of governmental authority. Or conversely the lack of this consideration can be used as evidence for the fact that what could kindheartedly have been interpreted as a mere supplementation must in fact be seen as obfuscation.

Let me adduce some additional evidence for this (restricting myself to the "Theory of Justice"):

(1) Rawls renders moral norms, which in Kant are considered valid irrespective of any wishing and desiring, commensurable with norms which make sense only in view of people's wishes and desires. Thus ([47], p. 176) people in the original position are said to "choose" (in view of the strains of commitment) both: basic rights and some insurance against poverty. Or ([47], p. 396) people are said to "want" more liberty in a sense not different from their "wanting" more wealth. Or ([47], p. 476) acting justly is said to be expressing one's nature, and thus to constitute one's "good". Or ([47], p. 434) goodness as rationality is used as accounting for all the "primary goods", with liberty again among them.

(2) On closer look ([47], p. 30), Rawls has defined his theory as deontological only by default, because he found utilitarianism already occupying the "teleological" box. But then he immediately lumps "not considering consequences" under the heading of craziness. That is, he does not distinguish between two crucially different ways of considering consequences, namely, one excluding this regard from the question of validity and including it only in view of the problem of motivational force (which is Kant's option), and the other of including this regard already in the validity issue (which is both Rawls' and the utilitarians' option, with the latter adding the goal of maximizing the good, the former abstaining from that addition).

(3) Rawls believes he needs to provide a naturalist basis for moral norms ([47], p. 438). He thus replaces what in Kant is a philosophical decision to adopt a view towards what can be considered "moral" about humans by a merely factual reference to "man's moral capacity" ([47], p. 505). But such a ground will not do. On the one hand, it involves Rawls in insurmountable difficulties when considering all those humans in whom that capacity is not – as a matter of fact – sufficiently developed ([47], pp. 505–9). On the other hand, he cannot explain why higher degrees of such capacity ([47], p. 505) should merely count as natural assets, rather than as a ground for higher or more extensive respectability. Finally, and most importantly, it remains quite unclear how a natural property should provide a sufficient basis for that unconditional duty which is required in order to – among other purposes – effectively exclude utilitarianism. (His remark that we should respect only those who can respect in turn ([47], p. 511) is not very helpful. Perhaps indeed we should hate or greet only those who can hate or greet in return, but it does not follow that we should therefore hate or greet them all.) After all, Rawls also argues that if there exist "some persons for whom the affirmation of their sense of justice is not a good" ([47], p. 575) and who therefore will not recognize any deliberatively rational grounds for complying with just institutions, then the persons in the original position would "agree that stabilizing penal devices can be employed" (loc. cit.). Since, so Rawls goes on ([47], p. 576), "the general affirmation of the sense of justice is a great social asset", the deviant nature of such persons "is their misfortune". As a result, what on this level of arguing should have been counted merely as the unhindered self-expression of the natures of the few is sacrificed to what should have been counted as merely the unhindered expression of the natures of the many. Or by having rephrased the deontological issue of justice in utilitarian terms, Rawls is left with no safeguards against these terms being also applied to other kinds of deviance.

(4) Rawls' frequent references to Kant must now be reinterpreted in the light of certain profound misunderstandings.

(a) The "deliberative revision" of Kant is motivated by a reference to Sidgwick's critique according to which "Kant did not show that acting from the moral law expresses our nature in identifiable ways that acting from contrary [i.e., bad] principles does not" ([47], p. 255). This critique is hard to understand in view of noumenal rationality being distinguished by a regard to humanity as an end in itself (*Grundlegung*, A69f).

This revision is also motivated by reference to Ross's critique that the "purely conscientious act is irrational" ([47], p. 477, cf. 478). Such criticism disregards the Kantian distinction betwen acting "out of" respect and acting "because of" a motive, or, what amounts to the same difference, between "taking an interest" and "being interested" (*Grundlegung*, A102ff). On Kant's response to a similar criticism see *Gemeinspruch*, A 210ff; *Religion* A5ff.

This revision is, thirdly, motivated by the claim that Kant's view towards a kingdom of ends remains "transcendent" ([47], p. 256). Rawls here presumably means that Kant restricts this view to a mere hope, which employs the notion of god and of a future state. He also objects that such "purely transcendent notions" are "lacking explainable connections with human conduct" But it is exactly the securing of such connections for the sake of which Kant poses the question of how the categorial imperative can be thought to motivate the will (*Grundlegung*, A102ff) and answers it in view of certain necessary postulates of practical reason the objects of which he admits must remain transcendent, i.e., beyond theoretical knowledge (*Kritik*, A215ff).

(b) In particular cases Rawls even attributes a narrowly teleological reasoning to Kant himself ([47], p. 338). But here Kant's describing an act to be morally illicit on the ground that the maxim guiding the will could not be entertained consistently (*Grundlegung*, A57, A69 note) has been mistaken for a simple regard to reciprocity.

(c) Rawls misunderstands Kant's noumenal freedom (on the "transcendence" of the nature of freedom see *Metaphysik*, A18) as "independence from natural contingencies and social accident" ([47], p. 255). Or he confuses the shame we experience when "we have acted as though we belonged to a lower order" ([47], p. 256), that is when we are subject to base cupidity, with our being subject to unequal endowments. [Perhaps he mistook Kant's comments on the conditions for self-contentment (*Kritik*, A212)]. No wonder Rawls should believe that governments, since they are obliged to respect our freedom (not being quite able to free us from our pathological desires) would have to "free" us from the effects of such inequality.

[35] In blurring over the fundamental difference between what is owed to people (out of respect for their humanity) – for example, liberty – and what can only be wished for them – for example, wealth – Rawls' own distinction between people's right to equal liberty and people's having to accept the unequal worth of that liberty ([47], p. 204) is no longer convincing. Thus, again, it becomes unclear how Rawls can escape the egalitarian solution (A similar criticism is made by Daniels ([11], pp. 265f).

[36] Thus Rawls renders himself vulnerable to Lyons [40] playing his general conception of justice off against his special one and thus to the charge of utilitarianism.

[37] In this context one is inclined to some suspicion on seeing Rawls define a "good (beneficient) act" as one which "advances and is intended to advance another's good (his rational plan)". For even if the good advanced is indeed my own good, why is there no regard paid as to my desire or aversion of having someone else (or perhaps this particular person) meddle with its advancement? (Dealing with rational two-year-olds reminds one of the need to respect their resentment of being thus "maternalized".)

A very similar lack of regard characterizes Rawls' political philosophy as a whole.

Already with Rawls' belief in the state's moralizing function in levelling out deserved and undeserved advantages the distinction between "private" and "public" is blurred. Individuals' natural capacities are considered as common assets ([47], pp. 101, 179, cf. [46], p. 229, note), and the state is entitled to secure their appropriate use. (Rawls' analogy of a family is not helpful: one can leave obnoxious relatives, but one cannot cease being a member of some state. See also [57], p. 105.) Rawls wishes to see such appropriateness determined by the moral principle that no one should profit from the cooperation of others without doing his share in turn ([47], pp. 343, 585). But then it follows, since invididuals have not much control concerning the extent of such societal cooperation by others, that the ensuing advantages can be forced upon them even against their will and can then be used to extract the required reciprocity.

[38] The admission of such a realm into political theory (cf. [47], p. 278) is compatible with Kantian reasoning (see for, example, *Gemeinspruch*, A252f; *Metaphysik* A164f, A325, A369f), as long as it remains subject to the constraints imposed by the strict government obligations (cf. Dietze [15], pp. 65ff, who speaks of a side-by-side account of the state of enforcing the idea of justice and as an arbitrator between interest groups. Yet all those who do admit the necessity of such a prudential realm within the Kantian framework ([57], pp. 102, 114; [49], p. 54; and also [39], pp. 186ff) fail to distinguish it clearly from the realm of unconditional obligation, and hence, authority.)

[39] This does not involve any contradiction in view of Rawls' reminder that "might does not constitute right" ([47], p. 134). For it is here not the power struggle by itself, the outcome of which is claimed as determining a just state of affairs. Instead, that power struggle is influenced in such a way that its predictable outcome will come as close as possible to some state of affairs determined as "just" on independent grounds (cf. [38], p. 258).

[40] Nor could such restraint in his unimpeded self-realization be objected to on Kantian grounds. It does not keep someone from employing his talents and skills in order to improve his situation as such. It may, of course, present a disincentive in view of the expected gains or of the difficulties and uncertainties involved in securing these gains, but that is a different matter. A Kantian government is unconditionally obliged to provide opportunities but does not have to guarantee any fixed rate of success.

[41] There is no reason why every detail of such intervention would have to be determined by the contractors in advance once and for all. They can also be adjustable to changing situations in order to secure the desired outcome. But then perhaps even Daniels' [11] problem of economic inequalities finding ever novel ways to work against equal liberty can be solved. The protection of that liberty constitutes one of the strict (not merely prudential) government obligations. Nevertheless, a constitution accounting for such obligations could also justify the institution of certain ad hoc measures in order to counteract any economy-related invasion of equal liberties.

[42] This is a bit similar to John Moskop's suggestion to interpret the fairness principle as part of a social ideal. One may well concede that, once the unconditional duties of the state are accommodated within the "justice" part of the fictitious contract, within its "fairness" supplement the state is given the prudential function of also "enhancing the welfare of its citizens" ([43], p. 336). The difference lies perhaps in the fact that on my account the implementation of this principle can be termed a – prudential or deliberative – government obligation in a stronger sense than is implied in Moskop's suggestion. A regard to an ideal can easily be pushed aside in view of practical, political, and economic

constraints, whereas a regard to rational self-interest meets that view on its own grounds.
[43] This is reminiscent of Daniel's interpreting health care on the model of granting opportunities. Yet the difficulty involved in treating all health care on this model has disappeared, since we are here concerned with only one aspect of a very complex matter. [Thus Buchanan's criticism of this model ([8], p. 19) would no longer have to apply.]
[44] Thus, finally, Fisk's claim that Rawls' political theory is invalid because of its socially pernicious consequences can now be repudiated on factual grounds as well. Fisk (summarizing his criticism [21], p. 67) argues that the social theory which follows from this view a) has no safeguards against human bondage due to imposed consciousness, b) permits the coercive state to dampen but not to eliminate social conflict by supporting oppressive groups and discouraging the oppressed, c) does not allow for real community. Against this claim we can now point out that a) chances of social advance are granted to every one, and that thereby every one is able to effectively free himself from the effects of indoctrination, b) that the state is morally right in supporting those who are efficient, as long as it guarantees that social conflict may serve as a motor in rendering individuals more efficient, c) that the sort of respect for each individual's humanity as an end in itself which is incorporated in Rawl's Fairness Principle, is already sufficient for "real community".
[45] This aspect of health care agrees with Green's ([23], pp. 112ff, 117ff) interpretation.
[46] The administration of compensatory measures may, for example, be trusted to specific agencies, as the German "Berufsgenossenschaften" (cf. [6]).
[47] On a superficial level, the conclusions concerning the need of a balance between the principles of solidarity and subsidiarity and thus concerning the limits of state involvement in social issues reached by Ritter ([50], pp. 7ff) on historical grounds, Nell-Breuning ([45], pp. 382ff) on the basis of Catholic social theory, Wittkämper, Boas, and Herder-Dorneich ([4] in a sociological context) may appear similar to the endorsement of such limits presented here. The fundamental difference lies, however, in the fact that these theorists, in accepting the involvement of politics with value commitments, would not accept the separation between strictly moral (justice-related) issues of right and strictly prudential regards for values insisted on here.

## BIBLIOGRAPHY

1. Barber, B.: 1975, 'Justifying Justice: Problems of Psychology, Politics and Measurement in Rawls', in N. Daniels (ed.), *Reading Rawls*, Basil Blackwell, Oxford, England, pp. 292–318.
2. Barry, B.: 1973, *The Liberal Theory of Justice*, Oxford, England.
3. Benhabib, S.: 1982, 'The Methodological Illusions of Modern Political Theory: The Case of Rawls and Habermas', *Neue Hefte für Philosophie*, Heft 1, pp. 47–74.
4. Bogs, H. *et al.* (eds.): 1982, *Gesundheitspolitik zwischen Staat und Selbstverwaltung*, Köln, F.R.G.
5. Brakemeier, H.: 1985, *Die sittliche Aufhebung der Staates in Kants Philosophie*, Frankfurt, F.R.G.
6. Brandts, H.: 1988, 'Occupational Health and Preventive Medicine', in this volume, pp. 323–344.
7. Brehmer, K.: 1980, *Rawls' "Original Contract" oder Kants "Ursprünglicher*

*Kontrakt*", Meisenheim, F.R.G.
8. Buchanan, A.: 1981, 'Justice, a Philosophical Review', in E. E. Shelp (ed.), *Justice and Health Care*, Reidel, Dordrecht, Holland, pp. 3–21.
9. Buchanan, A.: 1988, 'An Ethical Evaluation of the U.S. Health Care System', in this volume, pp. 39–58.
10. Daniels, N.: 198, 'Fair Equality of Opportunity and Decent Minimums: A Reply to Buchanan', *Philosophy and Public Affairs* **14** (1), 106–110.
11. Daniels, N.: 1975, 'Equal Liberty and Unequal Worth of Liberty', in N. Daniels (ed.), *Reading Rawls*, Basil Blackwell, Oxford, England, pp. 253–282.
12. Daniels, N.: 1983, 'A Reply to some Stern Criticisms and a Remark on Health Care Rights', *The Journal of Medicine and Philosophy* **8**, 363–371.
13. Daniels, N.: 1985, *Just Health Care*, Cambridge, England.
14. De Lue, S.: 1980, 'Aristotle, Kant and Rawls on Moral Motivation in a Just Society', *The American Political Science Review* **74**, 385–93.
15. Dietze, G.: 1982, *Kant und der Rechtsstaat*, Tübingen, FRG.
16. Doppelt, G.: 1981, 'Rawls' System of Justice', *Nous* **15**, (3), 259.
17. Dworkin, R.: 1975, 'The Original Position', in Daniels, *Reading Rawls*, pp. 16–52.
18. Dworkin, R.: 1975, 'Non-Neutral Principles', in Daniels, *Reading Rawls*, pp. 124–140.
19. Feshbach, M.: 1988, 'Health in the U.S.S.R. – Organization, Trends, and Ethics', in this volume, pp. 117–132.
20. Fleck, L. M.: 1987, 'DRGs: Justice and the Invisible Rationing of Health Care Resources', *Journal of Medicine and Philosophy* **12**, 165–196.
21. Fisk, M.: 1975, 'History and Reason in Rawls' Moral Theory,' in Daniels, *Reading Rawls*, pp. 53–80.
22. Friedman, M.: 1962, *Capitalism and Freedom*, University of Chicago Press, Chicago, Illinois.
23. Green, R. M.: 1976, 'Health Care and Justice in Contract Theory Perspective', in R. M. Veatch and R. Branson (eds.), *Ethics and Health Policy*, Ballinger Publ. Co., Cambridge, Mass., pp.111–126.
24. Green, R. M.: 1983, 'The Priority of Health Care', *The Journal of Medicine and Philosophy* **8**, 373–380.
25. Hare, R. M.: 1975, 'Rawls' Theory of Justice', in Daniels, *Reading Rawls*, pp. 81–107.
26. Herder-Dorneich, P.: 1977, *Soziale Ordnungspolitik*, Stuttgart, F.R.G.
27. Herder-Dorneich, P.: 1979, 'Solidarität, Subsidiarität, Emanzipation', *Stimmen der Zeit*, 197, 395 ff.
28. Höffe, O. (ed.): 1977, 'Vorwort', in *John Rawls' Gerechtigkeit als Fairness*, München, F.R.G.
29. Höffe, O. (ed.): 1977, 'Kritische Einführung in Rawls' Theorie der Gerechtigkeit,' in *Über John Rawls' Theorie der Gerechtigkeit*, Frankfurt, F.R.G.
30. Hyde, G.: 1974, *The Soviet Health Service*, London, England.
31. Kant, I.: 1793, 'Über den Gemeinspruch: Das mag in der Theorie richtig sein, taugt nicht für die Praxis', in *Berlinische Monatsschrift*, (ed.) J. E. Biester, 201–84.
32. Kant, I.: 1785, *Grundlegung zur Metaphysik der Sitten*, Riga.
33. Kant, I.: 1788, *Critik der practischen Vernunft*, Riga.
34. Kant, I.: 1793, *Die Religion innerhalb der Grenzen der bloßen Vernunft*, Königsberg.
35. Kant, I.: 1795, *Zum ewigen Frieden*, Königsberg.

36. Kant, I.: 1797, *Die Metaphysik der Sitten in zwey Theilen*, Königsberg.
37. Kazer, M.: 1976, *Health Care in the Soviet Union and Eastern Europe*, London, England.
38. Kuehl, K.: 1984, *Eigentumsordnung als Freiheitsordnung*, München, F.R.G.
39. Lindsey, A.: 1962 *Socialized Medicine in England and Wales,* University of North Carolina Press, Chapel Hill, North Carolina.
40. Lyons, D.: 1975, 'Nature and Soundness of the Contract and Coherence Arguments', in Daniels, *Reading Rawls*, pp. 141–168.
41. Maynard, A.: 1988, 'The Public and Private Regulation of Health Care Markets', in this volume, pp. 133–166.
42. Miller, R.: 1975, 'Rawls and Marxism', in Daniels, *Reading Rawls*, pp. 206–229.
43. Moskop, J. C.: 1983, 'Rawlsian Justice and a Human Right to Health Care', *The Journal of Medicine and Philosophy* **8**, 329–338.
44. Nagel, T.: 1975, 'Rawls on Justice', in Daniels, *Reading Rawls*, pp. 1–15.
45. Nell-Breuning, S. J., O. von: 1977, 'Der Staat und die Grundwerte', *Stimmen der Zeit* **195**, 378 ff.
46. Nozick, R.: 1974, *Anarchy, State, and Utopia*, Basic Books Inc., New York.
47. Rawls, J.: 1971, *A Theory of Justice*, The Belknap Press, Harvard University Press, Cambridge, Mass.
48. Rawls, J.: 1980, 'Kantian Constructivism in Moral Theory', *Journal of Philosophy* **72**, 515 ff.
49. Reiss, H.: 1977, *Kants politisches Denken*, Bern, Switzerland.
50. Ritter, G. A.: 1983, *Sozialversicherung in Deutschland und England*, München, F.R.G.
51. Sandel, M. J.: 1982, *Liberalism and the Limits of Justice*, Cambridge, England.
52. Sigerist, H. E.: 1937, *Socialized Medicine in the Soviet Union*, Victor Gollancz ltd., London, England.
53. Smith, A.: 1776, *An Inquiry into the Nature and Causes of the Wealth of Nations*, R. H. Campbell *et al.* (eds.), Clarendon Press, Oxford, England.
54. Sneed, J. D.: 1976, 'John Rawls and the Liberal Theory of Society', *Erkenntnis* **10**, 1, 1–19.
55. Stern, L.: 1983, 'Opportunity and Health Care: Criticisms and Suggestions', *The Journal of Medicine and Philosophy* **8**, 339–361.
56. Veatch, R. M.: 1976, 'What is a "Just" Health Care Delivery?', in Veatch and Branson (eds.) *Ethics and Health Policy*, Ballinger Publ. Co., Cambridge, Mass., pp. 127–153.
57. Volkmann-Schluck, K. -H.: 1974, *Politische Philosophie*, Frankfurt, F.R.G.
58. Wolff, R. P.: 1977, *Understanding Rawls: A Reconstruction and Critique of A Theory of Justice*, Princeton University Press, Princeton, New Jersey.

# SECTION III

# MACRO-ALLOCATION AND MICRO-ALLOCATION

RUTH MATTHEIS

# MACRO-ALLOCATION IN HEALTH CARE IN THE FEDERAL REPUBLIC OF GERMANY

In the Federal Republic of Germany a great many actors are on the stage when the play 'Health Care and Health Management' is performed. It is therefore not always easy for the spectator to recognize and understand the structure of the plot.

The actors may be grouped as follows:

Group 1
The parliaments on federal, state, and community levels with competence in legislation and budgeting.
Group 2
Health insurance, instituted by law 100 years ago, and regarded as the economic switchboard of a health care program which covers at present more than 90% of the population.

Health insurance in the Federal Republic: a corporation under public law which is self-administered. (There is not one central health insurance fund but all in all about 1,300 health funds operating under the same law – the Federal Insurance Code or *Reichsversicherungsordnung*.

More than 50% of the costs of health care is financed by compulsory health insurance. At present, its total annual budget amounts to almost 100 billion DM (60 billion U.S. dollars) compared to 10 billion DM in 1960 ([4], p. 42); roughly 30% of the total is spent for hospital care, 20% for ambulatory care, some 15% for drugs and other so-called medical remedies ([2], p. 34). Health insurance is funded by contributions related to the wage scale, half of which is paid by employers, the other half by the employees.

Health funds are managed by elected boards; half of their members represent the employers, the other half the employees.

Group 3
The providers of health care – doctors, dentists and their assistants, hospitals, and the pharmaceutical industry. (The total number of doctors – now 190,000 – increased by 235% during the last 25 years. An interesting fact is that the number of those working in hospitals – now

*Hans-Martin Sass and Robert U. Massey (eds.), Health Care Systems,* pp. 201–212.

76,000 – tripled, whereas those giving ambulatory care increased by only 13% ([3, 5], p. 42).

A special feature of the organization of health care in the Federal Republic is the virtually complete separation of hospital and ambulatory care. With the exception of teaching hospitals and some mental hospitals, no hospital is permitted to run out-patient clinics. Physicians in hospitals are usually fully employed and paid by the hospital. Doctors in private practice who, by the way, have a powerful lobby, insist that ambulatory care is their privilege, and no other groups of colleagues may interfere in this. An exception, again, are physicians in high positions in the hospital – usually heads of departments – who, in order to utilize their ample experience, are frequently licensed to provide treatment to out-patients.

There is a strong tendency among physicians toward specialization; the specialists in private practice outnumber the general practitioners. The patient is free to choose the latter, or present himself directly to the specialist; thus, no "first line" exists as, for instance, in Great Britain or the Netherlands. Physicians in private practice are required to be members of corporations under public law (*Kassenärztliche Vereinigungen*), associations of doctors in private practice that negotiate with the health funds concerning the amount of money the latter furnish for ambulatory care.

As for hospitals: there exist roughly 3,000 with 680,000 beds, which averages 11 beds for every 1,000 inhabitants. More than 50% of the beds belong to municipal hospitals, roughly 35% to non-profit agencies, and the rest are controlled by private owners [9].

There is a federal law which directs the financing of hospitals under which the 11 states (*Länder*) have to establish "hospital-plans" that require the consent of their governments. Those hospitals which are accepted in the "plan" have the costs of investment paid by tax revenues; administrative costs are covered by health insurance. There are special arrangements for those hospitals which are not accepted into the plans, but the policy is to continue to reduce their numbers [6].

In discussing the providers and system of health care, the pharmaceutical industry must be included, as drugs constitute the most widely used therapeutic intervention. The pharmaceutical firms are part of the chemical industry and one of its very powerful branches. There are more than 70,000 drugs on the market due in large part to the wide variety of combinations; these might be reduced to a much lower number of basic drugs if generic captions were used.

Drugs prescribed by a physician are in greatest measure paid by health insurance, the patient contributing only 2 DM per prescription. Everyday (over-the-counter) drugs included on a so-called "negative list" are not paid with health funds as, for instance, remedies for headache, colds, etc. [1].

Now that the principal actors in health care are on stage, it is evident that there is a great variety of competence within a number of groups and individuals to make decisions and to provide services. There is no strict central planning, but rather decision making by participation and consensus; the state confines its role to enforcing the legal regulations and to making certain they are observed. One may, therefore, observe a high degree of autonomy in health care and its management in the Federal Republic.

This takes us to the question of how macro-allocational decisions in the field of health are made, and perhaps, how they should be made.

Because of the decentralized organization of the health care system, there is no comprehensive national health budget, nor is there at present a national health plan. The revenues come from different sources – partly from taxes, partly as contributions for health insurance – and from parts of many budgets. With all the difficulties of demarcation one may say that the costs for health care at present constitute about 10% of the gross national product of the Federal Republic of Germany, which means more than 200 billion DM (120 billion U.S. dollars) per year for 60 million inhabitants. These 200 billion DM are distributed as follows:

| | |
|---|---|
| (1) prevention | 6% |
| (2) treatment | 50% |
| (3) sickness-related costs (sick allowance, etc.) | 32% |
| (4) research | 2% |
| (5) other | 10% [8]. |

The cost for health care during the last decades has risen much faster than the GNP; this caused growing concern and directed extensive discussions to the problem of financing of health care and health planning.

In order to exert a strong influence on the rising costs of health care, the federal government (by legislation in 1977) brought into being the so-called "Concerted Action in Health" and mandated it with the requirement that all those involved in health care should jointly develop (1) marks of orientation in the medical and economics field; (2) recommendations for more effectiveness and efficiency in health care, keeping in mind a level of care consistent with the level of medical science and the aim of a well-balanced distribution of costs [7].

This, in itself, sounds a bit like consulting an oracle. The Concerted Action was composed of 60 members, who were called together by the Federal Minister of Labor and Social Affairs, and represented the 11 states (*Länder*), health insurance organizations, physicians, dentists, hospitals, pharmacists, the pharmaceutical industry, the unions, and the employers. Now it is convened twice a year and formulates recommendations, especially with regard to physicians' and dentists' fees, costs for drugs, number of hospitals, etc.

Opinions about the efficacy of this institution are divided. Whereas, especially in the beginning, certain results in cost containment were obtained, the outcome of the last meeting in March 1985 was very disappointing. The various self-administered groups and institutions were unable to agree on most of the items on the agenda and even expressed the opinion that the government should take action, whereas it is rather reluctant to do so.

All in all, heated discussions during the last few years revolved around the costs of health care and sometimes gave the impression that one was in a market surrounded by salesmen: "Cut down the number of hospital beds," cry the health funds, "hospital treatment is too expensive. Our members will not take it if contributions rise indefinitely!" "Reduction of hospital beds will destroy working positions, therefore keep the number up," say the unions. "Find ways to reduce the number of medical students," doctors in private practice plead, "only then shall we give our consent to a limitation of funds and prices for ambulatory treatment. Too many eaters of a cake that cannot be enlarged make too small pieces!" The pharmaceutical industry, on the other hand, is interested in large numbers of physicians who write many prescriptions. And amid all this shouting are rather quiet voices deliberating the possible effects of cost-transparency for and financial participation by the patient.

So the financial question at present is a complex riddle that waits to be solved. But are there really no other problems than economic ones? Is there no health planning, no official policy?

Before one can attempt to answer this question, one must dwell a bit on another one: What will become of health and the health care system in general? Are there any indications that changes and developments, other than those planned by politicians and health professionals, may take place?

We live in the so-called technological age. It therefore seems only natural that we should practice technical medicine, and we would be ungrateful if we underestimated the great successes thereby achieved. The question is whether technology in medicine will continue to pervade more and more areas, e.g., artificial organs in addition to artificial joints, "hospital on the wrist" monitoring bodily functions and administering drugs, or whether there will be a counter-movement. Women already oppose the extensive use of technology in obstetrics. They oppose the "medicalization" of what they view as a natural process – giving birth to an infant. The same question may be raised in geriatrics: are all manifestations of old age symptoms of poor health or illness? Are we not "medicalizing" the life context which does not need it?

Another fundamental question which may sound peculiar coming from a professional: Will licensed physicians, who for many centuries were the principal health care deliverers, remain in this position, or is it likely that deprofessionalization will soon begin to take place? There are already indications of the latter: self-help groups are mushrooming; an increased acceptance of personal responsibility for health may be observed especially among younger age groups. The necessity to change, i.e., to improve life styles, is discussed everywhere. The entire concept of health may be changing with an emphasis on wellness rather than on the treatment of diseases. Since all this is in full swing at the moment, it is by no means possible to predict which direction the process will finally take. Meanwhile, those who have the responsibility for the health care system and for people must act, in spite of all the difficulties and uncertainties involved.

At present we face the following situation: *No radical changes are planned* by the federal government with respect to the system in general. Because cost-containment is regarded as most important today, higher expenses in one domain are possible only if counter-balanced by restrictions in other areas, i.e., new services may only be introduced with the reduction of other established services. At this point, our century-old Reichs insurance law reveals its problems. The basic idea is, of course, risk-sharing. One hundred years ago the greatest risk for a workman's family was his sickness, with all the costs involved. Grandmother, if she needed nursing care, was nursed by her daughter or daughter-in-law. Today, she goes to a nursing home with monthly expenses of 3,000 to 5,000 DM (1,800 to 3,000 U.S. dollars). While health insurance will pay 100% for comparatively small risks, e.g., costs for medication, it will not pay for nursing care in a long-term care institution. There has already been a year-long discussion to determine who should pay for the "big" risk of nursing care. Even high old-age pensions will not be sufficient to cover long-term care, so that people become dependent on social aid with all the psychological difficulties involved. Therefore: *a method to redistribute risk-sharing is one of the problems to be solved*. It also means that health insurance must be freed of costs which are not strictly related to sickness, as, for instance, contributions for the funeral, interruptions of pregnancy for other than medical reasons, etc.

"*Ambulatory Treatment as Far as Possible – Hospital Treatment Only if Necessary*" is a frequently heard slogan. There is no doubt about the fact that many hospital admissions have no strictly medical reason but are on a more or less social basis. This is even more true for the duration of stay. In Bavaria, health funds a few years ago decided to pay a bonus to doctors in private practice if the number of patients they admit to a hospital is below average. But this had only limited results, for the patient himself could still go to the hospital and bring about what is called a "self-admission." The underlying assumption that ambulatory treatment or care is always cheaper than hospital care may, by the way, be questioned when put in this absolute form. Good ambulatory care, at least in certain cases, may be just as expensive as intramural care, but it is as a rule more agreeable for the patient, when one takes into account his individual needs.

*Home nursing care* is strongly promoted in many places, as for instance in Berlin, again with the aim of reducing the number of hospital patients. Health insurance therefore also makes contributions to this need.

"*Health education*" is one of the magical terms frequently used. Health education is supposed to enable and motivate people to practice healthy life styles, and thereby to prevent, e.g., cardiovascular diseases and those respiratory and digestive system failures caused by the abuse of nicotine and alcohol. All this is fine and certainly correct. The only problem is that the tools by which we try to bring about changes of behavior, especially in adults, are still rather inadequate. What we know for certain is that the possession of knowledge does not necessarily correspond to behavior. Efforts to train young children in good health habits are often counteracted by the bad examples of adults. That does not mean that health education is useless, but it is a tedious job and, with respect to results, certainly a long-term process.

Let us return to the accusation that the *providers of health* often contribute to higher health care costs.

To limit the *number of medical students* is a much discussed item for two main reasons: the first is the quality of training. Patients can "endure" only a certain number of students, with bottlenecks developing in gynecology and psychiatry training programs. But our federal constitution guarantees free choice of profession. The federal ministry in charge, therefore, restricts itself to an appeal to the universities that no more students should be admitted than can be properly trained, and that examinations must aim at a high level of qualification. But the number of students has not only to do with the quality of training.

Second, having finished their training they want to make a living. Right now there are 65,000 *physicians in private practice*. Their number is expected to rise to 95,000 by 1995. No wonder that there are voices which question the principle that a doctor should be free to begin a practice wherever he or she wants to, with the health funds obliged to pay for his or her work. Since the government has no intention to interfere with the medical profession's independence, the federal minister leaves it to the self-administration groups to take appropriate action in this area.

Physicians in private practice, general practitioners as well as specialists, are reimbursed for individual performances, the charges being the result of negotiations between the health funds and the association of doctors at the federal level. The same applies to dentists. There is no doubt that such a system will produce a certain number of performances and medical interactions which may not be strictly indicated, but simply help to fund the replenishment of technical equipment. The problem is presently before the responsible agencies, but no changes are to be expected in the near future. In sum, 20% of the 100 billion DM per year spent through health insurance goes for ambulatory treatment, with an additional 12% for dental care. But the biggest piece of the cake – 30 billion DM – still goes to the *hospitals*. This only accounts for the operating costs, the capital costs incurred for construction being covered by tax revenues. The federal law requires that health insurance reimburse the operating costs produced by an "economically well-run hospital," one that gives high quality care at reasonable prices. The prices are the result of negotiations between the health funds and the individual hospitals. Right now they range between 150 DM per day (geriatrics, psychiatry) and 400 to 500 DM per day (university hospitals) with the possibility of having "total costs per case" paid for special types of care as, for instance, heart surgery, transplantation of kidneys, etc. The absurd fact is that the law permits economical behavior of a hospital to be penalized and uneconomical management to be rewarded. In many cases the partners will agree on so-called "preliminary costs" per day; the definite costs are fixed only when the financial outcome of the year is known. That is to say, if the hospital has generated a surplus, it will receive less money for the following year, and *vice versa*.

In Berlin, this system was discontinued in 1981 for municipal hospitals – the first year under the Christian Democratic government. Since then, definite prices are agreed on beforehand for the year in question, so that each hospital is clear about its budget and has to carry the appropriate risks; but, on the other hand, it is allowed to keep a surplus, should there be one, with the following condition: if a favorable economic development goes along with (and is likely to be caused by) a prolongation of the duration of stay which cannot be explained on medical grounds, there will be a slight reduction of payment, whereas a shortened stay is financially "rewarded." There is no doubt that hospital care is and will remain the most expensive form of care. Its capacity, therefore, must be reduced as far as medical responsibility permits. On the other hand, the shorter the stay the higher the cost-per-day will rise. Therefore, health insurance must understand that the cutting down of hospital bed days will inevitably cause a rise of the costs-per-day, as the first days in hospital usually are the most expensive ones. The patient, by the way, for the past three years has had to contribute 5 DM per day for a maximum of 14 days of hospital stay per year, a small sum that increases administrative costs without obvious results in the length of hospital stay.

Finally, we look at the *pharmaceutical industry*, another important "provider" that presently "absorbs" 15 billion DM per year. As it stands, the principle of "free market" competition plays an important role; the fact that there are 70,000 drugs on the market speaks for itself.

There are no more than 2,000, by the way, in East Germany.

As was already mentioned, a so-called "negative list" excludes several groups of "everyday" drugs from reimbursement by the health funds, a much criticized attempt to reduce costs in this area. The result is really meager, partly because a well-meaning physician who does not want to "burden" his patient with self-payment (nor to lose him!) will find it prudent to prescribe another medication which yields a similar effect.

Another attempt at cost containment is the so-called "lists of transparency" – telling the doctor which of a group of generically coded medications is the least expensive. As this interferes with the principle of free competition in a free market, a recent court case argued that one needs a special law to establish such lists.

All these considerations aim at the *providers* of health care. As far as the *consumers* are concerned, there are two main deliberations which aim at cost containment: One is *cost transparency* – this time for the patient. Because a West German patient, in contrast to patients in other countries, never receives a bill to be remunerated by health insurance later, he generally has no idea how much his own health care costs. As people are typically more reasonable than they are thought to be, it seems quite possible that by making them aware of the high costs this would help to reduce them.

A more difficult issue is the patient's *cost participation*. As was mentioned, the patient presently makes only a small contribution for each prescription (2 DM), and a maximum of 70 DM per year for hospital stay. Cost participation is strongly opposed, for instance, by the unions, which insist that it may hinder patients from seeking early treatment. It therefore needs careful consideration. Here general experience is proven again: to interfere with a privilege which has been granted for a long time is a very difficult undertaking, one which politicians are most reluctant to approach.

Whereas all of the measures under discussion here are directed partly toward the providers and partly toward the consumers of health care, other deliberations concern the priorities to be set and the necessity to decide on *age-related "targets" of health*. Listening to health experts and politicians, one gets the impression that prevention is high on the list of priorities. Looking at the bill, the greatest part of the resources are expended on treatment, and this will probably continue to be the case. It is worthwhile mentioning that health insurance in West Germany in its century-long history only began to pay for preventive programs about 15 years ago, which policy required amendments to the Reichs insurance law. To establish targets of health for special groups of the population could form a basis for the evaluation of health programs as well as the money spent on them, and could, although not a quick or simple procedure, help to make reliable decisions to increase or reduce such programs. (At a recent meeting of the Concerted Action in Health the minister announced his intention to have independent experts advise the members on questions of this type.)

The four major *moral* goals, according to Engelhardt, for a health care system are:

(1) equal health care to all citizens;

(2) the "best" health care to all citizens;

(3) effective cost containment; and

(4) the maximum amount of free choice for both patients and providers.

These four goals surely equal the task of squaring the circle. Number (1) is to be included in (2); the best health care to *all* citizens also means equal care, in my judgment. But what is to be understood by "best" health care? Perhaps "adequate" care is the term; this means that there is a long way to go in each country. How to combine adequate care with cost containment needs greater genius than has been provided until now, and on top of all that, a maximum amount of free choice for patients and providers reminds one of the title of an old German movie, "If All of Us Were Angels!" Being human, most of us will need correction and certainly guidance in order to arrive at a series of prudent decisions.

The Netherlands, at this year's World Health Assembly, which convened in May in Geneva, reported on the importance of analyzing the "limits of health care" on a worldwide basis under the following aspects ([10], pp. 21–23):

First: Determine the restrictions which must be imposed on the provision of health care on ethical grounds.

Second: Since limits on health care cannot disregard considerations of effectiveness, a distinction between effective and non-effective (no longer effective) health care must be made.

Third: Relative scarcity of financial resources is a limiting factor! In most countries new services can only be introduced by reducing or eliminating others; growing requests for funds from national income by the health sector will not be fulfilled.

Fourth: There is an obligation on the part of governments to promote health care. In many countries, it is deemed to be the responsibility of government to ensure that health services are up to a given standard and available to all no matter what their income. In other words, the government is obliged to enable people to receive their "right to health care." The government, however, also has to guarantee other basic rights for its citizens, e.g., sovereignty over their own bodies, and their right to be secure in their physical integrity. (A strong appeal was made to the WHO to assist member states to develop international guidelines under which reliable decisions could be made at the national level.)

Before concluding, let me offer a bird's eye view of the health care system in the Federal Republic of Germany:

(1) There is a century-old health (and social) insurance system which covers the great majority of the population.

(2) 10% of the GNP is expended on health care.

(3) Financial decisions for ambulatory treatment are made by health insurance and associations of health care providers (physicians, dentists, etc.) which technically are corporations under public law practicing self-administration.

(4) Financial decisions for hospital construction are in the end made by governments following consultations; for running costs, decisions are made by health insurance and hospitals within the framework of the federal law.

(5) There is growing concern over the constantly rising costs of health care.

(6) In spite of that, there is a great reluctance on the part of government to interfere with the freedom of health professions and the self-administration of the responsible groups more than is absolutely necessary.

(7) In accordance with (6), no strict central planning or budgeting exists, but decision making by participation and consensus through a new instrument was created in the Concerted Action in Health group, instituted in 1977.

(8) There is an obvious effort to switch from hospital care, as the most expensive form of health care, to ambulatory care.

(9) One can now observe certain indications that a new understanding of infirmity and health is extant, as well as a tendency toward greater social responsibility for one's own well-being.

*Senat für Gesundheit und Soziales,*
*Berlin, Federal Republic of Germany*

## BIBLIOGRAPHY

1. Bundesanzeiger Nr. 185 – 5.10.1982.
2. Bundesarbeitsblatt 12/1984, p. 31.
3. Deppe, H. U.: *Gesundheitssysteme und Gesundheitspolitik in Westeuropa*, Campus Verlag, Frankfurt New York.
4. Deutscher Bundestag – Drucksache 10/3374 22.5.1985, p. 12.
5. Deutsches Ärzteblatt, 82. Jahrg. Heft 15 – 12.4.1985, p. 27.
6. Krankenhausneuordnungsgesetz – 20.12.1984.
7. Reichsversicherungsordnung § 405 a.
8. Statistisches Bundesamt Wiesbaden, Fachserie 12, Reihe S 2.
9. Statistisches Bundesamt Wiesbaden, Fachserie 12, Reihe 6.
10. World Health Assembly 1985, A 38/VR/5, pp. 21–23.

BARUCH A. BRODY

# THE MACRO-ALLOCATION OF HEALTH CARE RESOURCES

There is little doubt that there is greater concern today in most countries about health care policy than there has been for a long time. I think that the reason for this concern is fairly evident. On the one hand, the tremendous rise in the cost of health care has made every society look at ways to control the cost of that care. At the very same time, there is a great concern that this control of costs not interfere either with the quality of health care or with the growing access of the less fortunate to that health care. So the current discussion of health care policy reflects, I submit, a conflict between the value of controlling health care costs, on the one hand, and the values of maintaining a high quality of health care and of maintaining an appropriate level of access of all citizens to health care, on the other hand.

My claim in this paper is going to be relatively straightforward. It is that we are unlikely in the context of the current policy debate to find an appropriate way to balance our attention to these different values. The reason why we are unlikely to do so is because we have not paid enough attention to a fourth value, the value of maintaining the maximum amount of free choice for both patients and providers. It is only when we do that, I shall argue, that we will be able to balance appropriately the first three values. I shall conclude my argument with several policy recommendations, recommendations which embody this additional value, and which range from the concrete to the global.

## I. IMPORTANT PRELIMINARY POINTS ABOUT A COMMON PICTURE

It is not necessary, I believe, to review a great many statistics from the United States, statistics that could be provided for most other Western countries, to prove what is now an obvious and well understood truth, namely, that the percentage of the Gross National Product (G.N.P.) allocated to health care has grown drastically in the post-World War II period and is continuing to do so, although perhaps at a somewhat lesser rate, despite major attempts to control the cost of health care. The response to this type of growth is a tremendous public outcry. The

*Hans-Martin Sass and Robert U. Massey (eds.), Health Care Systems,* pp. 213–236.

government, we are told, cannot continue to support health care, particularly for the elderly and the indigent, and the most vulnerable will not be able to receive the health care they require. The cost of private insurance premiums is becoming intolerable, we are told, and is becoming a major problem for industries which must compete in the international marketplace. The picture is a dismal one. The suggestion is that we need to do some drastic things if we are to avoid future crises in the provision of health care.

This picture forms a background for most policy analysis. In this first section, I want to make several preliminary moral and conceptual points about this picture that will be crucial for the later sections of this paper.

The first of these points has to do with the question of whether or not the rising percentage of the G.N.P. devoted to health care is a bad thing. I shall use for my analysis American figures, but I believe that similar factors need to be taken into account as we look at figures from other countries. Consider the period 1971 to 1981, a period of tremendous growth in the percentage of the G.N.P. in the United States devoted to health care. We know from a recent Health Care Financing Administration (HCFA) analysis [9] that about 60% of the increase in costs in that period was due to general inflation and population growth. Only the remaining 40% was due to factors that are specific to the health care industry. And a major portion of that 40% represents more visits to physicians or hospitalizations per capita and more services per visit or per hospitalization. These factors probably represent around 30% of the remaining 40% in the growth of health care costs. Are those increases desirable or non-desirable?

Consider the following argument [6] that they are desirable. The United States, later than most counties, made a major commitment to equalizing access to health care in the 1960s with the adoption of the Medicare and Medicaid programs. By the early 1970s, the elderly and the poor were taking advantage of these programs to get the health care which they traditionally would not have received. That represents a considerable portion of the increase in visits to physicians and hospitalizations, and we should welcome that. Moreover, the period of the 1970s saw an increasingly sophisticated medical practice. That is represented by the increase in services per visit and services per hospitalization. Unless we believe that this increasingly sophisticated medical practice was unfortunate, we should welcome these increases in costs as well. So, the argument concludes, the increases in health care costs that everybody is complaining about may really be something that ought to be welcomed, welcomed not for its own sake but for what it represents.

I do not by any means want to suggest that there are no problems with the rise in health care costs. Nor do I want to suggest that there is not major waste in the delivery of health care in the United States. All that I am saying in this first preliminary point is that we need to be extremely careful about moving from the claim that we have seen a great increase in health care costs to the conclusion that we have seen a bad trend which needs to be reversed. That is the first partially conceptual and partially normative point that I want to make in this section.

I turn now to a second related point. The United States has seen in the post-World War II period a tremendous increase in the percentage of G.N.P. devoted to the consumption of fine white wines, some coming from Germany or France, and some coming from California. No doubt, the total dollars involved are much smaller than the total number of dollars spent on health care. Still, the increase is real and significant. Why is there no public discussion of this increase? Why is there no public discussion of many other areas of the economy which have also seen tremendous growth in expenditures in the post-World War II period?

There are many obvious answers that can be given to this question. I want to focus on one that seems to be the most crucial. It is that these expenditures are paid for by those who receive the goods in question out of their own funds. There is no occasion for a national policy debate about the level of expenditures on these items, precisely because there is no need for a national policy on expenditures on these items. The situation is very different when we get to the question of health care costs. There, those receiving the benefits of health care are increasingly not those who pay (at least directly) for that care. Returning to the American experience, it is important to remember that as late as 1929, patient direct payments represented 88.4% of all personal health care expenditures. In 1950, they were still 65.5%. By 1983, they were only slightly more than 27% [11]. The remainder of health care was funded by private health insurance and by public health expenditures. The premiums on the private health insurance were, moreover, subsidized through favorable tax treatment of employer provided health insurance. In short, then, although American health care continues to represent a system in which private funding of health care plays a larger role than in most other Western countries, it is increasingly a system in which the funds to provide health care come from tax dollars. That being the case, there is a need for a national policy on health care expenditures, and the question of what percentage of the G.N.P. should be devoted to health care expenditures becomes a question for national policy.

Why this change in the pattern of paying for health care in the United States? It is clear that this change represents a growing recognition that the indigent, the elderly, and even large portions of the working middle class cannot afford the health care they would benefit from, not even if insurance schemes were available, without some public help. A favorable tax treatment for health care insurance premiums represents an attempt partially to subsidize the health care for the working middle and lower classes. The rise of the Medicare and Medicaid programs represents an attempt to provide an even greater amount of subsidization to health care for the elderly and the non-working indigent.

In short, our second major conceptual and normative point is the following: we have a need for a national policy about health care expenditures precisely because of the national commitment to provide health care for those who cannot afford to pay for all of their health care on their own (even if private insurance schemes were available). If there were no such commitment, then we would not need to make any policy judgments about the level of health care expenditures and about how much of an increase in health care costs is desirable.

These then are the two preliminary points upon which much of the rest of my analysis rests. They come to the following: Most modern countries, the United States perhaps later than others, have made a commitment that health care should be provided to those who benefit from it regardless of their ability to pay (even taking into account any private insurance scheme). That has led, even in the United States, to a rapid increase in the percentage of health care costs which are funded through tax dollars. Therefore, decisions to provide certain levels of health care become public rather than private decisions. That requires society as an organized group to address itself to the question of whether the resulting increase in health care costs represents an acceptable provision of better health care on a more equal basis, or whether it represents an unacceptable level of national resources devoted to this one area.

## II. THE EMERGING STANDARD VIEW

In this section, I want to develop what I take to be a standard view that is emerging in the United States. Standard views are peculiar things. It is not that they are *in toto* articulated by large numbers of authors. Rather, they represent a set of views each of which has a large amount of support and all of which together make a coherent picture. It is this standard view which I will attack in the next section, and to which I shall present an alternative, based upon my preliminary points, in the final section of this paper.

Let me begin by listing what I take to be the major claims of the standard view:

*Eliminating waste by Regulation and/or Competition* A considerable portion of American health care dollars is being spent wastefully and could be eliminated by some appropriate combination of stronger regulations and more encouragement of competition;

*The Merits of Preventive Medicine* America puts too much emphasis on treating illnesses once they have developed and not enough emphasis on preventive medicine, where that is understood both as measures taken by individuals to improve their own health and measures taken by society to improve the health of its citizens. If this overemphasis on curative care were negated, it would produce significant health care savings;

*All of This Will Not Be Enough* Nevertheless, it is unlikely that America will be able to solve its problems of increasing health care costs simply by better regulation, more encouragement of competition, and a prevention-oriented approach. The reasons for this are demographic and technological. Like most other countries, the United States is increasingly an aging country. Not merely are Americans getting older, but there are more and more very old Americans. These people will require tremendous health care expenditures, considering their needs for custodial care and for the management of chronic illnesses. Moreover, medical advances will provide us with the opportunity to help these people meet their health care problems, but all of these advances are very expensive;

*The Need for Rationing* Even a country as wealthy as the United States will have no choice but to turn to the rationing of health care, as the British have already.

Having now briefly listed the major claims of this emerging standard view, let me elaborate upon them.

*Eliminating Waste by Regulation and/or Competition* Those most concerned with cutting health care costs have been the politicians who are reluctant to vote for taxes to fund health care and the industrialists who do not want to pay increasing health care insurance premiums for their workers. They have adopted two major approaches to curtailing waste. One of these approaches is the regulatory approach. Examples of this approach include the certificate of need program, which prohibits major capital expenditures, particularly the building of new facilities, unless a need for them can be demonstrated, and the new peer review program which mandates the review of decisions to perform certain procedures which are widely thought to be used in a wasteful fashion to insure that the procedure is actually required in a given case. The second major approach is the market approach which attempts to control health care expenditures by providing economic incentives for avoiding their growth. These incentives may be directed towards providers by limiting the reimbursements they will receive. A good example of that is Medicare's recently adopted prospective payment program. These incentives may also be directed towards recipients, requiring them to pay for more of their health care. It is this thought which explains the recent moves to require patients to pay greater deductibles and co-insurance payments.

There is an extensive policy debate going on in the United States about the merits and demerits of these approaches. Some commentators believe that economic incentives would be sufficient by themselves without the need for stronger regulation. Alain Enthoven [8] would like to see the emergence of a series of competing health plans which consumers can choose to join, in part because of their ability to offer better health care at lower costs. This would provide, he claims, economic incentives for providers to eliminate waste in health care. There are others [10] who are very mistrustful of this approach of economic incentives. They are concerned that the introduction of economics into health care will break down the strong internal ethos of medicine to do the best that one can for one's patient. Many of those who are opposed to this economic approach would prefer the reliance on more regulatory mechanisms of the type mentioned above. I shall not in this paper enter into the controversy as to which of these approaches is preferable. I only want to make one point about both of these approaches, a point which the emerging standard view requires us to understand. These approaches are designed, and seem most appropriate, for the confrontation with waste in health care expenditures. It seems reasonable to believe that specific forms of wasteful behavior, whether capital expenditures or unnecessary procedures, can be diminished by regulation. It also seems reasonable to believe that other forms of wasteful expenditures can be controlled by providing financial incentives either to providers or to recipients to avoid them. Neither of these approaches seems so helpful, however, to issues other than eliminating waste.

*The Merits of Preventive Medicine* There is no doubt that America needs more preventive medicine. Some of this relates to the question of individual behavior. Large numbers of Americans simply need to stop smoking and need to drink in greater moderation. (Please note the crucial difference between the changes in the two cases.) Americans eat too much and eat too much of the wrong kinds of food, so we need to change our diets. Many of us will benefit in many ways from less sedentary lives, lives involving appropriate forms of exercise. These are messages that most Americans today know and believe, even though they may have great difficulty in following them. Preventive medicine is not, however, just a matter of individuals changing their life styles. There is also the question of the way health care is provided. America's problem with excessive infant mortality, at least as compared with other advanced countries, seems in part due to our failures in prenatal care resulting in unnecessarily large numbers of low birth weight neonates. We do better at bypass surgery than at effective screening programs to pick up hypertension in its early stages. We need to do more by way of appropriate screening programs for early stages of breast cancer, for early stages of colorectal cancer, etc. Again, these are messages that are increasingly believed, and form part of the ideology of the emerging health maintenance organization movement in America. Again, these are messages that are harder to act on than to believe. There is a third form of preventive health care which needs to be stressed in this context. While we are a long way from being able to identify all of the environmental factors and work place factors that harm our health, we certainly can at this time identify a wide variety of changes in the environment in general and the environment of the work place in particular that will have a highly beneficial impact upon our health status. Again, this is a message that is widely believed, although it turns out once more to be so hard to act upon.

The crucial point that we need to note about all three of these types of preventive health care is that there are two sorts of arguments that can be offered for them. One is the argument from the intrinsic value of the better health that they are likely to produce. We do want to live longer and healthier lives, and these forms of preventive health care can help us do so. The second, however, is an argument more relevant to this study. It begins with the observation that it often costs a great deal more money to repair the damages caused to our health by failing to take these preventive measures than it would have cost to prevent them in advance. In many ways, the appeal for preventive medicine from this second perspective is another appeal to the need to abolish wasteful health care expenditures. It is wasteful, the argument runs, to spend a great deal of money fixing people's bodies (if we can do so) rather than spending a lesser amount of money at an earlier stage to keep them healthy.

*All of This Will Not Be Enough* There is at the moment an optimistic belief in certain circles that the combination of tougher regulations and economic incentives for stronger competition is doing the job in controlling health care costs. The average number of days a patient stayed in a hospital declined dramatically in 1984, and it is widely believed that the prospective payment program is responsible for this, since it provides a tremendous incentive for hospitals to discharge patients earlier. The earliest data on the new peer review program seem to indicate a significant rate of disapproval of expenditures, resulting in hospitals tightening up on the number of procedures performed. It is not hard, then, for many to see the combination of these techniques as working effectively. The emerging standard view cautions us against this excessive optimism. It agrees that the measures that have been adopted have done a significant job in abolishing waste and will continue to do so. It argues, however, that after all of this waste is abolished, we will continue to face a problem of escalating health care costs because there are factors responsible for the rise in health care costs that go beyond waste.

One of these factors is the continued growth of new expensive technologies. Consider the organ transplant program. Liver transplants and heart transplants are now joining kidney transplants as the treatment of choice for certain conditions. These are expensive procedures. For a while, the traditional third parties, both public and private, avoided assuming these costs by claiming that the procedures in question were experimental. That is no longer a viable option. We are rapidly seeing the emergence of third party responsibility for paying for such procedures. These procedures are not wasteful, so the mechanisms designed to eliminate waste will not in the long run enable us to avoid the heavy expenses of these transplant programs. Such examples are not confined to the area of transplantation. In short, the first reason why all of this will not be enough is that medical scientists are doing a good job in developing new efficacious techniques, but ones that are very expensive. This is why one of the claims of the emerging standard view is that we cannot afford all of the new high technology medicine.

There is a second factor that is extremely relevant here and which is an additional part of the reason why all of these cost control measures will not be enough. This factor is the aging of American society. Like Europe before us, the United States is becoming top heavy with elderly citizens. Maximum human life span has not changed significantly, and there is a well known debate as to whether it is likely to change at all. But there is a dramatic increase in average life expectancy that has allowed many more people to survive into old age. Average life expectancy in the United States has increased by more than 25 years in the 20th Century, from 47 years in 1900 to 73 years in 1980, and it is continuing to increase. It is projected [7] that one in five citizens will be over 65 years of age by 2040, and, more importantly, that we will have 13 million individuals over the age of 85 by that year. Already Americans over 65 account for more than three times the *per capita* health care expenditures of Americans under 65. So an increasing number and percentage of elderly suggest rapidly increasing health care costs. Of particular concern is the group over 75 and, to an even greater extent, the group over 85. Members of these very old groups are far more likely to need custodial nursing home care. Nursing home care costs are already growing rapidly, and that increase in costs comes before the major increase in the percentage of the very elderly in our population. Moreover, all projections are probably underestimates because they fail to take into account further medical developments that will enable us to extend the life expectancy of even more Americans. In some sense, the potential successes of preventive medicine have frightening implications in this context, for they may mean even more people reaching these very old ages with tremendous demands on the health care system. In short, the second reason why it is thought that all the measures currently adopted will not be enough is the demographic factor of an aging population.

We have here then the crucial argument of the emerging standard picture. It is that successful medicine, one that combines preventive medicine with advances in high technology curative medicine, will result in a level of health care expenditures that continues to rise rapidly no matter how much we use regulation and competition to eliminate waste. The conclusion is that we cannot afford high technology medicine for all who will benefit from it.

*The Need for Rationing* One response to all of this would be simply to accept the inevitable, to say that once we have abolished as much as we can of the waste in American health care, we simply have to accept the remaining continuing increases in health care expenditures. If that increase represents not waste but beneficial health care on an equitable basis, what is wrong with that? The emerging standard view has a powerful response to that suggestion. It is simply this: there are many things that are desirable and upon which our society can expend its resources; health care is only one of them. No matter how desirable the health care in question, there comes a time when the increasing allocation of resources to health care means the inability to spend money on other desirable goals. At some point, we simply cannot accept the ever increasing percentage of the G.N.P. devoted to health care, no matter how beneficial that health care. The great policy debate, so this argument runs, has to be about what is that point and what to do when we reach it.

What then can we do when we feel we have reached this point? The emerging standard view says that we will have to ration [14]. What is the difference between rationing and eliminating waste? A wasteful health care expenditure is a health care expenditure that produces no benefits for the patient. Unnecessary surgical procedures are wasteful precisely because they are unnecessary, precisely because the patient can obtain all of the benefits of restored health without the procedure (in many cases, these wasteful procedures are worse than that, for each carries its own risks). Eliminating wasteful health care expenditures is eliminating expenditures which provide no benefits to patients. Rationing is another story. It is identifying health care services which provide some benefit but which are very expensive, and it is deciding not to provide the services in question even though they produce some benefits for the patients. Naturally, in any rationing policy, attention turns first to those health care services which involve the heaviest expenditures and the least benefits. Nevertheless, the crucial thing about rationing is that it means withholding from patients something that will be of benefit to them. This is what distinguishes rationing from merely eliminating waste. This is also what makes the last of the claims of the emerging standard view such a strong claim. What it is saying is that the only solution to the problem of rising health care costs in a world of high technology medicine and an aging population is to deny, for economic reasons, some health care to those who would otherwise have received it and benefitted from it.

The classic example used in these discussions of rationing is the British policy [13] of denying dialysis to patients suffering from end-stage renal disease who are above a certain age. It is not that these people would not benefit from dialysis. The truth of the matter is that without dialysis, they die from end-stage renal disease, and with dialysis, they live for some period of time. It is rather that, in a world of limited resources, a decision has been made not to expend significant resources for this form of care on people after a certain age, perhaps with the thought that it is more effective to expend the resources on providing health care to the younger who will live more years with a higher quality of life. Advocates of rationing might point to this example with considerable pride, claiming that it shows how a civilized country can control health care costs by rational thinking about the allocation of health care resources, or they might just accept this example as indicating the inevitable prices we will pay for a rationing policy. Whatever attitude is expressed, however, the crucial point the emerging standard view wishes to make is that rationing is what will be required if we are honestly to confront the problem of meeting the health care needs of our population without depriving our society of the ability to meet other needs.

## III. RESISTING THE EMERGING STANDARD VIEW

I shall now argue that there are three very good reasons to resist the emerging standard view. The first is that its policy of rationing is necessarily a policy carried out independently of the wishes of the patient and involving implicit and/or explicit deceit. Such a policy runs counter to the whole tendency towards honest discussion and shared decision making that is so important in modern medical ethics and jurisprudence. The second is that there is no coherent basis for decision making concerning either how much of society's resources should be allocated to health care as compared to other goods, or how much of the resources allocated to health care should be allocated to various aspects of health care. The introduction of rationing would lead to total chaos because of the lack of a principled basis for making such decisions. Finally, in a significant number of cases, the introduction of a rationing policy would result in a violation of our fundamental commitment to the value of human life. Let me elaborate upon each of these criticisms.

How does rationing actually work? How do British physicians, for example, manage to avoid dialyzing patients whose lives would be saved by dialysis. Let me quote from Aaron and Schwartz's ([1], pp. 36–37) recent study of rationing in the British health care system:

> British physicians are candid about the way they discourage patients from insisting on dialysis. Asked how he would explain to her family the prospects of a 65 year old woman with kidney failure, one general practitioner first told us that he did not think that it was up to him to decide whether she should be dialyzed, that he would leave the decision to the consultant . . . When pressed on whether he might save everyone time and anguish by discouraging referral, he described how he would talk to the family. 'I would say that mother's or aunt's kidneys have failed or are failing and there is little that anyone can do about it because of her age and general physical state, and that it would be my suggestion and my advice that we spare her any further investigation, any further painful procedure, and we would just make her as comfortable as we can for what remains of her life'. Remarkably few of the criteria for rejection are explicitly stated. Age, for example, is not officially identified as an obstacle to treatment . . . Because of the respect that most patients have for physicians, the recommendation of the doctors is usually followed with little complaint, particularly when the disease does not manifest itself in a way that is recognizable to the patient or when it is only one manifestation of a multifaceted disorder like diabetes.

What is wrong with such an approach? It seems obvious that rationing works because the patients are passive and accept, out of respect for the physician's authority, the claim that nothing more can really be done. Moreover, it seems clear that these physicians are at best deceiving their patients and at worst are just lying to them. It is just not true about these elderly patients that "there is very little that anyone can do about it." Patients can be kept alive, even if they are elderly and diabetic, for considerable periods of time with careful aggressive management of kidney failure through dialysis and relevant support. So Aaron and Schwartz's study suggests, as we might have suspected anyway, that rationing requires patient passivity and a fair amount of deceit.

It is worth noting, as a parenthetical remark, that there have been some recent British court decisions [2] which have explicitly rejected the

American doctrine of informed consent, arguing that British medicine rests much more upon trust and it is better off because of that. It is interesting to put that British decision into the context of the rationing of health care resources. If British patients are less informed, it may well be that the purpose of their being less informed is not to maintain a beneficial trust relation but rather to enable the system of rationing to proceed with less tension. In fact, Aaron and Schwartz report elsewhere:

> One leading nephrologist told us, for example, that a patient who sits down in the waiting room will probably be seen and a slot found for him in the dialysis program ([1], p. 107).

In other words, patients who understand can beat the rationing system, so it is crucial that large numbers of patients do not understand how the system works.

It is easy to see why such an approach of rationing is unlikely to work in America and moreover is something that Americans should resist. It is unlikely to work in America because American patients are increasingly knowledgeable and non-passive and have the threat of malpractice suits as a powerful weapon to force physicians to provide them with the care from which they could benefit. Moreover, there are powerful reasons why Americans should resist the suggestion. One of the most positive developments in recent years has been the emphasis in American medical ethics and medical jurisprudence on patient involvement in decision making and on honest communication between physicians and patients. A world of rationing, a world in which patients are denied health care from which they would benefit because society does not want to expend its resources upon it, is a world which necessarily involves decisions made independently of the patient and which are best facilitated by techniques which minimize the patient's awareness of what is going on. Those of us who believe in patient involvement and in physician-patient honesty can hardly welcome a policy of rationing.

There is one response to this issue which needs to be considered further. Defenders of rationing might claim that rationing would not necessarily involve deceit if the issues were discussed publicly in advance of a particular patient being ill so that members of society perceived that the decisions to ration care were fair rational decisions made socially and not prejudicial decisions of individual physicians in particular cases. But I think that this is unlikely to work for several reasons:

First, the affluent will, presumably, continue to be able to purchase at extra cost the care otherwise rationed. A public understanding of the situation will include, therefore, a public understanding that some will die or suffer serious illness because they did not receive the care that others who are more affluent do receive. This is a hard message for people to accept, and the difficulty will therefore lead to a desire to keep rationing decisions private rather than public.

Second, it is one thing to agree in advance that certain care be rationed. It is another thing to accept the rationing when one or one's family is ill and is having care rationed. The difficulty of obtaining patient and family acceptance *at that point* is what drives providers to deceit or at least to minimize awareness.

To my understanding, it is these two factors that help explain why, contrary to the defenders of rationing, British physicians behave as described by Aaron and Schwartz.

I turn then to a second reason for rejecting the recommendation of rationing, namely, the lack of a reasonable basis for rationing policy. In order to see why this is such a difficult problem, we need to remind ourselves that there are two issues that we will need to confront if we adopt a rationing policy. The first is the question of what percentage of our G.N.P. we allocate to health care expenditures as these have traditionally been understood, as opposed to other good things (including food, shelter, clothing, education, etc.) which are also positively correlated with better health status. The second is the question of how to decide, within the funds allocated for health care in the narrower sense, which forms of health care get higher priority and which get lesser priority. Once we keep in mind the nature of both of these choices, we shall see how unlikely it is that a rationing policy could rest upon any principled basis.

There are those who talk about a right to health care independent of the individual's ability to pay for that health care. There are those, like the recent President's Commission [12], who prefer to talk of the social obligation to provide health care, regardless of the individual's ability to pay for the health care. Whichever approach is adopted, we need to keep in mind that a rationing policy means accepting the point that the right to health care is only a limited right, is only the right to some of the health care from which an individual can benefit, or the corresponding point that the social obligation to provide health care is only a limited obligation to provide some of the health care from which an individual can benefit. A rationing policy forces us to confront the question of the extent of the right to health care, and it forces us to confront that question keeping in mind that there are these two aspects which we have identified.

Let me use two examples to help illustrate why I think that we will be unable to adopt a rational rationing policy that addresses the questions I have just mentioned. There recently appeared an important study [3] involving a cost-benefit analysis of weekly viral cultures in pregnant women with recurring genital herpes. Pregnant women with recurring genital herpes have to face the difficult question of whether or not to deliver by caesarean section to avoid the possibility of transmitting herpes during the birth process to their newborn infant. The trouble is that a caesarean delivery increases both the risk of maternal morbidity and mortality and the cost of child birth. It is standard practice therefore not to deliver all such women by caesarean section but rather to screen for recurrent infection and to deliver by caesarean section only those women with evidence of a recurrent infection. One way to screen would simply be to inspect for lesions. Women with subclinical infections pose a problem for such a simple screening program, and it had been widely recommended that such pregnant women undergo viral cultures during the last 8 weeks of pregnancy. The authors of the study suggest that the adoption of such an aggressive screening program may be a very expensive way of avoiding neonatal herpes. Based upon various assumptions, they concluded that it costs between $400,000 and $5,000,000 to avoid one case of neonatal herpes, with the most likely estimate being $1,840,000. Now is this form of more aggressive screening something which should be provided under an appropriate rationing policy?
policy?

I have chosen this example for two reasons. The first is that it reminds us that once we decide that we are not going to spend more than a certain percentage of our G.N.P. on health care, we will need to examine a very large number of routine health care activities and ask ourselves where they stand on the list of priorities for health care expenditures. It is not just the question of deciding whether or not we are going to provide kidney dialysis for the elderly and heart transplants for those in heart failure. The second is that it makes it easy to understand why no approach to the question of rationing or to the question of the extent of the right to health care (those two really being the same question) even begins to help us address these types of concrete day-by-day questions.

Let me explain. Calculations of this sort are what is meant by cost-benefit analysis in health care. A cost-benefit analysis is an analysis which identifies costs and benefits, assesses their magnitude, and determines which forms of health care can be provided at costs that are justified by the benefits. Now if we are comparing two forms of health care, and trying to see which provides the identical benefit at a lesser cost, this is unproblematic. But in order to ration, we need to do more than just that. We need to be able to decide at the end whether the cheapest way of obtaining particular medical benefits obtains them at a cost which is justified in terms of the benefits. If it is not, then we might well ration the care. If it is, we will still have to compare it with other ways of obtaining other benefits. The point of this first example is that the defenders of rationing have offered us no principled basis for deciding whether the benefits are worth the cost, so they have offered us no principled basis for even the simplest form of rationing. In the case of this study, lacking a basis for deciding whether the prevention of neonatal herpes is worth that cost, how can we begin to decide whether to ration that care?

There is an even more difficult type of example which we will need to confront. Consider a public hospital that has a large number of indigent patients. For a variety of reasons, the women in question are at high risk for a low birth weight delivery with all the complications associated with it. Even better prenatal care does not seem to resolve all of this problem. The neonates born often require expensive neonatal intensive care. Would that form of health care be provided under a rationing policy? Would we be making a better decision if we failed to provide that form of care and if we used the money to provide these families with better nutrition, housing, and education for their normal children? I am not raising this simply as the question of better preventive care. I am making a different point. How do we decide in a principled fashion whether it is more appropriate to provide health care to those who cannot pay for it themselves, or to spend the same resources in providing other goods to the very same people? Which allocation of resources would be better for them? Which do we have a greater moral obligation to provide?

I come then to the final set of questions which I wish to raise about a rationing policy. In many cases, the ones that have been most often discussed, a rationing policy is a policy which denies life-saving care to the elderly and the terminally ill, a policy that results in the death of these patients. Note that this is not like the elderly patient who says it is not worth spending all of the money that I have saved for retirement to keep me alive for a short time longer, so please let me die so that my spouse can have something to live on for the rest of his or her life. It is a question of our letting them die because of our judgment about their life and its worth. The belief in the sanctity of human life is a belief that all human life is of infinite value and that any amount of money should be spent to save a human life. It is not necessary to hold such an extreme belief in the value of human life to be troubled by a rationing policy, although obviously anybody who believes in the sanctity of human life must be opposed to rationing. It is sufficient to hold weaker but substantial views about the value of human life to be greatly concerned about the evaluations that would be presupposed in these cases by a rationing policy. Those who have advocated a rationing policy, particularly as it addresses itself to the question of life-preserving care of the elderly or the terminally ill, will need to tell us much more explicitly their view of the value of these human lives. Short of their doing so, I urge that we be very skeptical about their ability to so in a satisfactory fashion and that we be skeptical about the rationing policy which they are recommending.

## IV. SOME MODEST PROPOSALS

How then ought we decide how much high technology medicine we can afford to provide to those who can benefit from it in a world with an increasingly aging population? In this final section, I would like to present three suggestions. I believe that all of them deserve careful consideration.

All three of these proposals have one fundamental theme in common. It is the fundamental theme that was already implicit in our rejection of rationing. It is the theme which emphasizes the individual choices of patients rather than collective social choices. All three of these proposals postulate as an essential value the value of individual choice and freedom.

Let us begin with the first proposal. Suppose that all of the following three hypotheses are valid:

(a) There are several large classes of patients about whom there is general (although not complete) agreement that various levels of aggressive and expensive health care could be withheld and/or withdrawn, providing that the patient, if competent, or the family or guardian, if the patient is incompetent, concurs.

(b) Health care that could be limited with that concurrence is presently often provided, primarily because the question of its provision is not raised by any of the parties involved.

(c) There are substantial costs incurred in the provision of this care and instituting measures to limit it could provide considerable savings in our national health care budget.

If all three of these hypotheses are valid, then we would have a proposal for meeting at least part of the problem of rising health care expenditures which would avoid the objections we raised to rationing. It would involve patients and their families choosing to limit care, rather than physicians rationing, so it would continue the tradition of patient involvement and honest communication. The basis for the withholding of care would be a principled basis, namely, the wishes of the person from whom the care is withheld or the wishes of those who speak for him. Finally, since these cases usually involve patients who are terminally ill and/or have a very low quality of life, and since we have their concurrence to limit care, there are many plausible views about the value of human life compatible with such a policy of limiting care.

What are some of these classes of patients? I would suggest at least the following: (1) persistent vegetative patients; (2) patients with a documented history of persistent severe dementia; (3) patients with very severe motor disabilities despite rehabilitative efforts; (4) patients with chronic progressive recurrent illnesses who have been hospitalized extensively and who despite aggressive management are severely impaired in their daily activities; (5) cancer patients with documented metastatic disease for whom effective curative therapy is not available and who are severely limited in their remaining capacities; (6) patients with documented failure of three or more major organ systems. The details of these classifications could be spelled out more clearly, but it seems to me that we do have a widespread consensus that such patients need not be managed aggressively if they undergo acute crises and if they or those who speak for them wish no further aggressive therapy.

Would the withholding of further aggressive medical care in such cases make a significant dent in national health care expenditures? It is impossible at this point to answer that question definitively. But there are reasons for being optimistic. The very factors that are going to cause a crisis, no matter how much we avoid waste, namely, an increasingly aging population which can be kept alive with high technology medicine, are the very factors which are involved in so many of these cases. So many of the patients in these classes are patients who are old and who are kept alive by highly aggressive measures. So, a policy of agreeing to withhold care at the request of these patients coupled with a policy of always honestly providing that option to the patient might by itself be enough to make a major impact upon rising health care costs. If used as a supplement to techniques designed to avoid waste, it may be the missing ingredient that will enable us to avoid rationing.

This then is the first of my proposals for dealing with the problem of rising health care costs. It is obvious that the theme of patient choice is central to that proposal. And it is, of course, this theme which differentiates my first proposal – and the others to follow – from rationing proposals. This recognition of the centrality of patient choice in my first proposal suggests the possibility of still further ways of dealing with our problem by employing the theme of patient choice.

We need to return at this point to remarks made in the first section of this paper, viz, that the reason why we have a social problem is because society is increasingly paying directly for the health care that individual citizens are receiving. We are doing that because we judge that there are many patients whose financial status does not allow them to receive the health care from which they could benefit. But we choose to meet this need either by providing the care directly (in public hospitals or in the Veterans Administration system) or by offering them a publicly funded insurance program (the Medicare and Medicaid programs). Having done that, we find ourselves compelled to ask the question: What sort of care should we provide in the public hospitals or fund in these programs? The inability to answer this question in a principled fashion is part of what made us so skeptical about rationing policies.

We noted above that there were two major difficulties that needed to be addressed. One is the question of which health care to provide once we decide how much we want to spend in providing health care for the indigent. The other is the question of how much traditional health care we are going to provide as opposed to other things we might provide to those who are indigent. Remember that we do not have to worry about the question of what health care to provide to the non-indigent (where that is understood in the broad sense required in the medical context) because that decision is something they can address themselves.

My next two proposals allow one or both of these crucial questions to be answered by the recipients and not by society in general. Suppose we have decided that there is a certain amount of money that we are prepared to allocate to health care expenditures for those who cannot pay for health care themselves. The question that was posed for any rationing policy was which health care to provide. The obvious suggestion is that we do not need to answer that question. We could simply provide that level of funding in the form of a voucher to those who are medically indigent and allow them to choose. As Enthoven [8] has suggested, they would choose among insurance schemes for the provision of health care. They might choose to join some scheme which emphasizes more preventive medicine and better and more pleasant primary care and which provides that by not providing certain expensive high technology last ditch efforts. They might choose a more catastrophe oriented policy, one which downplays primary care and preventive medicine. Why do we need to make that choice as opposed to their making that choice? Why not develop our system around the theme of liberty?

Notice again that this scheme avoids the major difficulties that we raised against rationing. This is a scheme that involves patient decision-making and patient honest understanding rather than passivity and ignorance. The health care that the patients receive and the health care that they do not receive will be determined on a principled basis, namely, their choices given their budget constraints, and not by some rationing decision made by others. Finally, to the extent that patients choose better health care on the primary level at the cost of foregoing certain expensive high technology life-saving efforts, our sense of the sanctity of life is less offended when they do not receive those efforts. We see this not as our withholding care from these people, but rather our allowing them to make and follow their value choices.

Note that this second proposal can incorporate within it the first proposal. One scheme which many might well choose to purchase is one which offers better health care at earlier stages of life by avoiding costs at later stages of life through a policy of aggressively questioning people about the use of life preserving technologies in certain contexts. Many, but by no means all, patients would probably opt for such a scheme. So the adoption of the second approach would, in fact, be a continuation of the first approach as well.

Still, there is a crucial question unresolved by the adoption of the second more comprehensive approach. That is the question of how much of our G.N.P. we wish to allocate to subsidizing health care for those who are unable to pay for health care from which they could benefit. Why is this such a hard question? In part it is because, as we suggested in Section III, there are other things we may wish to provide to the indigent (food, housing, education for the young, etc.) and we do not see a principled basis for deciding how much health care to provide as opposed to these other things. This leads to the third and most radical of my proposals. It is one that I have advocated elsewhere [4], and I just wish to review it briefly here.

Suppose that we recognize that what we really are interested in doing is helping the indigent, but that we have no particular desire to provide them with a prioritization of things that are valuable. Suppose we recognize that what justice really demands of us is that we aid the indigent in confronting their indigency rather than with some specific aid. This would suggest providing the indigent with money rather than with health care (or any other) vouchers. They would then use the money available to meet needs and desires according to some system of priority that they adopted. We would not need to answer either the question of what percentage of our G.N.P. should be devoted to helping the indigent meet their health care needs or the question of what form of health care to provide. All we will need to deal with is the general question of how much help we want to give to the indigent in our society. The choice of how to use that money will be made by the individuals who receive the funds and not by us. That is why this proposal, like the previous two proposals, is based upon the value of liberty.

Naturally, this still does not answer all of the questions. There remains the background question of what percentage of our G.N.P. should be devoted to helping the indigent. I have elsewhere [5] attempted to provide a principled answer to that question, and I am working on a book, tentatively titled "The Redistributive Budget," which will attempt to develop that answer more fully. I will also be arguing there that the reasons for having a redistributive program actually suggest that the redistributed funds belong to the indigent, and that gives us still another reason for thinking that they should be able to choose how the funds are used.

I have made three proposals, each of an increasingly broad nature, of how to deal with the question of what forms of health care should be socially funded for those who cannot afford to pay for their own health care even with the help of private insurance schemes. Each of these proposals does not attempt to answer the question directly. What each does is create decisional processes which enable those who will receive the health care to answer the question. In the first proposal, they would simply be given an option, which they probably should have anyway, about whether they should receive certain care. The other proposals would provide funds to them, allowing them to make choices rather than imposing the choices upon them in a rationing process.

When I make these last two proposals, there is an objection that is always raised, the objection of the patient at the hospital-door. It runs in its most dramatic fashion as follows: suppose that one of the recipients of these vouchers (or these funds) makes choices, even reasonable choices, that do not cover insurance for some esoteric life-saving therapy for some unusual illness. Suppose now that he or she contracts that illness and arrives uncovered by insurance at the hospital. Would we, and should we, turn him away and allow him to die? And if not, does that not mean that both of your suggestions about patient choice collapse?

I think not, but the answer is complex. At least the following points need to be made:

(a) If we have in place an adequate redistributive program, and if this patient had the opportunity to make informed choices about the use of the funds earlier on, we would be under no obligation to provide that care and he or she would have no right to it.

(b) We might nevertheless do so, either out of a sense of compassion or a sense of respect for human life.

(c) We should not do so if we become convinced that it would contribute to setting an example which led others to neglect health care expenditures with the expectation that extra social funding for medical emergencies would always be available.

I also find that when I make these proposals, there is considerable resistance to them. The resistance often comes from those who are involved in the process of providing health care, or other goods, to the indigent. They argue that those who are indigent, those whose health care needs we worry about in a social decision process, are incapable of making decisions. Against that, I would make only two points: First of all, we will never find out whether they are capable of making decisions until we give them a chance to do that. Moreover, perhaps we are creating an incapacity in them to take responsibility for themselves by the process of not allowing them to make decisions.

Let me conclude, then, with the following thought which helps summarize what I have argued for in this paper: We have a social problem about the growing percentage of the G.N.P. devoted to health care only because of the decision that we have made, a decision which is quite appropriate, to help those who are indigent receive the health care from which they will benefit but which they cannot afford. Rationing, which is a morally objectionable policy, is emerging as a fashionable suggestion because we are so used to thinking about aiding the indigent in terms of deciding what to provide them. If we thought more and more about allowing them to make most of the decisions, we would have less of a problem of how to deal in a principled fashion with the question of rising health care costs. It may be that the value of liberty is the value that we need to emphasize if we want to have an affordable, high quality, just system of health care.

*Center for Medicine, Ethics and Public Issues*
*Baylor College of Medicine*
*Houston, Texas, U.S.A.*

## BIBLIOGRAPHY

1. Aaron, H. and Schwartz, W.: 1984, *The Painful Prescription: Rationing Hospital Care*, Brookings, Washington
2. Annas, G.: 1984, 'Why the British Courts Rejected the American Doctrine of Informed Consent', *American Journal of Public Health* **74**, 1286–88
3. Binkin, N. *et al.*: 1984, 'Preventing Neonatal Herpes', *Journal of the American Medical Association* **251**, 2816–2821
4. Brody, B.: 1984, 'Health Care for the Haves and Have-nots', in Shelp, E. (ed.), *Justice and Health Care*, Reidel, Dordrecht, Holland, pp. 151–159.
5. Brody, B.: 1983, 'Redistribution Without Egalitarianism', *Social Philosophy and Policy* **1**, 71–87.
6. Brody, B.: 1987, 'Wholehearted and Halfhearted Care: National Policies vs. Individual Choice', in Spicker, S., Ingman, I., and Lawson, I. (eds.), *Ethical Dimensions of Geriatric Care*, Reidel, Dordrecht, Holland, pp. 79–93.
7. Bureau of the Census: 1984, *Projections of the Population of the United States, by Age, Sex, and Race: 1983 to 2080*, Washington: U.S. Dept. of Commerce.
8. Enthoven, A.: 1980, *Health Plan*, Addison-Wesley Publishing, Reading, Penn.
9. Freeland, M. S. and Schendler, E. C.: 1983, 'National Health Expenditure Growth in the 1980s', *Health Care Financing Review* **4**, 1–58

10. Fuchs, V. R.: 1984, 'The "Rationing" of Medical Care', *New England Journal of Medicine* **311**, 1572–73.
11. Gibson, R. *et al.*: 1984, 'National Health Expenditures: 1983', *Health Care Financing Review* **6**, 1–29.
12. President's Commission for the Study of Ethical Problems in Medicine: 1983, *Securing Access to Health Care,* U.S. Government Printing Office, Washington D.C.
13. Prottas, J., Segal, M., and Sapolsky, H.: 1983, 'Cross-National Differences in Dialysis Rates', *Health Care Financing Review* **4**, 91–104.
14. Thurow, L. S.: 1984, 'Learning to Say "No"', *New England Journal of Medicine* **311**, 1569–72.

STUART F. SPICKER[1]

# RIGHTS, REASONABLE EXPECTATIONS, AND RATIONING: A COMMENTARY ON THE ESSAYS OF RUTH MATTHEIS AND BARUCH BRODY

## I. MEDICAL COST INFLATION AND THE RIGHT TO HEALTH CARE

Ingredient in the very structure of capitalist democracy is the necessity for so-called "trade-offs," a term in ever-increasing use these days in the context of the economics of health care. In fact, the very idea of trade-offs in the health care context tends nowadays to plague us all; even the affluent are following the economics of health care and wondering where it will all lead. The furrowed brow is now commonly worn on the heads of physicians and patients alike. The worry is that with respect to medical and health care (for those who draw this distinction) equality and efficiency, democracy and capitalism, humanitarianism and rational spending, even ethical principles and economic propositions are now in conflict; this can also be expressed as a conflict between rights and dollars. There is one way to resolve the conflict: namely, provide a powerful argument to show that there is a *right* to health care for each and every citizen. After all, it is also a maxim of democratic capitalism that equally distributed (essentially cost-free) rights should not be given a pecuniary value – ought not to be bought and sold for money. U.S. citizens do not tend to be required to pay for their right to free speech, for example; however, since we do not pay for cost-free rights out-of-pocket, so to speak, we often lack the incentive to restrain ourselves economically when we insist on a right to health care – clearly a "costly right." (I tend to subscribe to two fundamental psycho-economic maxims: if you give each of us total freedom to choose our "goods," we tend to eat at the Ritz; second, if you do not provide each of us with psycho-economic incentives in the way of continuous motivations, we tend to grow lazy and fail to exercise even our rights. The founders of democratic capitalism, it seems, were as good at understanding human behavior and the human psyche as they were at understanding the value of individual freedom [2] devoid of side constraints.) To repeat: rights, like the purported right to health care, are usually universally distributed and thus equality is emphasized, sometimes even at high economic cost. Since rights are not to be bought and sold, a society which

*Hans-Martin Sass and Robert U. Massey (eds.), Health Care Systems,* pp. 237–253.

defends the proposition that every citizen has a right to health care is compelled to work to provide every citizen with all that the right to health care entails, and that includes not only the material goods of pharmaceuticals, medical equipment, and the latest high technological apparatuses, as Dr. Mattheis illustrates so well ([15], p. 203), but also the time and talent of health care personnel, especially physicians and nurses.

In 1980, Edmund D. Pellegrino, M.D. – the Director of the Kennedy Institute of Ethics at Georgetown University – asserted that "Undeniably the physician, all other health professionals, and health care institutions have a crucial role in the equilibration of economics and medical morality" ([19], p. 183). Furthermore, he was surely correct when he added that "Ultimately, our health economic policies must reflect what it is we expect in the way of health and medical care in an advanced democratic society" ([19], p. 183). Indeed, it appears that Dr. Pellegrino agrees with so many who claim that we all have a right to health care, though he does not explicitly state this in his eloquent remarks on the subject [19]. I shall maintain, however, that this claim is surreptitiously or tacitly included not only in Dr. Mattheis' essay ([15], p. 211), but also in such deontological pronouncements as " . . .the physician cannot reject the person in need and remain faithful to the obligations he/she incurs in agreeing to treat" ([19], p. 179); or " . . .it would be a violation of the obligation of the covenant between physician and patient not to place this patient's concerns and needs above economic, social, and political considerations" ([19], p. 176). Where, one might ask, is the argument for such rights and obligations? Dr. Pellegrino might rightly appeal to the Anglo-American medical tradition, but precisely the values of that tradition are challenged and called into question these days. Put positively, an argument which sustains the proposition that 'there is a right to health care' would indeed generate an obligation on the part of all health professionals; this is built into the asymmetry between rights and duties (or obligations). That is, though a claim to a duty often does not entail that anyone has a right, a just claim to a right to health care would indeed entail an obligation on the part of health professionals; or, in legal parlance, the formulation is not 'The citizen has a right to health care' but rather, 'The citizen has a right to health care such that health professionals shall . . . .' My point here is that virtually all the noble claims with regard to patients' rights and physicians' obligations fail to consider the *loss of autonomy for health professionals*, not to mention the danger which enforcing a so-called 'right to health care' entails.

Permit me to provide the premises which lead to the valid inference that there is no right to health care:

(1) there are natural (inalienable human, essential) rights – like the right to life and liberty;

(2) each person has a right to pursue values and carry out *peaceful* actions that sustain his/her life;

(3) health professionals, as persons, also pursue values which sustain their lives;

(4) medical and health care includes, necessarily, physicians' and health professionals' time and services (they choose to render health care and service in order to support their lives);

(5) rights can be and are protected by law, that is, they should, in principle, be enforced if they are to have any social significance; this includes the legitimate force of government, e.g., by way of regulation and mechanisms to protect citizens against force and fraud;

(6) one expression of the legitimate authority of government is force of arms; whereas the use of illegitimate force is the antithesis of freedom;

(7) freedom is fundamental to rational choice and the individual's peaceful pursuit of his/her values;

(8) if medical and health care are rights, they entail a government's use or threat of force against health professionals;[3]

(9) however, the use of force to restrict liberty and coerce the lives of health professionals who engender healing acts is illegitimate (though not incompatible with taxation) and therefore inconsistent with the second premise: that persons have a right to pursue values and carry out peaceful actions that sustain their lives;

(10) moreover, if one claims that there exists a right to health/medical care for all citizens, then one denies that health professionals have a right to live in peace and freedom.

QED: there is no right to health/medical care.

I have taken this slight digression for an important purpose: to show that the central issues which confront us in the conflict between economic and pecuniary decisions in health care *and* certain ethical propositions depend upon more than defending, however virtuously, the "traditional moral obligations of physicians to individual patients" ([19], p. 175). Such appeals to the moral preeminence of physicians' concern for their patients will not suffice, however much we agree with the values of our traditions. That is, since there is no right to health care justifiably to be claimed by any citizen, we are thrust into difficult times in which dollars do indeed speak very loudly. By this I mean that in the absence of any valid claim to a right to health care, we are compelled to (1) confront costs that we hope would promote efficiency and economizing; (2) make hard choices that require comparing quality of care with health and medical costs; (3) create and display incentives that would produce socially productive efforts; and (4) make those necessary compromises (perhaps too vulgarly called "trade-offs") that would mostly benefit patients and health professionals. Having said this, I should immediately add that I personally reject any political paradigm that defines *all* relationships in terms of the marketplace; but the history of the economics of health/medical care has inevitably led us to challenge medical professionals as gatekeepers and to reject the long-standing and entrenched tradition to expect the medical profession to monitor itself. To be sure (I agree with Professors Brody and Mattheis), we have to resist defining all relationships in terms of the marketplace (should anyone so claim), while at the same time we respect the power of market forces to control the ever-rising costs of health care, even if we happily persuade our representatives to allow the present percentage of the real GNP (U.S.=10.6%; F.R.G.=10%) for health care to increase quite considerably – a policy Professor Brody has shown is not inconsistent with our present national (indeed, international) options. A recent study has shown that "Paradoxically . . . Americans are not particularly concerned by the increasing share of the nation's resources that is devoted to health care" ([2], p. 613). But we *are* all concerned about the compelling condition of infirmity which, as Dr. Pellegrino rightly observes, tends to affect our attitudes toward the costs of medical/health care: He remarks: "When we become ill and are in immediate need, we are far less likely to think about the economic impact of care *versus* cure and prevention." ([19], p. 179). It is for this very reason, on my account, that we shall have to solve our current crisis, and radically reconstitute the present health care system, *prior* to the full range of self-serving, individual patient choices which arise when we are ill. For the question at issue, as always, is, "What *now* are the *legitimate expectations* of those who are sick when they consult a health professional and approach the access point to care?" I believe it is less than helpful to look to the traditional values and simply to assert that "A distant threat to the economic well-being of society cannot compete morally with the immediate and urgent need of the patient" ([19], p. 176). In short, we must challenge the long-standing *legitimate expectations* of physicians and health professionals in general. In the U.S., we shall have to confront our population with the debate over health/medical policies in order thereby to determine how much we value health care, how we should modify the health care system and, more importantly, to what extent we are willing to *compromise* or trade off our traditional, legitimate expectation that, for example, in most clinical situations, when we fall ill, everything possible, whatever the cost, will be done in our interest. We would all appreciate the opportunity to live in a world where patients' needs are preeminent and only the best care provided. Our compatriots in the United Kingdom and other European nations would like to join us in this, I believe, but the full story of the selection and treatment of patients with End-stage Renal Disease (ESRD) in the U.K. [12], for example, suggests that such ideals do not dictate policies, and in the U.S. such ideals are no longer possible to achieve, at least not on the basis of our present health care system and rational economic principles that take further scarcity and even rationing seriously.

## II. THE SHARED PREMISES OF RUTH MATTHEIS AND BARUCH BRODY

I now turn to our two contributors to this section of the volume. On reading their initial essays, it is striking to note the formidable series of premises and assumptions which the authors embrace. I shall attend to the most salient and important ones for our purposes:

(1) There exists a scarcity of and burden on economic resources – pecuniary, human, and material – given the competing demands of the ever-increasing number of very elderly patients on the present and future health care rolls ([4], pp. 216–217), ([15], p. 206), as well as the future high technological medical prospects of (1) "artificial organs in addition to artificial joints, (2) 'hospital on the wrist' monitoring bodily functions" ([15], p. 205) and (3) natural and artificial heart transplantation for patients of all ages [11] – neonates as well as the elderly, decisions at the very edges of life.

(2) The present health care system is significantly inefficient and wasteful ([14], pp. 213, 217) ([15], pp. 203–204; 211), in that too often

there is no significant benefit, following treatment, to functioning human life. Hence, with regard to the provision of health services and the increased demand for medical resources it has become necessary to (i) contain the overall medical/health costs by a prospective payment system ([15], p. 208) (often crudely reflected in the present and rising GNP), (ii) avoid "production losses," such as the additional cost when workers are removed from the labor force due to illness, which might have been avoided by preventive programs ([4], p. 217), or the enormous time and material required for children and others who are chronically ill, (iii) continue efforts (already underway) to reduce expensive institutionalization, and (iv) reduce, wherever possible, the number of medical graduates and the costs of medical education and training ([15], pp. 206–207).

(3) They agree that it is essential to assume a *scarcity* of resources and place limits on health care ([15], p. 211), an assumption if challenged with regard to the present that will surely be true in the not-too-distant future in the U.S. Hence it is a prudent assumption on the basis of which conflict between ethics and economics of health care makes eminent sense. *Rationing* – that is, to apportion or reapportion and distribute resources at a level below market prices – is necessary critically to consider now more than in the past ([4], pp. 217; 228–233), due principally to the enormously increasing societal costs for health care ([15], pp. 211–212). The corollary to this proposition, noted by Dr. Mattheis, is the requirement that we begin to consider in detail which use of health care resources directed to certain purposes will therefore not be spent on or directed to other purposes ([15], p. 211); or, with respect to ESRD in the United Kingdom, for example, which patients with ESRD will be selected for renal haemodialysis treatment and who will be denied this treatment and thereby suffer a premature death.[4]

(4) The authors also appear tacitly to agree that economic arguments taken in isolation cannot in themselves justify ethical propositions, although they do not directly argue for any particular ethical theory and adopt it to justify what we *ought* to do; but they do strive to have all economic decision making in the health care arena cohere with the principle of individual autonomy: the view that every citizen should retain the maximum amount of free choice as a moral agent in his or her own health care and economic decision making ([4], pp. 230–233), ([15], p. 203), though Dr. Mattheis does question the ideal of "a maximum

amount of free choice" for patients and providers ([15], p. 211).

## III. RATIONING THE "GOODS" OF HEALTH CARE

A recently published comprehensive Master Plan Report to the University of Connecticut Health Center and teaching hospital opens by pointing out that rapid changes are occurring in the external environment in which academic medical centers must function, including increasing competition among health care facilities for a shrinking number of inpatients; reimbursement for inpatient care on a fixed per-case, prospective payment (pricing?) basis,[5] rather than on a per-patient-day basis; limitation of patient choice through leverage by employers and insurers; and competitive systems which will effectively exclude certain patient populations, e.g., the indigent, the elderly, and other high risk groups.

It is almost commonplace, then, to observe significant ambivalence in public attitudes toward ever-changing health care economic factors. (Dr. Mattheis also observes that attitudes of West German citizens are difficult to change, e.g., patients are not used to paying too much out-of-pocket; it has been an old privilege ([15], p. 209) Yet, on the one hand, there are documented references to patterns of continued scarcity of health care resources; on the other, some of the very same writers who profess allegiance to this judgment, perhaps even Professor Brody, seem to disregard its serious implications, e.g., the necessity for new standards of rationing, though Brody is well aware of the fact that rationing of care has often occurred as have decisions to terminate treatment. Furthermore, they seem to agree that scarcity is likely to continue, since it is now clear that no country in the world appears willing to provide state-of-the-art medical care for all its citizens. Yet daily rhetoric prevails: one Senior Vice President and Treasurer of a Connecticut hospital remarks: "We're not sure what the long-term impact of the new regulations will be. We may have to manage our resources differently. But we will always meet our responsibilities to the community. If people are sick, they will get the level of care they need. That is why Mount Sinai Hospital [Hartford, Connecticut] exists" ([16], p. 9). Do such pronouncements really convince? Do they jibe with present economic reality and trends? I think not.

## IV. REASONABLE EXPECTATIONS: A DEFENSE OF GOVERNMENTAL RATIONING POLICY

Scarcity of resources generates difficult economic and ethical trade-offs, and even suggests rationing [13], ([4], p. 229). Whereas in the past the wealth of the U.S. had in many instances permitted it the nearly exclusive use of medical criteria for making explicit health care utilization decisions from a moral point of view, we are now faced with economic conditions which radically revise the basis on which these utilization decisions will henceforth be made, as well as the type of decisions themselves; for now we must control rising costs while (1) maintaining quality of care, (2) not intentionally inhibiting the less fortunate in their need for access to and receipt of health care, and (3) retaining the citizens' maximum amount of free choice ([4], p. 213).

There can, then, be little dispute with our authors' formulations of the portended difficulties in the ever growing health care requirements for the ever increasing number of aged in the U.S. and West Germany due to a significant rise in the average life expectancy.[6] I readily acknowledge with them: (1) our society's bias against the old-old; (2) the data on the increasing demographic momentum and costs for various life-sustaining programs; (3) the data on the ever increasing demand for chronic care; as well as (4) the general thesis that the elderly will be very expensive to care for in the years ahead ([4], p. 221), since there will surely be a greater number of dependent elderly in our society than ever before, and they will not be illness free, either.

But such references and facts are virtually useless in determining what governments *ought* or *ought not* to do. Further, there is little solace in appealing to reasonably reliable cost benefit analyses (CBAs), though it is true that CBAs necessarily entail allocation decisions since they lead to recommendations that require that we select some persons for screening, testing, and treatment, and deny these services to others. We can properly criticize the United Kingdom, for example, for failing to initiate their policy of financial "incrementalism" on the basis of a comprehensively conducted CBA [12]; as Professor Brody quite correctly points out in citing J. H. Aaron and W. B. Schwartz's *The Painful Prescription* (1984; pp. 36–37), the "selectors" for treatment of End-stage Renal Disease are the general practitioners, since they serve as the gatekeepers for renal hemodialysis ([4], pp. 223–225). Unfortunately, these decisions are not made outside the clinic, as perhaps they should be if we hope to eliminate the tendency on the part of physicians to deceive their patients.

I should like now briefly to recapitulate Professor Brody's position with regard to the moral permissibility of rationing by government intervention. Although Brody's view of rationing policies is not restricted to the health care context, I shall restrict my discussion, as he does, to this particular context. It appears that Brody not only maintains that the U.S. "will be unable to adopt a rational rationing policy" ([4], p. 227), he argues for a stronger claim in his concluding paragraph: "Rationing is morally objectionable" in the context of the health care needs of the indigent ([4], p. 235). I take it that his stronger claim is based upon the fact that when a government establishes a rationing policy it necessarily restricts otherwise free citizens from deciding for themselves what they wish to pay for out-of-pocket. Furthermore, rationing the indigent restricts their freedom, as well as the freedom of the taxed middle class and well-to-do. This quasi-libertarian position which values highly the freedom of citizens to choose and pursue their view of the good consistently insists on maintaining the freedom of the indigent (who are offered assistance through some voucher system) by not restricting or "earmarking" a subsidy when it is derived from others by taxation, or charity, or skimming off a percentage from the "inflated" bill of the non-indigent hospital patient to serve as "revenue" for the poor patient.

Brody's view is that *the individual should ration himself* – patients and families should choose to limit care, rather than respond to physicians' judgments ([4], p. 214). I would also prefer this state of affairs, and for much the same reasons Brody does; but unless individuals pay for their own care there is really little incentive to ration themselves. Yet we do not (in our day of stressing the autonomy of patients and health professionals) need to encourage physicians to deceive their patients about costs and revenues. It is bad enough that they are too often the gatekeepers and rationers of care. Worse yet, we do not need to undermine clinical judgment entirely and thereby deprofessionalize the medical professions, as Dr. Mattheis has warned in her remarks ([15], p. 205).

Before addressing Brody's line of argument, which leads to his conclusion that government rationing of health care is "morally objectionable," let me summarize the standard arguments *against* rationing policy:

In a condition of scarcity, once the waste and inefficiency are under control (e.g., spending the same but utilizing 20–30% of the budget more effectively), we either (a) opt for *universal treatment*, an unfeasible option given our premise of scarcity and the fact that people in the U.S. do not want to pay for all that good quality health care costs, or we (b) opt for *universal non-treatment*. But this is only permissible if the selection of some patients for treatment and others for no treatment is morally arbitrary. Hence, both universal treatment and universal non-treatment are rejected options. This leaves us with the allocation problem at *both* the macro and micro levels. Those who argue *against* rationing policy, then, maintain that the criteria for selecting some patients are not only medical, as perhaps they should be; rather, they are "social" and made only by individual physicians. Hence these sorts of rationing decisions are arbitrary and morally suspect. In short, random selection (which, of course, is a systematic method of selection) is necessarily imprecise and can only lead to inequities. Some maintain that this is also an irrational strategy or policy. Moreover, rationing policy fails to protect patients equally. In sum, rationing is invidiously discriminatory and unfair, since physicians have the power to decide who receives and who does not receive treatment. Finally, rationing challenges our original commitment to the value of life ([4], p. 224), and a rationing policy (among other factors) tends to lead physicians to deceive their patients which may infringe on tort, if not criminal law.

I am certainly in agreement with Brody that rationing policy generates certain worries and even some dangers. But his objections fail to convince me that rationing policy is intrinsically "morally objectionable." Or, to put my view forward more directly – arguments which are proffered to reject rationing policy, as undertaken by governments, are, while rationally defensible, *useless* and *impractical*. My view is that governments, in order to survive (and I challenge Dr. Brody to cite a case where a nation state has survived, absent rationing policy) must be permitted to coerce tax dollars from its citizens for selected and specific purposes – like advancing the well-being of all citizens by seeking to provide minimally adequate health care especially for its currently uninsured indigent citizens. Reliance on a system of charity fails and will continue to fail; it is hardly an option given the needs of millions of U.S. citizens who have no health insurance of any kind.[7] Arguing against rationing policy is, to me, like the arguments philosophers have proffered to defend solipsism. They too are rationally defensible, but useless and impractical. Hence my argument in defense of rationing policy in a time of great scarcity (scarcity, of course, not due primarily to general economic hard times, but to the long-term unappeasably voracious appetite of the health care industry) is as follows:

Since resources are scarce, governments are compelled to carry out CBAs and cost effective analyses (CEAs) in order to discover ways to curtail expenses and contain costs via macro-allocative decisions. Allocative mechanisms are possible to establish equitably, especially when great care is taken to exclude only those persons from treatments and health care who would not derive a reasonably positive outcome from these treatments. A set of criteria for selecting individuals to receive treatment, and others not to receive treatment, can be rationally and equitably established; such allocations need not infringe upon a physician's right to practice, though at times it does require restricting the movement of physicians to the more lucrative private services so as not to endanger the indigent by dangerously restricting the availability of health professionals in their region. Moreover, such rationing policy need not be invidiously discriminatory but may be established on reasonable grounds. A rationing policy, after all, will seek to save as many lives as possible, knowing it cannot save all.

Although I have argued that there is no right to health care, it is not in my view inconsistent to defend rationing, however undesirable at times this alternative may be. We may, indeed, have to deny life-saving care to some as a matter of policy; it will be, as Tristram Engelhardt has remarked, unfortunate but not necessarily unfair.

I simply do not understand Brody's assertion that, "We could simply provide that level of funding in the form of a voucher to those who are medically indigent and allow them to choose" ([4], p. 232). Where, if one takes the premise of scarcity seriously and at least debates the significance of rationing policy, are the *additional* resources to enable us to provide this "level of funding"? I really believe Brody wheels them in behind him on a stretcher, e.g., in the form of new Medicaid dollars. In truth the stretcher is empty; no additional funds will be redirected from other U.S. "priorities" to the GNP for health care, nor, perhaps, should they be. (This is a critical assumption of my position on rationing.) The recognition of this essential fact compels us to face the need for a rationing policy squarely. We may wish to keep government interference to an absolute minimum, as Dr. Mattheis has observed regarding the West German people ([15], pp. 204, 211), but can we continue to do do so and discharge our self-imposed *duty* to the poor? Unfortunately, Brody's proposal assumes we can realistically aid the indigent with (1) money, or (2) "unearmarked" vouchers, or (3) other financial stratagems. I seriously question these options. I do not here wish to question the *capacity* of the indigent to make their own decisions in terms of their own view of their economic priorities. Indeed, it is precisely because I am inclined to respect the judgment of the indigent that I worry over Brody's passion to stress their autonomy in deciding for themselves what they shall do with a voucher *not "earmarked"* for their family's health care. Poor people will of pressing necessity spend supplementary vouchers on what is needed at the next moment. To establish a voucher program which enables people to purchase health care is to discover that in a short time the monies are expended – even if well spent. That is, these same people will appear at the doors of hospitals and clinics! (I have no confidence in the claim that, in time, the indigent will reallocate their subsidy and purchase health care insurance or join a health care plan for their family's needs and wants [8, 9].) Moreover, we shall not be willing or able to turn them away because their "credit" vouchers for health care are no longer available. If one attempts to counter this view by arguing that the expending of the vouchers for other necessities – food, shelter, gasoline, etc. – is the proper use of the voucher, then one is begging the question and really supporting a supplementary income not "earmarked" for health care. That is, one is defining the totality of everyday needs as in the service of "health" maintenance, and thus the original policy has been modified. I repeat: I am opposing the policy of not earmarking funds for health plans should they be offered by voucher to the indigent. If we must ration and allocate by a special formula for redistribution of our limited resources, then the indigent will have to live with a modest restriction of their liberty – they will indirectly receive aid from others that is directed to specified purposes as outlined in the initial rationing requirement. One need suffer no moral umbrage here.

I do not mean to oversimplify the mechanisms required of a rational, rationing policy, since things are very complex, given that the supply-side, the health professionals, is confronted by multiple constituencies on the demand-side: not only patients, but employers, unions, insurance carriers, and government. In sum, I see nothing "morally objectionable" in a rationing policy when undertaken at the federal level for the express purpose of advancing the well-being of all citizens. Rationing does not entail either an excessive infringement on the freedom of or venal paternalism toward those receiving rations.

I was pleased to note that as late as August, 1983, Brody himself argued, in a very creative article, that there are "two types of legitimate infringement upon the rights of others. The first type of case," he writes, "involves a major threat to the well-being of an individual or group, where that individual or group can nullify the threat only by infringing . . . upon the rights of others . . . . The second type of case," he continues, "involves a major opportunity for advancing the well-being of an individual or group, where that opportunity can be exploited only by infringement . . . upon the rights of others" ([3], p. 83). Does not the second case serve quite well here – in the context of the millions of U.S. citizens who have no health insurance today? Must not the U.S. government "exploit" and tax the rest of us? (Please note that this argument does not depend upon any claim that citizens have a right to health care, only that citizens have a right, generally speaking, not to be infringed upon unnecessarily. And here we can argue back and forth about the right of the government to tax us and our right not to be taxed.) In short, we must be careful here, given my opening remarks and argument, to distinguish a "national commitment to provide health care for those who cannot afford to pay for all of their health care on their own" (Brody's point, [4], p. 216) from the fact that the U.S. government has no national stance such that any individual citizen can claim that he or she has a right to health care. Governments by their policies can signal that they have obligations without their citizens laying claim to a right. The present U.S. Administration, as the reader is no doubt well aware, pursues a policy of (1) containing taxes and (2) restraining Federal social spending, including spending on health care. Hence we in the U.S. can anticipate little increase in the percentage of the G.N.P. for health care beyond the present 10+ percent; indeed, at this writing it is beginning to show a modest decline.

It would be interesting to pursue, should we have time, the issue of whether the indigent are entitled to receive health care resources based on a theory of *just compensation for past debts incurred by citizens who are no longer living*. After all, many have been the beneficiaries of extended average life expectancy, thanks to all sorts of advances in public health and medical research. Could the present indigent claim a right to present resources, not on the basis of a right to health care, but on the basis of a *right of transfer* from citizens who lived in the past?

Professor Brody has, I know, given thought to this problem and line of reasoning, and I, for one, would like to see him further explore this interesting suggestion.

## V. CONCLUDING WORD

I have indirectly reviewed various options: (a) we could *spend less by improving productivity*, thereby saving as much as 20–30% to create a more efficient health care system. But if we accomplish this, we shall still have to face a second phase – the need to manage with less than an efficient and frugal system will have available in the form of revenues. In this difficult phase we will by definition have to (b) *spend less by doing less*. That is to take scarcity, not efficiency, seriously. Then the difficult decision point will have been reached: (1) will we, by spending less, continue to disenfranchise those who have no health insurance (12–15% of the U.S. population) or (2) will we, to spend less, redistribute our inadequate resources in such a way that the indigent (as well perhaps as some of the middle class) can endure by way of having not the best, but at least "decent' or "minimally adequate" care which a government deems itself duty-bound to provide? If the latter course, it may be best to establish a rationing policy and continue to *tax* all citizens equitably (proportionally?) in order to provide the necessary revenues for discharging this obligation. In any event, political and economic reality will obstinately demand their day. I believe that in the future, illness will make harsh demands on us which the reality of economic and political factors will cause us to fail to fulfill. Regrettably, all we can hope to avoid, since we cannot in my judgment avoid rationing and misfortune, is *gross* inefficiency, inequity, and unfairness.

*University of Connecticut Health Center,*
*School of Medicine,*
*Farmington, Connecticut, U.S.A.*

## NOTES

[1] I am especially grateful for the excellent critical comments provided by my friends and colleagues, Thomas Halper and H. Tristram Engelhardt, Jr.; the remaining inadequacies are clearly the fault of the author.

[2] See [21], Bk. IV, Chap. V, p. 508: "The natural effort of every individual to better his own condition, when suffered to exert itself with freedom and security, is so powerful a

principle, that it is alone, and without any assistance, not only capable of carrying on the society to wealth and prosperity, but of surmounting a hundred impertinent obstructions with which the folly of human laws too often incumbers its operations; though the effect of these obstructions is always more or less either to encroach upon its freedom, or to diminish its security."

[3] An established right to health care would generate not an *in personam* right against individual physicians, who would then be duty-bound to provide care under coercion, but an *in rem* right against *all* health professionals taken collectively; to the "world" of them at large (as they have the knowledge and expertise others do not possess). Health professionals would incur a duty to provide health care; it would be owed to all injured or sick persons. In sum, an established right to health care would be a positive, *in rem* right against the *community of health professionals*. Duties of the community of health professionals would be correlated with patients' positive *in rem* rights to receive health care. This supports the conclusion that there is no right to health care. See [10], pp. 131 *et passim*.

[4] "*Allocative efficiency* is obtained when one person, department or service cannot be made better off without making another person, department, or service worse off" ([7], p. 16). Allocative efficiency is illustrated when, for example, one considers the value of preventive medicine in cost containment: At what point, one may ask, are resources better employed in the prevention of illness than in curing it?

[5] "Pricing" is not quite the same as the *actual* "*payment by*" the third-party payer; discrepancies here can often leave the payee (hospital) with less than it believes it is due. Therefore, we are more precise if we speak of the "prospective pricing system". I owe this point to H. Thomas Ballantine, M.D.

[6] One should be careful not to confuse "life span" with "average life expectancy"; it is the latter that has been increasing in this century. The human "life span", on the other hand, has actually little changed from the time of our ancient ancestors. The important point is that more and more of us are living three score and "thirty," thus increasing dramatically the average life expectancy ([4], p. 221).

[7] Accurate information on the number of U.S. citizens without either private or public health insurance is difficult to ascertain. For example, from 1950–1976, some 25–50% of the increase in physician service expenditures was attributed to the *growth* in insurance coverage [6]. According to the U.S. Census Bureau, in 1983, the percentage of uninsured U.S. citizens was fifteen [5]. Other sources claim that "In 1983, about 29.2 million people – 12.6 percent of the population – had no health insurance" ([14], p. 59). Mary O'Neil Mundinger's and Uwe E. Reinhardt's ([20], pp. 20–28) estimates differ only slightly: Mundinger remarks that "Thirty-five million people, or 15 percent of the population, are without [private or public] health insurance today – an increase of 10 million, or 25 percent, since 1977" ([17], p. 4).

Wilensky and Walden remark that the problem of the uninsured is often overstated, for data tend to fail to note that (1) many uninsured persons are uninsured only on a temporary basis, i.e., less than a year, (2) a large proportion of uninsured are relatively healthy younger persons between the ages of 6 and 24, (3) less than 1% are over 65 years, and many belong to the middle class or above and are presumably uninsured by choice. (4) Even the difference between the proportions of uninsured and insured who rated their health as poor was so small that it may well have been explained by normal sampling error [22]. Moreover, (5) although increase in coverage has been accompanied by even higher charges for insurance, this has not actually depressed service utilization, since these charges are not close enough in time to the use for consumers to link the two [18]. Again (6) since over three-fifths of the population's insurance coverage is employer provided, the percentage of uninsured is necessarily quite volatile and fluctuates with employment trends.

Aday and Anderson are closer to the truth, perhaps, in their remark that in 1984 only 9 percent of the total population was uninsured ([1], p. 1335). In any event, individual citizens and other interest groups will have to determine what is to be done with respect to these uninsured citizens, especially those who truly do not have the ability to pay for their own health care or health insurance. This problem was raised early on by Professor Brody ([4], p. 216).

## BIBLIOGRAPHY


1. Aday, L. and Anderson, R.: 1984, 'The National Profile of Access to Medical Care: Where Do We Stand?' *American Journal of Public Health* **74**, 1331–1339.
2. Blendon, R. J. and Altman, D. E.: 1984, 'Special Report: Public Attitudes About Health-Care Costs', *New England Journal of Medicine* **311** (9), 613–616.
3. Brody, B.: 1983, 'Redistribution Without Egalitarianism', *Social Philosophy & Policy* **1** (1), 71–93.
4. Brody, B.: 1988, 'The Macro-allocation of Health Care Resources', in this volume, pp. 213–236.
5. Census Bureau: 1985, *Survey of Income and Program Participation*, U.S. Government Printing Office, Washington, D.C.
6. Dyckman, Z. Y.: 1978, *A Study of Physicians' Fees*, U.S. Government Printing Office, Washington, D.C.
7. Eastaugh, S. R.: 1981, *Medical Economics and Health Care*, Auburn House Publishing Company, Boston, Massachusetts.
8. Enthoven, A. C.: 1978, 'Consumer-Choice Health Plan: Inflation and Inequity in Health Care Today', *New England Journal of Medicine* **298** (12), 650–658.
9. Enthoven, A. C.: 1978, 'Consumer-Choice Health Plan: A National-Health-Insurance Proposal Based on Regulated Competition in the Private Sector', *New England Journal of Medicine* **298** (13), 709–720.
10. Feinberg, J.: 1980, *Rights, Justice, and the Bounds of Liberty: Essays in Social Philosophy*, Princeton University Press, Princeton, New Jersey.
11. Graven, D., *et al.*: 1984, *The Price of Life: Ethics and Economics: Report of the Task Force on the Affordability of New Technology and High Specialized Care: Life at Any Price?*, Minnesota Coalition on Health Care Costs, Minneapolis, Minnesota.
12. Halper, T.: 1985, 'Life and Death in a Welfare State, End-stage Renal Disease in the United Kingdom', *Milbank Memorial Fund Quarterly* **63** (1), 52–93.
13. Heckler, M. M., Schwartz, W. B., DeVries, W. C., Hofmann, P. B., Lamm, R. D., Capron, A. M., and McIntyre, R. L.: 1985, 'Ethics, Rationing and Economic Reality', *Federation of American Hospitals Review* **18** (3), 14–43.
14. Iglehart, J. K.: 1985, 'Medical Care of the Poor – A Growing Problem', *New England Journal of Medicine* **313** (1) (July 4), 59–63.
15. Mattheis, R.: 1988, 'Macro-allocation in Health Care in the Federal Republic of Germany', in this volume, pp. 201–212.

16. *Mount Sinai Hospital Annual Report*: 1984, 'No Easy Answers: The Issue of Medical Ethics', published by Mount Sinai Hospital, Hartford, Connecticut.
17. Mundinger, M. O.: 1985, 'Health Service Funding Cuts and the Declining Health of the Poor', *New England Journal of Medicine* **313** (1) (July 4), 44–47.
18. Pauley, M. V. and Langwell, K. M.: 1982, *Research on Competition in the Market for Health Services: Problems and Prospects*, Applied Management Systems, Silver Springs, Maryland.
19. Pellegrino, E. D.: 1980, 'Medical Economics and Medical Ethics: Points of Conflict and Reconciliation', *Journal of the Medical Association of Georgia* **69**, 175–183.
20. Reinhardt, U. E.: 1985, 'Economics, Ethics, and the American Health Care System,' *The New Physician* **34** (9) 20–28, 42.
21. Smith, A.: 1776, *An Inquiry into the Nature and Causes of The Wealth of Nations*, Edwin Cannan (ed.), (1965), Modern Library, New York.
22. Wilensky, G. R. and Walden, D. C.: 1981, *Minorities, Poverty, and the Uninsured*, Department of Health and Human Services, Hyattsville, Maryland.

WOLFDIETER THUST

# POLITICAL-MEDICAL ALLOCATIONS IN THE COMPULSORY HEALTH INSURANCE PROGRAM IN THE FEDERAL REPUBLIC OF GERMANY

Compulsory Health Insurance (CHI) in Germany provides medical care benefits and interacts in various legal and functional ways with the social insurance system of which it is an intrinsic part.

The accompanying diagrams illustrate the most important relationships of the statutory health insurance program; these diagrams help to explain the complexity of the sytem.

Diagram 1 shows the structural relationship between the insured person, the Statutory Health Insurance Fund, the Association of CHI Physicians, and the physicians admitted to CHI practice. The interdependence of these four structural elements is regulated by the RVO (*Reichsversicherungsordnung*) or Reich Insurance Code, which in turn is subject to the economic situation, labor market, political situation, and health and social factors.

Compulsory Health Insurance benefit expenditures are distributed as follows: 32% for inpatient benefits, 21% for drugs, medicines, dressings and corrective prescriptions, 14% for dental benefits, 15% for other services, and 18% for outpatient benefits.

The National Master Agreement regulates benefits and settlements, and defines the relationship between the CHI fund and CHI physicians at the national level. The National Master Agreement is a contractual agreement autonomously administered but subject to governmental requirements through various state ministries. The National Master Agreement defines the relative worth of each professional service, expressed either in points in a relative value system or as a specific medical fee. The Federal Insurance Code, although promulgated by the Federal Republic of Germany, returns the administration of contracts to the Associations of CHI Physicians and to the CHI Funds in the various states or regions, thus fulfilling the micro-regulative requirements of the National Master Agreement. The interests of the CHI physicians and the interests of the CHI funds are thus balanced.

Each Association of CHI physicians distributes the combined payments among its member physicians. Distribution of payments for professional services is determined by the number of cases and the cost

*Hans-Martin Sass and Robert U. Massey (eds.), Health Care Systems,* pp. 255–265.

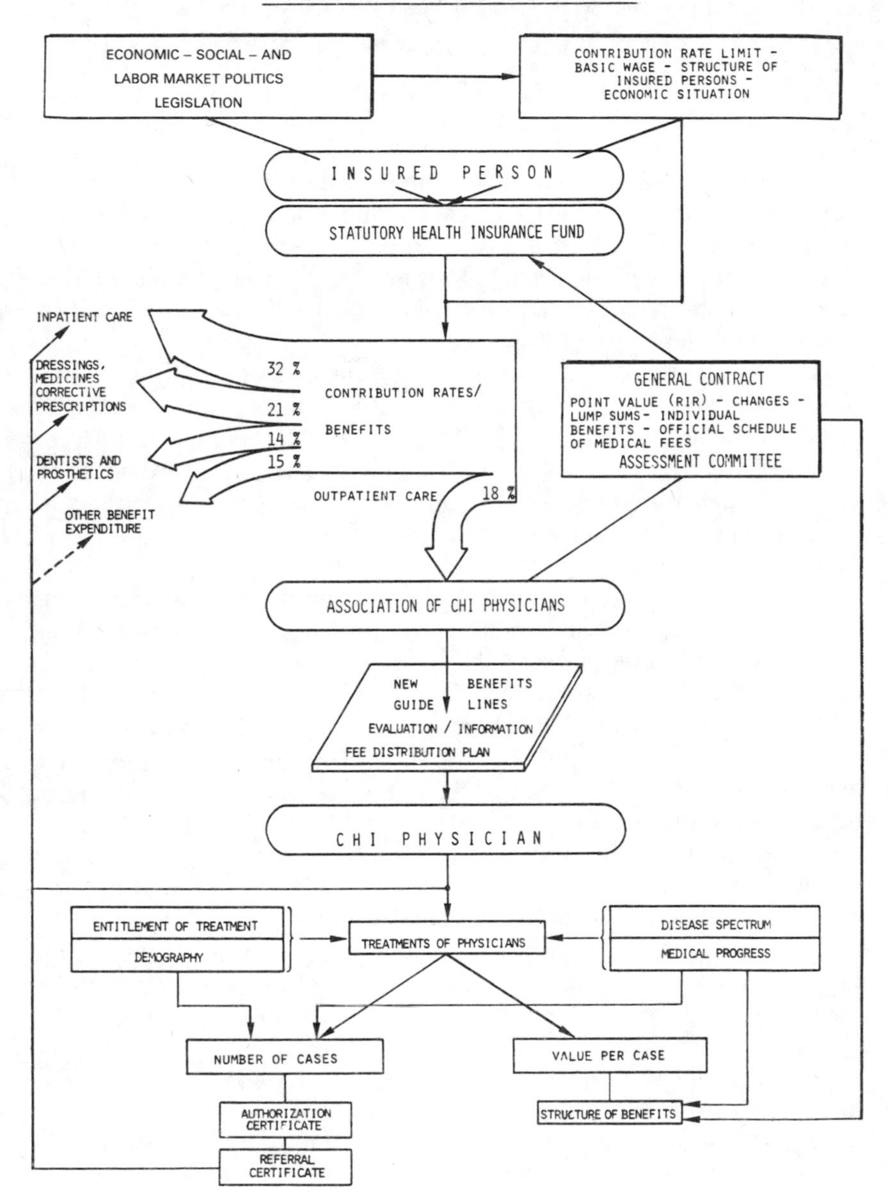
COMPULSORY HEALTH INSURANCE SYSTEM IN GERMANY
ECONOMIC – SOCIAL – AND LABOR MARKET POLITICS LEGISLATION
CONTRIBUTION RATE LIMIT - BASIC WAGE - STRUCTURE OF INSURED PERSONS - ECONOMIC SITUATION
INSURED PERSON
STATUTORY HEALTH INSURANCE FUND
INPATIENT CARE
DRESSINGS, MEDICINES CORRECTIVE PRESCRIPTIONS
DENTISTS AND PROSTHETICS
OTHER BENEFIT EXPENDITURE
32 %
21 %
14 %
15 %
CONTRIBUTION RATES/
BENEFITS
OUTPATIENT CARE
18 %
GENERAL CONTRACT
POINT VALUE (RIR) - CHANGES - LUMP SUMS - INDIVIDUAL BENEFITS - OFFICIAL SCHEDULE OF MEDICAL FEES
ASSESSMENT COMMITTEE
ASSOCIATION OF CHI PHYSICIANS
NEW BENEFITS
GUIDE LINES
EVALUATION / INFORMATION
FEE DISTRIBUTION PLAN
CHI PHYSICIAN
ENTITLEMENT OF TREATMENT
DEMOGRAPHY
TREATMENTS OF PHYSICIANS
DISEASE SPECTRUM
MEDICAL PROGRESS
NUMBER OF CASES
VALUE PER CASE
AUTHORIZATION CERTIFICATE
STRUCTURE OF BENEFITS
REFERRAL CERTIFICATE

per case, taking into consideration the severity of illness and intensity of care, costs of hospitalization, length of illness, and the costs of the medications, dressings, drugs, and health aids.

## I. THE ALLOCATIONAL STRUCTURE OF THE COMPULSORY HEALTH INSURANCE SYSTEM

Ninety-two percent of the population of the Federal Republic of Germany, including West Berlin, is covered by Compulsory Health Insurance; a considerable number of those covered by CHI have supplemental protection determined by personal preference.

The institutions administering CHI (i.e., the funds) consist of the statutory health insurance funds which include the unemployed, self-employed, artisans, farmers, housewives, and others not on regular salaries; the compulsory health insurance funds of salaried employees; and funds for trade associations (the Seamans Insurance Fund and the National Society of Miners). This historically evolved system influences today the varied contract forms between the CHI funds and the National Association of CHI physicians. The values of services covered under the RVO funds have a point value whereas the substitutional (private) insurance programs show DM values.

Parallel to the CHI and private health insurance, there is a third category of 1.3 million insured which includes those persons who are either legally entitled to free medical care (e.g., in the National Armed Forces or Border Patrol), or are members of independent social institutions such as the National Railroad Men's and Mailmen's health insurance funds.

In 1984 in the Federal Republic of Germany there were over 1,200 separate Compulsory Health Insurance Funds with nearly 40 million insured, and 15 supplementary or substitute health insurance funds with nearly 16 million insured. The CHI funds are local or regional; the supplementary or substitute insurance funds are national.

By law the local associations of CHI physicians and the National Association of CHI physicians must guarantee outpatient services for those insured by the CHI funds, substitute health insurance funds, the social funds Seaman's Health Insurance Funds and Miner's Guild Health Insurance Funds. Diagram 2 depicts the scope of outpatient medical services provided, including maternity care, early detection, and curative care.

If a CHI physician plans to care for patients in the hospital he must apply to his association of CHI physicians for registration as an external hospital physician, and must submit a supporting letter from the hospital indicating the willingness of the hospital to grant him external physician status and indicating the total number of beds which will be made available to him. Inpatient medical care must not constitute the main portion of his medical activities.

CHI outpatient medical practice involves prescription of drugs, dressings, medicine, and aids, certification of sick leave, and hospitalization, all regulated by the guidelines of the National Committee of CHI physicians and CHI Funds (National Master Agreement).

The total amount of CHI insurance in 1984 was 103.3 billion DM. The expenditure per type of benefit was as follows:

| | |
|---|---|
| Hospitalization | 33.1 billion DM (32.0%) |
| Medical treatment | 19.0 billion DM (18.4%) |
| Prescribed drugs | 15.5 billion DM (15.0%) |
| Dental treatment including dental protheses | 13.9 billion DM (13.5%) |
| Disability income benefits | 6.3 billion DM ( 6.1%) |
| Dressings, medicines, and aids | 6.0 billion DM ( 5.8%) |
| Miscellaneous benefits | 9.5 billion DM ( 9.2%) |

Inpatient (hospitalization) benefits include the costs of prescribed drugs. If one includes the costs of prescribed drugs to outpatients, the costs or benefits of the CHI would then total roughly one third for outpatient care and one third for inpatient care.

## II. ALLOCATION OF MEDICAL SERVICES AND COSTS

As of December 31, 1984, there were 192,000 physicians in the Federal Republic of Germany including West Berlin. Thirty-five thousand physicians were not engaged in medical activities. These physicians, all of whom have completed their medical education and are licensed, are non-working, mostly women, inactive, or retired. A small but increasing number of physicians are unemployed.

The 156,000 active physicians are distributed in their activities as follows: 48% inpatient (hospital) services, 42% in outpatient services (private practice), 10% are engaged in other areas such as public health, the governmental medical services, military, administration, industry, or research.

Outpatient CHI medical benefits for those insured within the social insurance system were guaranteed, as of December 31, 1984, through the services of 72,000 physicians. These physicians are divided into three groups. The largest group is the "*Kassenärzte*" or CHI outpatient physicians themselves, who are the basic providers of CHI medical services. These physicians are self-employed and therefore have full financial responsibility for their practices. This group includes 62,300 physicians or 87% of all active CHI physicians. Of all physicians in private practice, 95% are licensed to render CHI medical benefits. Five percent are in private practice outside the social insurance system, although outpatient medical practice is generally not feasible without participation in CHI medical practice for obvious financial reasons.

Every licensed physician is entitled to establish a CHI practice in a location of his own choosing. Nevertheless, certain requirements must be fulfilled in order to be admitted to CHI outpatient service. Entrance in the registry of physicians, which is a prerequisite for membership in an association of CHI physicians, requires official licensing as a physician, 18 months' preparatory time for the CHI physician activity, and an introductory course jointly sponsored by the Association of CHI Physicians and the CHI funds.

In 1984, 5,600 physicians (7.8% of CHI practicing physicians) were authorized to render services reimbursable by CHI funds. This authorization itself does not provide admission to the Association of CHI Physicians, but is granted in order to provide such outpatient medical services as early detection and prophylactic medical care.

Health Insurance Funds and the Associations of CHI Physicians are autonomous self-administered organizations. The Compulsory Health Insurance funds, the substitute Health Insurance funds, the Seamen's Health Insurance funds, and the National Society of Miners are independent, self-administered public corporations under public supervision analogous to the federative system of the Federal Republic of Germany. Therefore, the various Health System funds represent a three-level system: (1) health insurance fund; (2) state association of the health insurance funds; (3) national association of Health Insurance funds consisting of state associations.

The substitute health insurance funds are organized on a national basis represented by the salaried employee health insurance funds and the workmen's health insurance funds.

The organizations administering CHI funds have joined together to achieve certain goals which are in the interest of the general public. For example: establishment of guidelines for uniform implementation of certain procedures; participation in the legislative process; negotiation of contracts; collection and analysis of statistics.

The Associations of CHI Physicians, with few exceptions, are state organizations performing their functions under autonomous self-administration, subject to supervision. In some states there is more than one autonomous association. Membership in an Association of CHI Physicians is not an automatic legal right, but, as indicated above, each interested physician must apply for admission to CHI practice as a registered, full-time, or part-time CHI outpatient physician or external hospital physician.

The primary functions of the Association of CHI Physicians are to (1) represent the economic interests of CHI physicians by negotiating the regional agreements with CHI funds for the provision of CHI medical care; (2) distribute remuneration among the physicians in CHI practice; (3) assume, on behalf of the CHI funds, the mandate of assuring the provision of outpatient medical care for the entire population covered by CHI funds; and (4) fulfill the obligation of monitoring the compliance of physicians in CHI practice with legal and contractual stipulations.

The associations of CHI physicians together form the National Association of CHI Physicians as a corporation of corporations, in which the registered CHI physician is an indirect member through his regional association. The National Association of CHI Physicians is subject to supervision by the Federal Ministry of Labor and Social Affairs.

The primary function of the National Association of CHI Physicians (*Kassenärztliche Bundesvereinigung*) is the protection of the interests of the CHI physician at the national level in dealing with legislative bodies in (1) concluding the national Master Agreements with the National Associations of the Health Insurance Funds, which are designed to define the general content of the regional agreements; (2) participation in the National Committee of CHI Physicians and CHI Funds for adopting guidelines on the provisions of CHI medical care; (3) participation in the preparation of the medical service assessment schedule in

SYSTEM OF ASSESSMENT FOR MEDICAL SERVICES

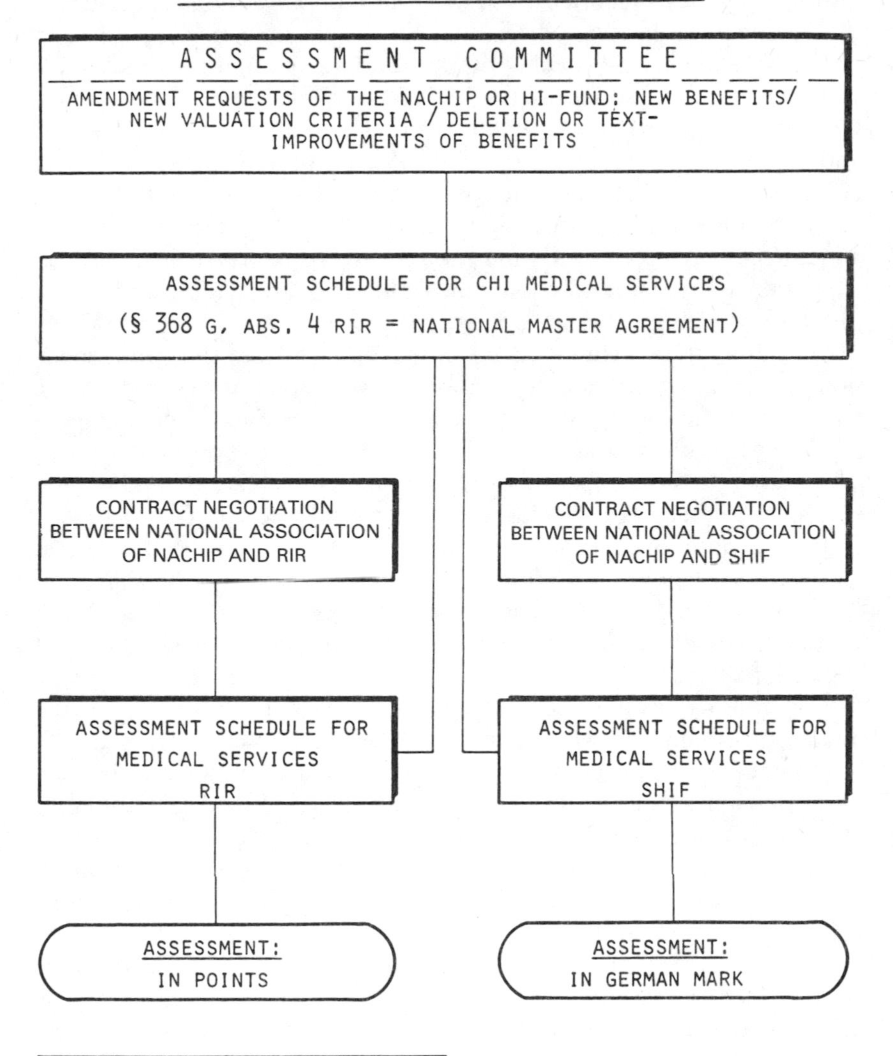


NACHIP = NATIONAL ASSESSMENT CHI PHYSICIANS
RIR = REICH INSURANCE REGULATION
SHIF = SUBSTITUTE HEALTH INSURANCE FUND
Source: NACHIP

the Assessment Committees; and (4) appointment of representatives to the National Arbitration office.

The settlement of accounts of the CHI physicians is between the contract partners (Association of CHI Funds and the Association of CHI Physicians) at the national level. These payments are defined within the official schedule of medical fees of the National Master Agreement agreed upon by the contractual partners (RVO, § 386 i, 8). The Assessment Committee stipulates the type of medical services for which fees may be charged, and the relative worth of such services as expressed by a point value system or in D-Marks, representing the micro-orientation within the system.

The Committee includes equal representation from the two associations: seven members from the National Association of CHI physicians, and seven from the National Association of CHI Funds, including the substitute health insurance funds. It is also the responsibility of the Committee to adjust periodically the assessment scale, taking into consideration new developments in medical science and technology.

The adoption of certain contractual billing provisions has led to two contractual schedules of CHI medical fees as follows: (1) the substitute health insurance funds schedule of medical fees has a DM value; (2) the Statutory Health Insurance Funds schedule of medical fees has a point rating schedule.

The transactions and legal structure of the CHI services within the system of Compulsory Health Insurance are illustrated in Diagram 3.

A beneficiary of CHI who requires medical attention receives an authorization certificate, which entitles him or her to medical treatment, and the physician renders medical services to the patient as required.

The Substitute Health Insurance Fund contract provides that physicians under contract to a fund be remunerated for individual outpatient services according to the current fee schedule. The remuneration process is as follows: The physician bills his Association of CHI funds for the services he has provided; the association of CHI funds in turn examines the appropriateness and the medical justification for the claim submitted by the CHI physician. If the charge is approved, the substitute health insurance fund pays the Association of CHI Physicians an amount equal to the monetary value of the services as indicated in the schedule, deducting an administrative charge; the Association in turn pays the physician.

The CHI funds system for remuneration of services is the Unified Medical Assessment schedule; the point monetary value varies from one CHI physicians association to another. The National Association of CHI Physicians distributes collective remuneration to the local Association of CHI Physicians in accordance with the Unified Medical Assessment schedule, after medical justification of the services provided. The physician in turn receives his share of the total assessment from his Association of CHI Physicians. A physician thus has a claim against his association and not against the CHI funds directly. An individual physician does not have a legal right to payment for all the services he has submitted but to a proportion of the total remuneration. In this manner the system guarantees the patient competent benefits, and the CHI physician his entitled proportional share of the monetary value for services rendered.

The balance between funds available and medical services authorized by CHI fund and the demand for cost containment has top priority in current political discussions.

An authorization certificate itemizes the medical services the physician provides, validating his claim for payment.

The vouchers are examined by the Association of CHI Physicians for mathematical correctness and medical accuracy; reimbursements are made quarterly.

The Association of CHI Physicians has also been given the responsibility for monitoring the efficiency of the provision of medical care, checking the services rendered, the prescriptions, and the medical certificates which are produced by CHI outpatient physicians. To fulfill these legal requirements, the Association of CHI Physicians maintains auditing and review committees.

The usual auditing procedure compares the relationship of the charges of an individual doctor to the average of others in his field, and to his own charges. This procedure, which has been developed in practice and shaped by judicial action, is based on the premise that the majority of CHI physicians work efficiently. Thus statistical averages, when based on sound statistical practice, can be of qualitative significance. The auditing committees are therefore empowered to adjust fees submitted for payment.

CIRCULATION OF AUTHORIZATION CERTIFICATES AND FEES IN THE CHI SYSTEM

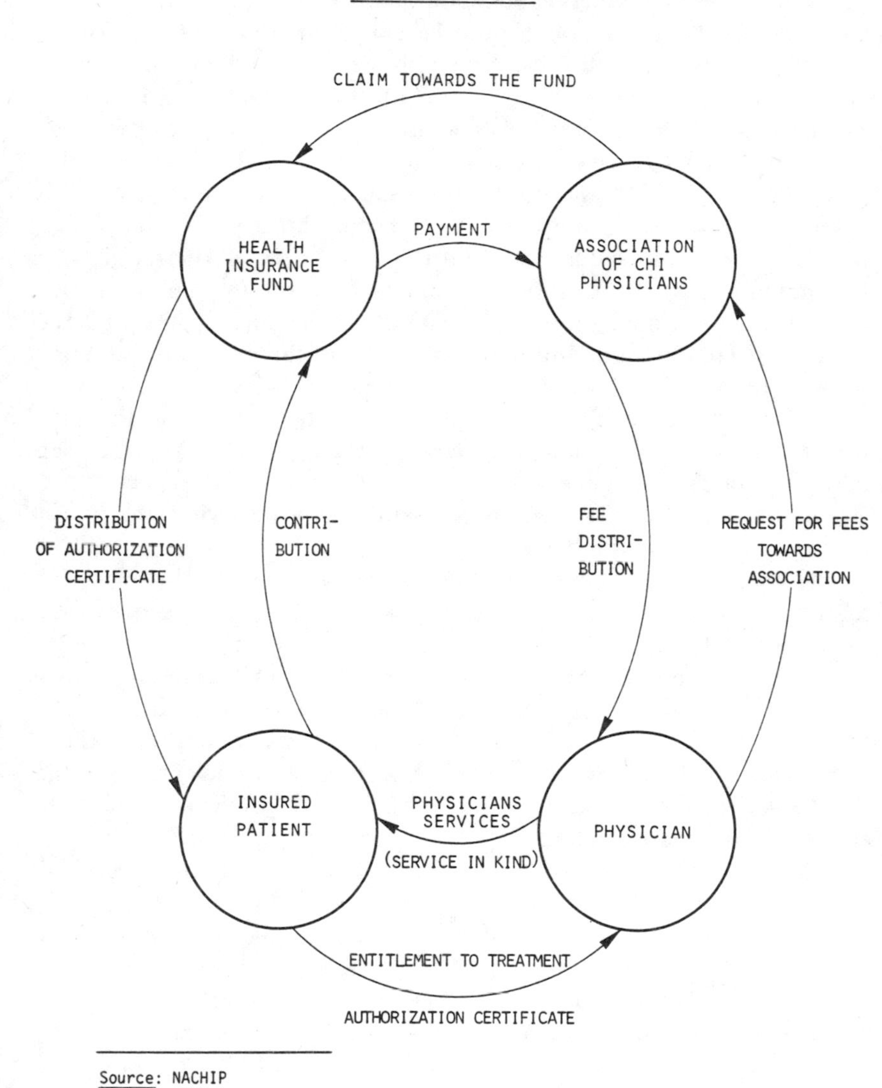


Source: NACHIP

SUMMARY

The provisions of the Social Insurance System, originating over a hundred years ago during the Bismarck era, represent social justice and guarantee adequate medical services for the citizens of the Federal Republic of Germany including West Berlin. The medical benefits provided by the physician, through the Association of CHI physicians, should therefore be considered as optimal.

Because of the increasing costs of health insurance within the last few years, the governing mechanism of the Compulsory Health Insurance System has been questioned. The Compulsory Health System is not a financial system in the true economic sense; expenditures have evolved from the medical fee schedule. These expenditures do not have a market price value within the economic system and probably never can have.

The valuation standards form only a part of the governing or controlling mechanism; social value judgements, the resources of the economy, the comfort and security of the quantity and quality of individual medical benefits must also be considered.

*Kassenärztliche Bundesvereinigung,*
*Köln, Federal Republic of Germany*

BIBLIOGRAPHY

1. Arnold, M. *et al.*: 1982, *The Medical Professions in the Federal Republic of Germany*, Deutscher Ärzteverlag, Köln.
2. Blüm, Norbert *et al.*: 1984, 'Ordnungspolitik im Gesundheitswesen', *Bundesarbeitsblatt*, H. 12, 1–33.
3. Brandt, A. *et al.*: 1980, *Cost-Sharing in Health Care*, Springer, Berlin, Hamburg, New York.
4. Bundesärztekammer:1985, *Tätigkeitsbericht 1985*, Deutscher Ärzteverlag, Köln.
5. Federal Republic of Germany, Press and Information Office:1980, *Social Security for All: Social Services in the F.R.G.*, Walter Bogs, Stuttgart, Berlin.
6. Federal Republic of Germany, Federal Ministry for Youth, Family and Health:1980, *Daten zum Gesundheitswesen,* Kohlhammer, Stuttgart, Berlin.
7. Federal Republic of Germany, Federal Ministry of Labor and Social Affairs:1981, *Insurance Code* (*Reichsversicherungsordnung 1880*), transl., Intern. Labor Office, Geneva.
8. Kassenärztliche Bundesvereinigung (NACHIP):1984, *Die ärztliche Versorgung in der Bundesrepublik zum 31.Dezember 1983*, (Blaue Reihe 34–84), Deutscher Ärzteverlag, Köln.

MARTIN PFAFF

# MICRO-ALLOCATION IN THE HEALTH CARE SYSTEM: FISCAL CONSOLIDATION WITH STRUCTURAL REFORMS?

## I. INTRODUCTION

This paper takes Thust's overview of the system of health care in the Federal Republic of Germany (F.R.G.) and De Wachter's list of moral concerns of today's health care system as points of departure, in order to pose – and attempt to answer – the following fundamental question: What are the scope and limitations of a micro-reallocation in the health care system which would achieve a greater degree of fiscal consolidation of the burgeoning health care costs without violating ethical norms?

In his overview of F.R.G.'s health care system, Thust's interest focusses only on the "18% of the benefit expenditure . . . namely the expenditure which is being spent on outpatient medical care (ambulatory care)" ([18]). While the observer may surely limit his attention to any aspect of a system being studied, an undue narrowing of the focus would not do justice to the central role of the doctor in well-nigh all aspects of health care: after all, medical doctors are the main hospital resource allocators; furthermore, they play a central role in prescribing drugs and medical implements. Thus, the role of doctors in F.R.G.'s health care system is far more encompassing than what would be reflected in the 18% of costs cited by Thust. The doctor's scope for ethical judgments in and about the health care system thus must be broadened to encompass well-nigh the entire system reflected in Thust's overview diagram (No. 1). That doctors do not only play a fundamental role in individual but also in social choice processes is tacitly implied by Thust's observation that almost all doctors working in free practice are providing CHI (compulsory health insurance) medical care as it would not be easy for a doctor to do without the CHI patient. This is why social choice enters the picture in such a significant fashion, since CHI patients are part of the system of social insurance which is being financed partly by individual contributions and partly by tax-financed transfers out of the public exchequer.

*Hans-Martin Sass and Robert U. Massey (eds.), Health Care Systems,* pp. 267–291.

Undoubtedly, public interventions are not limited to the macro-level but affect also the micro-level, particularly in the context of attempts at cost containment. This is a point that should perhaps be considered when interpreting Thust's diagram (p. 264).

Finally I can only applaud Thust's concluding citation that social valuations on the financial budget, on the extent of medical security to be accorded to an individual, and on the extent and quality of structural resources to be provided, must take precedence over the management of the health care system via prices and fees. This conclusion surely highlights the relevancy of ethical choices – at the individual and collective level – for the governance of the health care system.

De Wachter's stimulating paper enters fairly and squarely into the very substance of these ethical judgments. In his introductory statement he reminds us of the four moral concerns in today's health care system:

(1) providing equal health care to all citizens,
(2) providing the best health care to all citizens,
(3) controlling health care costs, and
(4) maintaining the maximum amount of free choice by both consumers and health providers ([1], p. 1).

These differ from what were traditionally seen as the *goals of medicine*:

(1) to hold back death,
(2) to relieve suffering and pain,
(3) to overcome disability and
(4) to restore function.

Indeed, the goal specification formulated by the organizers of this international colloquium already reflects the perspective that health care is not only of individual concern (to the patient and his doctor) but equally of social concern, as it entails issues of equality in health care and of controlling health care costs.

De Wachter's discussion of the Dutch Secretary of Health's enunciation of five types of limits to health care – to my mind – gets at the heart of the choices with which public decision makers are confronted today:

(1) The limit of what does and what does not belong to health care;
(2) the limit of what is or is no longer ethically permissible;
(3) the limit of what is or is not (no longer) effective therapy;
(4) the limit of what can and cannot be done financially;
(5) the limit of state authorities interfering in health care.

In every country similar questions will have to be posed and answered, given the cultural and social values, the political priorities and the economic resources of each country. The answer may in fact differ somewhat from country to country but the general lines of approach presumably would be similar.

De Wachter calls for a "change of mind" away from "today's curative health care services" (with their Cartesian overtones) and toward a "general systems" view of the interdependencies between ill health and "the person's ability to succeed at the level of human relations." I shall turn to this theme in my subsequent observations.

De Wachter furthermore cites Musschenga's priorities of health care: to treat first

(1) diseases which can be successfully treated;

(2) diseases which occur frequently;

(3) diseases which appear early in life [1].

On intuitive grounds, I would not wish to quarrel with this listing. However, when Musschenga argues that a consistent application of cost-benefit analysis to health care policy is not possible, I find myself in less agreement:

(1) If the emphasis is on the word "consistent," then it becomes a matter of definition!

(2) However, if we substitute the term "cost-effectiveness analysis" for the term cost-benefit analysis, a broader scope for economic choice appears possible.

(3) Even if we do not place a monetary value on a person's life, or on the man-months of productive life made possible by different treatments, can we not measure in real terms – say, in terms of man-months of productive or functional life – the relative effectiveness of different types of medical services costing identical sums of money?

(4) Does not Musschenga imply a similar concept of cost-effectiveness when he arrives at the following list of health care priorities:

(a) life-threatening diseases among the young;

(b) quality-threatening diseases of all ages;

(c) life-threatening diseases of old age?

Problems of ethical choice are perhaps less cumbersome when we pose questions about the third type of limit, i.e., "the limit of what is or is not (no longer) effective therapy." I wish to address my remaining comments and questions to this point:

(1) Let me start with some general statements on the scope of the problem, and then

(2) Proceed to some perspectives on a policy of fiscal consolidation

with structural reforms which may avoid some of the ethical pitfalls pointed out time and again in the papers of this colloquium.

## II. THE NEED FOR FISCAL CONSOLIDATION

### *The Scope of the Problem*

Prior to the worldwide recession starting 1974/75, Western industrialized countries placed their faith in mixed-economic systems on the possibilities generated by the integration of economic growth with social progress. Indeed, the shaping of purposeful social policy in general and of health policy in particular could, to a considerable extent, be accomplished by redirecting a part of income growth. Severe conflicts over the distribution of income could generally be avoided. Undoubtedly this phase of economic and social expansion covered up a whole gamut of structural problems, including outdated and ineffective patterns of the provision of health care, of remunerating suppliers, of covering some risks but not others within the Statutory Health Insurance System (*Gesetzliche Krankenversicherung*) of the Federal Republic of Germany.

The fiscal crisis of the welfare state in general, and of the system of statutory health care in particular, has cast a glaring light onto these structural deficiencies. By making it impossible, simply on grounds of severe financing bottlenecks and ceilings, to continue established patterns, it forced a more fundamental reexamination of the system.

Such a review, undoubtedly overdue even before the recession, made it impossible to ignore fundamental defects; similar to the situation found in most other industrialized countries, in the Federal Republic the share of health expenditures in G.N.P. has increased continually during the past three decades. The rate of increase declined in the mid-1970s, and sped up again in the early 1980s. In Canada, such a decline occurred starting in 1970, in Italy and Switzerland starting in 1975, and in the U.K. starting in 1977.

Taking even a longer-run (secular) view, health care expenditures have increased over the past 100 years; they were generally linked to the expansion of health personnel (3.2 physicians per 10,000 population in and sped up again in the early 1980s. (See Table 1 for some pertinent 1876 may be contrasted with 19.8 physicians or more in the late 1970s).

During this time the average life expectancy of the population has increased; the mortality of certain age groups accordingly has declined. At the same time, the incidence of some illnesses, particularly of chronic degenerative diseases, has increased together with the higher life expectancy and the changed age composition of the population. A decline in average mortality thus appears to be accompanied by an increase of average morbidity.

Undoubtedly, the increase in average life expectancy was due to a whole range of factors – from the increased standard of living, better nutrition, better housing, more time for recreation, to better medicine.

What gives cause for doubt about the efficacy of the further expansion of health expenditures is that during the past decades such increases have not always led to a commensurate improvement in the health status of the population, as reflected in indicators of mortality and morbidity. Indeed, when judged by an international comparison of ten leading industrialized countries, in 1975 the F.R.G. ranked first in health expenditures when measured by the share of health expenditures in G.N.P., only to be followed by the U.S.A. When 17 indicators of mortality differentiated by age and sex – ranging from perinatal mortality to death prior to age 65 – are considered, the F.R.G., however, ranks last (that is, it shows the worst health status of all), only to be preceded by the U.S.A. ([4], p. 52).

Health expenditures and health status thus do not necessarily show a positive association; on the contrary, they often exhibit a negative correlation!

In a similar vein, experts have noted that a further increase in health service resources – from doctors and nurses to hospital beds – will not necessarily improve the health status of the population. On the contrary, in the case of acute illness, such an expansion is likely to be counterproductive, and is likely to have negative effects on the health status of the population ([16], p. 17).

A fundamental reexamination of the efficacy of the health care system thus appears to be called for, ranging from methods of financing and control to the types of risks covered. In particular, the F.R.G.'s Statutory Health Insurance System covers well-nigh all "old risks" – illness and the costs resulting from the consequences of illness – but excludes the so-called "new risks," such as nursing care and psychiatric care, which can lead to catastrophic situations for individuals and families. In the course of a structural reform, the question of the balance in the coverage of old and new risks has to be answered as well.

The recession has affected both sides of the budgets of health insurance funds: In so far as it leads to smaller increases in wages than would have occurred during periods of higher growth, it has slowed down the rate of increase of revenues from contributions. And in so far as it has led to an increase in unemployment, a growing share of the potentially active population ceased to pay contributions at all. (An increase of employment by one percent would have increased the revenues of social security institutions by about DM 4.4 billion in 1981.) At the same time, expenses of these institutions increased, exerting pressuress on an already overworked system from both the revenue and the expenditure sides.

As revenues could not be raised and expenses continued to mount, the call for measures of cost containment could no longer be ignored. Starting in 1977, a series of measures were enacted, consisting mainly of consisting mainly of the

(1) Health Care Cost Containment Act (*Krankenversicherungs-Kostendämpfungs-Gesetz*) of July 27, 1977;

(2) Hospital Cost Containment Act (*Krankenhaus-Kostendämpfungs-Gesetz*) of December 22, 1981; and

(3) Cost Containment Extension Act (*Kostendämpfungs-Ergänzungs-Gesetz*) of January 1, 1982.

Common to these measures is the intention of curtailing the rates of increase of statutory health expenditures via various macro- and micro-level instruments, ranging from budgeting to intervention on resources and technology, without, however, changing the basic structure of the existing system. Although these measures did lead to a short-run reduction in the rates of increase of major components of health expenditures, the long-run forces causing the expansion seemingly have been affected slightly but have not been fundamentally curbed.

Indeed, the government has had to make the following admission: "Through the . . . aforementioned laws (only) measures which appeared to be of greater urgency were undertaken. A series of problems, especially those of a structural type, continue to remain unsolved" ([2], p. 10).

The present system could thus be termed "cost containment *without* structural reforms," and be contrasted with a policy of "fiscal consolidation *with* structural reforms" which I shall advocate as one possible perspective for health policy for the 1980s and 1990s.

As the German system differs greatly from that of other countries, particularly from that of the U.S.A., I shall sketch the brief outlines of the present system of cost containment, and only thereafter I shall turn to the longer-run perspectives.

*The Statutory Health Insurance System*

At present, the Statutory Health Insurance System is financed mainly from contributions of its members. Membership is obligatory to all wage and salaried workers (except civil servants) with monthly wages and salaries of more than DM 420 but not in excess of DM 4275 (in 1987). This ceiling does not apply to blue collar workers, however. Presently, the contribution rate amounts to about 12% of income, half of which is paid by the employee and half by the employer.

White collar workers with monthly income of over DM 4275 (in 1987), civil servants, and self-employed persons may opt for membership. The major part of the contributions of the retired, disabled, the aged, and the unemployed are paid by the old-age and unemployment insurances, respectively. Starting in the middle of 1983, pensioners have to pay part of their contributions themselves (three percent, and starting in 1985, five percent of their old-age income). Students are covered for very low rates.

In 1985, the Statutory Health Insurance System consisted of 25.6 million members, plus 10.6 million pensioners. In addition to the paying members, there are about 20 million non-paying (but also insured) dependents of members. The Statutory Health Insurance System covered over 92% of the population, with 8% covered by private insurance.

Only in an euphemistic sense, however, may we speak of a "community of the insured." Apart from civil servants and high-income individuals who generally are not part of this "community," several types of lower risk groups have been able to obtain their own insurance system within the overall Statutory Health Insurance System. Thus we find seven different types of Health Insurance funds:

| | |
|---|---|
| Local health insurance funds | (*Ortskrankenkassen*), |
| Craftsmen's health insurance funds | (*Innungskrankenkassen*), |
| Company health insurance funds | (*Betriebskrankenkassen*), |
| Farmers' health insurance funds | (*Landwirtschaftskassen*), |
| Seamen's health insurance funds | (*Seekrankenkassen*), |
| Miners' health insurance funds | (*Knappschaftskrankenkassen*), |
| Special health insurance funds For specific groups of workers (such as technicians) | (*Ersatzkassen für Arbeiter und Angestellte*) |

Within these seven groups, we find 1299 individual health insurance funds.

By law, all health insurance funds are obliged to grant identical benefits. Nonetheless, the special health insurance funds project the image of being a superior type of health insurance fund. The local health insurance funds, in juxtaposition, are also obliged to accept all those persons who are not covered by any other type of health insurance. Sometimes they pose higher risks; therefore, their contribution rates are somewhat higher than those of the special insurance funds. Furthermore, since employers are obliged to pay the same contribution rates for employees insured within the local and special health insurance funds, the individual employee covered, say, by the technicians' health insurance fund (which belongs to the special health insurance funds) pays a lower contribution rate than his colleague covered by a local health insurance fund.

Proponents of this system argue that it tends to enhance competition among funds for membership. In reality, this claim is essentially meaningless since the insured individual cannot choose freely the fund to which he wishes to belong. Furthermore, over 90% of benefits are laid down by law such that competition becomes well-nigh impossible. At best, it is only operative in marginal areas, such as in counseling and preventive care, in which instances "competition" may lead to questionable results.

Some special health insurance funds use their privileged position to reduce the contribution rates to be paid by their members. One further advantage enjoyed by them is the ability to negotiate as one bloc at the federal level with organizations of suppliers.

The associations of health insurance physicians can thus divide the different types of health insurance funds, in order to negotiate individually and at a decentralized level and to obtain more favorable terms. The potential for "countervailing power" is thus severely curtailed on the part of the health insurance funds.

Thus, "competition" among health insurance funds presently works to the advantage of physicians. In the early 1960s, their incomes were about five times the incomes of the average wage and salary recipient; in the early 1980s they approached almost the ten-fold value! Not surprisingly, the number of new entrants into the medical profession has increased; the total number of doctors was 154,000 in 1985, and it is likely to increase to over 250,000 by the year 2000. Since each doctor has the power to create the demand for his own services, the costs of health insurance are bound to go up commensurately. Little wonder, therefore, that a continuous increase of health expenditures is being forecast for the future!

## III. THE PRESENT SYSTEM: COST CONTAINMENT WITHOUT STRUCTURAL REFORMS

The various measures of cost containment can only be understood against the background of the organizational setting of the health care system. Since "costs" entail the product of prices and quantities, and since the latter are determined largely by the suppliers (physicians) and partly by the patients, a purview of the pricing mechanisms offers a good starting point for our inquiry. As is evident from Table I, per diem charges, fees-for-service, "market prices," and mark-ups fixed by the government all co-exist within the system. They are arrived at in complex negotiations between health insurance funds, hospitals, states, the federal government, associations of health insurance funds, and associations of health insurance physicians, or in routine interactions between physicians and patients (as in the case of pharmaceuticals).

The complexity of the process is exemplified by the ambulatory medical care delivery system (Table II): it is based on a negotiated process, whereby federal (and state) associations of health insurance funds interact with the federal and state associations of health insurance physicians, in order to determine the delivery of medical services,

TABLE I
Pricing mechanisms of the German health care system

| Medical service | Pricing mechanism | Parties involved |
|---|---|---|
| Hospital care | Per diems, fixed by Federal States, covering "approved" operating costs; | Federal States |
| | voluntary negotiations of hospitals and h.i. funds in advance; if they come to terms, the States authorize the result | Hospitals, h.i. funds |
| Ambulatory medical and dental services | (a) *statutory health insurance:* fee-for-service; fee-schedule negotiated between h.i. funds ass. and h.i. doctors association at Federal level; mark up negotiated at State level | Associations of h.i. funds and of h.i. physicians |
| | (b) *private health insurance:* "market pricing" within margins of official fee-schedule. | Physician (patient) |
| Pharmaceuticals | (a) price of producer: "market pricing" | Producer (physician) (patient) |
| | (b) markup of dealer: fixed by the Federal Government | Federal Government |
| | (c) markup of pharmacies: fixed by the Federal Government | Federal Government |
| Public health services increase no pricing mechanism | | |

Source: [3], p. 45.

the system of economic monitoring, the fee schedule (since 1977), the system of remunerations (since 1977), and the ceilings.

Since 1977, a National Health Conference has been added, in which the federal and state governments, health insurance funds, hospitals, physicians, the pharmaceutical industry, employers, and unions are represented. Its main task is to issue guidelines in advance which should serve as points of reference for the forthcoming year. In 1986, a seven-member Council of Health Advisors has been established, as well, to advise the National Health Conference.

The main features of the major pieces of legislation pertaining to cost containment, the Health Care Cost Containment Act of 1977 and the Hospital Cost Containment Act of 1981, are shown in Tables III and IV.

(1) In the field of medical-technical aids and of dentures, a greater

TABLE II
Scheme of ambulatory medical care delivery under health insurance

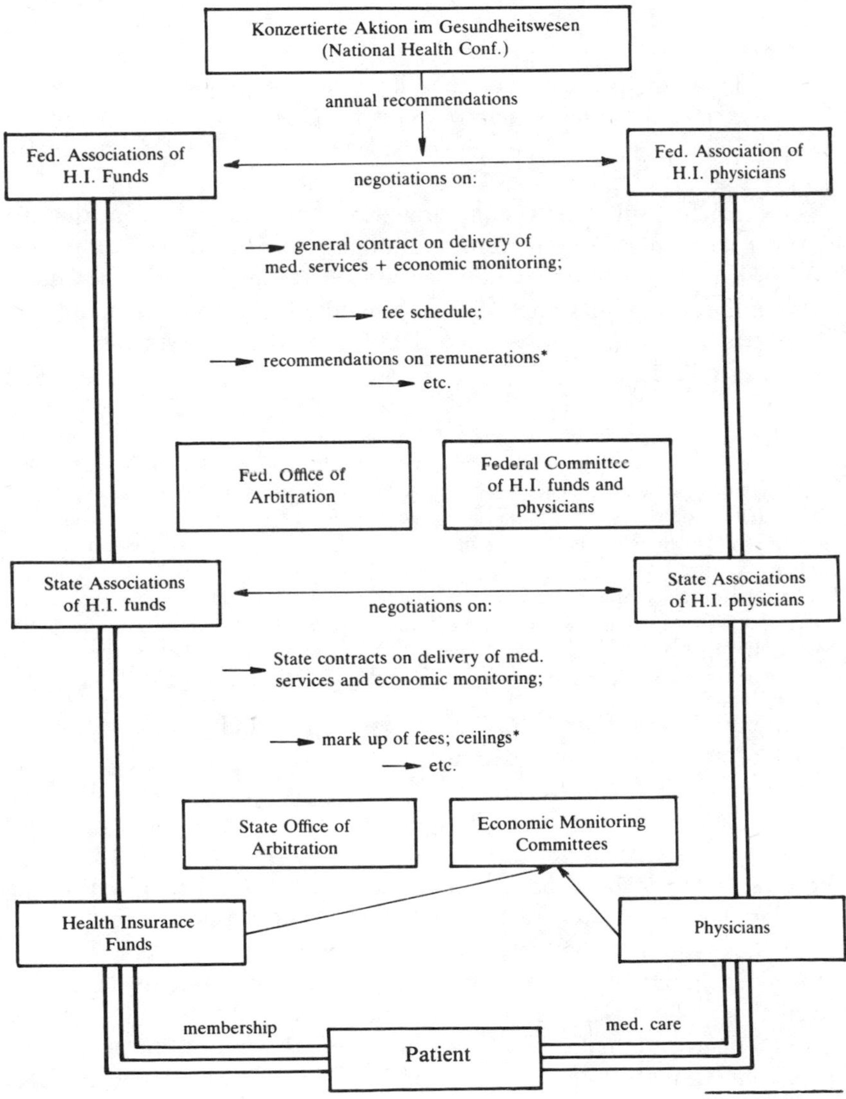


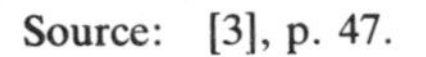
Source: [3], p. 47.

efficiency is to be attained (for example, by changing benefits as regards dentures).

(2) Price competition is to be enhanced among suppliers of dental services of a technical nature, of medical-technical aids, and of spectacles.

(3) Finally, the process of eliminating unnecessary and uneconomical benefits (started by the earliers laws) is to be strengthened; this provision pertains particularly to pharmaceuticals, aids, and treatment in health spas.

Undoubtedly, the laws summarized above contain a considerable array of instruments of intervention, ranging from health care budgeting and pricing mechanisms to intervention on resources and innovation, both at the macro- and micro-levels. Nonetheless, a growing number of observers concludes that the potential of these regulatory measures has not been fulfilled:

In view of the general rejection by the medical practitioners, the competition for members prevailing between health insurance funds, and so on, the budget limits envisioned by law were not implemented in practice (as was the case with pharmaceuticals). This experience shows that the presence of regulatory instruments in itself does not suffice, as long as the political will for their use is not present, and as long as the government is not able (or willing) to assert itself.
– Furthermore, the long-run increase in the number of doctors, dentists, pharmacists, and medical-technical assistants was not curbed sufficiently. Since supply is able, in the case of the health care system, to create its own demand, the long-run failure (of the cost containment effort) is thus preordained ([8], p. 151).

But perhaps we should also look at the increase in health personnel in terms of its employment creation effects.

In my view, the final judgment about the effectiveness of these measures to curb the *overall* expenditures of the Statutory Health Insurance System has yet to be made. Indeed, it is possible that – for other reasons, such as the decline in reported morbidity noted during the past few years because of the impact of the economic recession – overall expenditures will stagnate, while the degree of effectiveness of the system in terms of broader medical goals remains at a low level. But even this possibility appears to be remote. More likely than not, the present degree of medical ineffectiveness is likely to continue even while cost containment efforts run afoul simply because of a waning of the political will. Against these prospects, the alternative of fiscal consolidation *with* structural reforms appears to hold considerable appeal.

TABLE III
Strategies of health care cost containment and the H.I. cost containment act of 1977

| Strategic approaches via . . . | Provisions of act |
|---|---|
| I. *Health care cost budgeting* (macro-intervention) | |
| (a) overall | (a) – National Health Conference issuing recommendations or guidelines (not binding) |
| (b) sectoral | (b) – annual negotiations in advance on increase of *totals* of remunerations of *physicians* and of *dentists*, taking into account expected growth of general incomes etc.; annual negotiations in advance on maximum increase of volume of prescriptions |
| II. *Utilization and pricing mechanisms* (micro-intervention) | |
| (a) target actors: physicians and dentists | (a) – uniform structure of fee schedule (mark ups still variable) – parity on economic monitoring committees, *casting vote alternating* – information on prices and effectiveness of drugs |
| (b) target actors: consumers | (b) – 1 DM per item of prescription – exclusion of drugs for minor illnesses – 20% of costs of dentures (co-payment) – more information on costs of care |
| III. *Org. structure of health care system* (organizational intervention) | – limited outpatient privileges for hospitals – more staff privileges for office-based physicians – obligation of H.I.-physician associations to provide for efficient medical-technical services |
| IV. *No. of providers and facilities* (intervention on resources) | |
| V. *Advance of medical technology* (intervention on innovation) | |

Source: [3], p. 56.

TABLE IV
Strategies (and instruments) of the hospital cost containment act of 1981

| Strategic approaches via . . . | Provisions of the act (instruments) |
|---|---|
| I. Macro-interventions | |
| (a) global | (a) – Inclusion of hospitals in the National Health Conference<br>– Specific guidelines for hospital requirement plans<br>– More active participation of H.I. funds and hospitals in the process of planning hospital requirements, taking the form of a close cooperation with the State authorities |
| (b) sectoral | (b) – Classification of hospitals into 4 groups based on the services rendered |
| II. Micro-interventions | |
| (a) to influence quantities | (a) – Consultation among hospitals and private practitioners regarding the purchase and utilization of large-scale medical equipment<br>– Economic utilization of large-scale equipment<br>– Assistance in reducing unnecessary hospital beds, by granting rights for compensatory payment, etc. |
| (b) to influence prices | (b) – Agreement on per diems negotiated between H.I. funds and hospitals (in lieu of their being fixed by States)<br>– General orientation of per diems in terms of future time periods |
| (c) to influence quantities and prices | (c) – Obligation of the German Hospital Association (Deutsche Krankenhausgesellschaft) and the association of H.I. funds to arrive at joint recommendations as regards<br>– the principle of incurred costs coverage ("Selbstkostendeckungsprinzip")<br>– the principles of efficient and effecttive resource use<br>– Formulation of contracts between the State H.I. funds and the hospitals to ensure efficient hospital care<br>– Institution of economic monitoring committees through the State H.I. funds and the German Hospital Association |

Source: [8], p. 177.

## IV. THE ALTERNATIVE: FISCAL CONSOLIDATION WITH STRUCTURAL REFORMS?

### *Regulatory versus Market-Type Strategies?*

Particularly in the context of political discussions, it is often argued that two fundamental types of antithetical strategies of structural reform may be implemented: a regulatory *vs.* a market-type strategy.

A closer examination will reveal, however, that the so-called market-type strategy requires for its implementation a host of regulatory measures, including administered prices and the requirement to grant insurance protection to lower income groups. Thus, the distinction between the two strategies becomes a matter of degree rather than of kind; both entail major interventions into ongoing processes of the supply, utilization, and financing of health services.

They differ, however, in terms of the groups on whom the burden of reform is to be placed:

(1) The market-type strategies focussing on the "demand" side, that is, the patients, attempt to curb "excessive" utilization by instituting coinsurance and other price-equivalent instruments.

Its supporters argue as follows:

(a) The shifting of costs to consumers relieves the fiscal pressure experienced by health insurance funds.

(b) By instituting variable coinsurance rates, the structure of demand for health services can be influenced, and priorities in health policy can be implemented.

(c) Price elements tend to curb excessive demand. Indirectly, they also influence the behavior of suppliers of health services, and thus they tend to curb the rise of health expenditures.

The critics of coinsurance, however, point to the following:

(a) Coinsurance is generally ineffective in curbing demand. In those cases where it can be made to be effective, its consequences are undesirable on the grounds of social policy: they place burdens particularly on the lower income groups which often lead to an underutilization and postponement of necessary treatment.

(b) The curtailment of demand via price-equivalent instruments is incompatible with the basic philosophy of the Statutory Health Insurance concept: to supply medical services solely on the basis of need, and to finance the system solely on the basis of ability to pay (that is, on the basis of the insured persons' income).

There is little doubt that coinsurance (and other types of pricing schemes) places the burden of structural reform mainly on the shoulders of patients rather than on the insured as a whole. This applies in particular to those types of goods and services which have been excluded (not been included) in the catalogue of benefits covered by the health insurance funds: in their case, the patient experiences a 100 percent coinsurance.

(2) Market-type strategies have also been advocated for the *supply-side*. They include a strengthening of competition, not only among health insurance funds, with the aim of reducing inefficiencies and unnecessary costs, but also among suppliers of pharmaceuticals and medical-technical aids. Thus far, the competition for membership among statutory health insurance funds has tended to increase rather than decrease costs since competition focussed on the granting of benefits rather than on the reduction of contributions. Knowledgeable observers of F.R.G.'s health services system, however, doubt that anything like the health maintenance organizations (HMOs) found in the U.S.A. is likely to be instituted in the F.R.G. They pose several questions about the HMOs:

(a) Have they utilized the taxpayers' money (especially Medicaid funds) more efficiently than other types of institutions?

(b) Have they attempted to pick out the "good risks" among potential members?

(c) Has the quality of treatment been equivalent to other types of providers ([17], pp. 401–420)?

Furthermore, has the Enthoven Plan been able to avoid some of the shortcomings of the earlier HMOs ([17], pp. 321–323)?

Apart from the limitations to competition among health insurance funds imposed by present institutional arrangements, there appears to be ample scope for enhancing competition among suppliers of pharmaceuticals and medical-technical aids (I shall return to this theme later).

(3) Regulatory strategies consist of budget limitations, rationing of health goods, administered prices, and so on. Undoubtedly, the measures of cost containment enacted thus far contain a host of such regulatory measures. As was pointed out before, they have not been as effective as expected, simply because the various measures were not enforced sufficiently. There is no doubt, however, that a strategy of budget limitation is most successful in curbing expenditure growth, as was exemplified convincingly by the experience of the U.K. and Canada. However, these very examples indicate that budget limitations generally lead to rationing of health services by individual medical practitioners.

Budgetary regulation alone, however, is no guarantor of structural reform. For the latter, a more conscious effort has to be made to influence not only the volume of expenditures and thus the input side, but rather, the very composition of the output should change simultaneously to enhance the efficiency and effectiveness of the system. At the same time, such a policy must place the burden of reform in a more balanced fashion upon suppliers and users of health resources.

*Some Perspectives on a Policy of Fiscal Consolidation with Structural Reforms* [16]

Any attempt at fiscal consolidation must grapple with two fundamental questions:

(a) Are health services being offered at an excessive cost?
(b) Are too many health services being offered?

The first question aims at the price and the second at the quantity of health services. Structural reforms of the existing system of provision must focus on both dimensions.

*Structural Reforms of Price Systems*

Consumers generally have a very weak position *vis-à-vis* suppliers. Thus, competition among private sector suppliers is often severely curtailed (as is exemplified by F.R.G.'s pharmaceutical industry), leading to excessive prices and incomes.

Similar effects occur also for health goods supplied by the public sector. Because of structural elements present, for example, in the form of an extreme separation between the ambulatory and inpatient sectors of health care, inefficiencies occur on the supply side which lead to higher costs.

Indeed, some observers estimate that structural changes can yield substantial savings even without reducing the effectiveness of the health system: "Schäfer estimates that 30 to 40% of total expenses can be saved without reducing benefits. Jahn estimates potential savings to amount to over DM 10 billion in the inpatient sector alone which amounts to about one third of present outlays for the hospital sector. Other experts do not quantify the potential savings, but their qualitative judgment of the situation and their long list of proposed reforms indicate that they envisage substantial magnitudes (of savings)" ([20], p. 20).

*Structural Reform of the Quantity Dimension*

If one proceeds from the assumption that (with some notable exceptions in the genetic field) the main determinants of public health are generally found in the living conditions and in the environment, then the main emphasis of health policy must be placed on controlling working and living conditions: "I believe that for most types of illnesses, their prevention by combating their causes is less costly, more humane, and more effective than the curative treatment after their occurrence" ([5], p. 11).

Unfortunately, the information required for such a conception of health policy is thus far severely lacking in the F.R.G. What is needed, for example, is detailed information on costs and benefits by type of illness and the evaluation of medical services, especially of diagnostic and therapeutic processes.

A policy of fiscal consolidation with structural reforms can focus on several dimensions:

*Savings from more efficient resource use* result if and when (1) the control shortcomings of the public sector are reduced, leading to a more efficient transformation of inputs into health outputs; and (2) the functioning of private markets is improved. As was pointed out before, the potential savings are enormous.

*Savings from curtailing the excessive provision of curative services* focus solely on the output side of the health system. They result if and when benefits and services are eliminated which either do not improve the health status of the population or which even lead to its deterioration. There is a considerable list of such ineffective treatments:

(a) The treatment of *coronary diseases* in hospitals, especially in intensive care units is generally less effective than home treatment, prevention, and timely counseling of prospective victims.

(b) The antibiotic therapy of *chronic bronchitis* is often superfluous, partly even damaging, since the infection of the bronchial system is generally only a secondary phenomenon; the primary determinants are found rather in the form of chronic damage due to chemical causes such as smoking, air pollution, and other factors.

(c) The antibiotic therapy of *interstitial nephritis* is partly superfluous, partly even damaging, since the presence of bacteria in the urinary tract plays only a minor role for the development of chronic renal insufficiency.

(d) The excessive consumption of *pharmaceuticals* is partly superfluous when judged by their effectiveness, partly even damaging, when side effects are considered; cases in point are the risks of an "adverse episode in the hospital" of which about one half are due to treatment with pharmaceuticals, and of "adverse reactions" to medicines. (In the F.R.G., 20 of the 50 most widely sold pharmaceutical drugs were found to be of doubtful effectiveness – placebo effects, only partially effective drugs, combination drugs of limited effectiveness. Cases in point are oral hypoglycemic agents for the treatment of diabetes; the improper treatment of liver disease; drugs for treatment of heart conditions which were found to be largely ineffective; the adverse effect of multiple prescriptions on patient compliance; drugs for combating excessively high cholesterol levels which are ineffective; excessive prescription of Vitamin $B_{12}$, half of which is adjudged to be ineffective; the use of pharmaceuticals for dieting; and the use of tranquilizers, fever-reducing drugs, etc. which may be ineffective.
drugs, etc. which may be ineffective.

The potential savings from more effective use of pharmaceuticals have been estimated for the F.R.G. to amount to about DM 8 billion (in 1982), of which about DM 5.6 billion pertain to publicly financed pharmaceuticals. Furthermore, it is estimated that one-seventh of all hospital days result from adverse effects of medicines; the potential savings would amount to an additional DM 4 billion per year (in the F.R.G.).

*Hospital infections*, particularly infections of the urinary tract, of wounds, bronchial passages, and sepsis often result from the excessive stay in hospitals and from excessive consumption of antibiotics. If the U.S. experience (between 2% and 15% of patients, $1 to $10 billion) applies to the F.R.G., a corresponding reduction of the duration of stay in hospitals could lead to potential savings between DM 2 and 2.5 billion per year.

*Savings by enhancing preventive measures*

(a) The failure to treat *hypertension* results in about 200,000 cases of work disability in the F.R.G., or to 5 million work days lost. The causes are found in inadequate counseling, care, and control by the medical institutions, as well as the lack of discipline of some patients.

(b) *Inadequate prophylactic* care leads to dental problems; for example, fluoridation of drinking water would lead to a substantial reduction of caries and to potential savings of DM 4 to 7 billion per year in the public health insurance system alone!

(c) Inadequate prevention of accidents also causes substantial loss of days worked, deaths, and disability. "The greatest uncontrolled epidemic," namely traffic accidents, leads to the "almost largest group of illnesses." Based on the experience of Austria, the introduction of speed limits on the Autobahn and other roads would lead to a reduction of traffic accidents by 49% and of deaths due to traffic accidents by about 80%, yielding potential savings between DM 1.5 and 3 billion per annum (in the F.R.G.).

A more forceful program of primary prevention thus could lead to substantial savings in health expenditures.

*Filling Gaps in Curative Medicine without Increasing Costs*

It is very likely that the high degree of professionalization of medicine and existing systems of remuneration lead to a situation where too little information, counseling, and individual care of patients are combined with too little control of the therapeutic process.

It is difficult to estimate the savings potential from filling existing gaps in curative medicine. It is highly probable, however, that the added costs resulting from improved information, counseling, and individual care are likely to be more than compensated by savings in costs, as is exemplified clearly by the case of hypertension.

We may conclude this broad overview of potential savings from structural reforms (see Table V for a summary), by pointing to the necessary changes in the behavior of suppliers and consumers of health services, as well as of their environment, which are a prerequisite for realizing such savings. Indeed, such changes are likely to be forthcoming only if and when systems and instruments of regulation are created by law which would help to steer human behavior in the direction necessary for a more rational health system.

## IV. CONCLUSION

An effective curb to the further increase of health insurance expenditures must combine several measures encompassing, for example,

(a) obligatory health insurance budgeting;
(b) limiting the number of health insurance physicians;
(c) eliminating the privileged status of special health insurance funds;

TABLE V
Experts' Estimates of potential savings in the statutory H.I. system

| Potential savings | | Necessary measures/assumptions |
|---|---|---|
| DM | year | |
| 5.6 billions | 1982 | Substitution of pharmaceuticals through less expensive products of equal or greater effectiveness |
| 1.5 billions | 1978 | Increased deductible for drugs from the present (approx.) 5 to 15 percent |
| 6 billions | 1982 | Limiting the excessive use of medicines |
| 375 millions (DM 150,000 per patient) | 1981/82 | Treatment of hemophiliac-patients as in the USA (cheaper, equally effective prescriptions) |
| 10 billions | 1982 | Savings in the inpatient sector |
| 2 billions | 1981/82 | Removal of trade barriers in the case of pharmaceuticals |
| 25 to 34 billions (= 30–40% of the total) | 1980 | Reducing technical-diagnostic measures, control of illness, less double diagnostics, patient's own payment for trivial illness, less use of hospitals |
| 1.5 to 3 billions | 1981/82 | Reducing the risk of accidents by lowering the speed limits |
| 4–7 billions | 1980 | Fluoridation of drinking water |
| 2 to 2.5 billions | 1982 | Reducing the risk of hospital infections by half by reducing the duration of stay |
| 4 billions | 1982 | Reducing the harmful side-effects of pharmaceuticals by half by reducing the length of inpatient care |

Source: [16].

(d) improving the information of patients regarding the expenditures charged by individual health insurance physicians for individual treatments;

(e) improving the control by health insurance funds of the types of treatment prescribed by the individual health insurance physician;

(f) analyzing the expenses charged by individual health insurance physicians more thoroughly and critically.

In short, since the major cause of increased health insurance expenditures is found in unnecessary, i.e., ineffective, medicines and treatments, fiscal consolidation in general, and cost containment in particular, must be combined with structural reforms aimed at increasing the efficiency and effectiveness of the whole system. At the same time, such reforms cannot simply consist of the shifting of burdens to the shoulders of patients: the health service suppliers, and the larger community of the insured, must carry an overwhelming share of the burden of structural reforms.

After all, the social quality of measures of structural reform should be adjudged by the burdens they place on the shoulders of the strong rather than on the backs of the weak.

*Department of Economics,*
*Universität Augsburg*
*and*
*International Institute for*
*Empirical Social Economics, Augsburg*

*References Pertaining to Table v:*

*(ad 1):*

Abholz, H. -H.: 1980, 'Welche Bedeutung hat die Medizin für die Gesundheit?', in Deppe, H. -U. (ed.), *Vernachlässigte Gesundheit.* Zum Verhältnis von Gesundheit, Staat, Gesellschaft in der Bundesrepublik Deutschland, Köln, p. 29.

Bauer, J. and Lehman, H.: 1981, 'Zur Entstehung, Behandlung und Rehabilitation von Herzinfarkt', in Badura, B. (ed.), *Soziale Unterstützung und chronische Krankheit.* Zum Stand sozialepidemiologischer Forschung, Frankfurt/Main, pp. 185–257 (esp. pp. 185, 191, 194).

Losse, H. and Loew, H.: 1977, 'Chronische Pyelonephritis – Wandel eines Krankheitsbegriffes', in *Medizinische Kritik* 72, pp. 1610–1615.

Systemanalyse 1978/2: 1978, 'Systemanalyse des Gesundheitswesens in Österreich', Eine Studie über Entstehung und Bewältigung von Krankheit im entwickelten Kapitalismus,

Vol. 2, Institut für Höhere Studien und Wissenschaftliche Forschung, 2nd ed., Wien, p. 203.

*(ad 2):*

Geißler, U.: 1981, *Statement V: Probleme der Mittelumverteilung*, Symposium 'Pflegebedürftige in der Bundesrepublik Deutschland', Bad Boll, p. 5.

*(ad 3):*

Grönert, J.: 1982, 'Die Lücke im System der sozialen Sicherung: Ungenügende Absicherung der Pflegebedürftigkeit', in *Sozialer Fortschritt*, 8, pp. 186–188 (esp. p. 187).

*(ad 4):*

Jahn, E.: 1981, 'Expertise über die Möglichkeiten einer Besserung des Gesundheitsstandes und Hebung des Leistungsniveaus der sozialen Krankenversicherung in der Bundesrepublik Deutschland nach Verlust des finanziellen Spielraumes der sozialen Krankenversicherung', in Wissenschaftliches Institut der Ortskrankenkassen (ed.), *Leistungssteigerungen im Gesundheitswesen bei Nullwachstum*, WIdO-Materialen 14, Bonn, pp. 57–87 (esp. p. 78).

*(ad 5):*

Reher, R.: 1981, *Arzneimittelimporte, Möglichkeiten und Wettbewerbsbelebung auf dem Arzneimittelmarkt*, Bundesverband der Ortskrankenkassen (ed.), p. 6.

*(ad 6):*

Schaefer, H. (no title): 1981, in: Wissenschaftliches Institut der Ortskrankenkassen (ed.), *Leistungssteigerungen im Gesundheitswesen bei Nullwachstum*, WIdO-Materialien 14, Bonn, pp. 47–56 (esp. p. 51).

*(ad 7):*

Bierau, D.: 1983, 'Straßenverkehrsunfälle 1982', in *Wirtschaft und Statistik*, 3/83, pp. 247–250.

Kaprio, L. A.: 19, 'Gesundheitsprobleme in Industriegesellschaften', in: *Aufgabe Zukunft, Qualität des Lebens*, Vol. V, Frankfurt/Main, p. 17.

Schaefer, H. and Blohmke, M.: 1972, *Sozialmedizin*, Stuttgart, p. 141.

Systemanalyse 1978/1: 1978, 'Systemanalyse des Gesundheitswesens in Österreich', Vol. 1, Institut für Höhere Studien und Wissenschaftliche Forschung, Wien, p. 355.

Systemanalyse 1978/2: 1978, 'Systemanalyse des Gesundheitswesens in Österreich', Vol. 2, Institut für Höhere Studien und Wissenschaftliche Forschung, 2nd ed., Wien, p. 354.

Unfallverhütungsbericht: 1982, Bundestagsdrucksache, pp. 63.

*(ad 8):*

Büttner, W. *et al.*: 1979, 'Zahnmedizinische Prophylaxe (Epidemiologie und Ätiologie der Karies, Kollektive und individuelle Prophylaxemöglichkeiten)'. Wissenschaftliche Kommission "Zahnmedizinische Prophylaxe", WIdO-Schriftenreihe 4, Bonn.

Schuchart, F. and Kübler, G.: 1983, 'Zwei Jahre Karies-Prophylaxe, Modellvorhaben im Bonifatius Kindergarten, Lörrach', in *Die Krankenversicherung*, Vol. 35, pp. 80–84.

Schicke, R. K.: 1978, *Präventiv-Medizin: Kosten und Nutzen*, Hartmannbund, Bonn. p. 18.

*(ad 9):*

Köpcke, W. et al.: 1976, 'Maßnahmen gegen Hospitalinfektionen', in *Medizinische Klinik* 71, pp. 1057–1061.

Müller, H., Sepsis: 1974, in: *Medizinische Klinik* 69, pp. 1632–1633.

## BIBLIOGRAPHY


1. De Wachter, M. A. M.: 1988, 'Public Policy and Moral Issues: The Netherlands', in this volume, pp. 97–116.
2. Federal Republic of Germany, German Bundestag: 1982, *Report of the Federal Government according to Article 2 §6 of the Health Insurance Cost Containment Act*, 9th Electoral Period, Printed Matter 9/1300 of 2 February 1982.
3. Geissler, U.: 1980, 'Health Care Cost Containment in the Federal Republic of Germany', in Brandt, A., Horrisberger, B., and von Wartburg, W. P. (eds.), *Cost-Sharing in Health Care*, Springer Publishing Co., Berlin, Heidelberg, New York.
4. Maxwell, R. J.: 1981, *Health and Wealth: An International Study on Health Care Spending*, Lexington, Kentucky, Toronto, Canada.
5. McKeown, T.: 1982, *Die Bedeutung der Medizin: Traum, Trugbild oder Nemesis?*, Frankfurt/Main, F.R.G.
6. Müller, W.: 1983, 'Ausgaben für Gesundheit 1981', in *Wirtschaft und Statistik*, pp. 726–731.
7. Pfaff, M.: 1983, 'Finanzierungs- und Ausgabenströme im Gesundheitssektor: Eine realtypische Betrachtung fiskalischer Systeme', in Häuser, K. (ed.), *Finanzsysteme. Ideal- und Realtypen - Gesundheitswesen und Hochschulbildung*, Duncker & Humbolt, Berlin, pp. 57–173.
8. Pfaff, M.: 1983, 'Steuerungsinstrumente in der GKV – Das Beispiel der Kostensteuerung', in Wissenschaftliches Institut der Ortskrankenkassen (ed.), *Strukturfragen im Gesundheitswesen in der Bundesrepublik Deutschland*, WIdO-Materialien No. 21, WIdO, Bonn, pp. 133–179.
9. Pfaff, M. and Asam, W.: 1978, 'Distributive Effects of Real Transfers via Public Infrastructure: Conceptual Problems and Some Empirical Results', in Griliches, Z., Krelle, W., Krupp, H.-J., and Kyn, O. (eds.), *Income Distribution and Economic Inequality*, Frankfurt/Main, F.R.G., New York, Toronto, Canada, Chichester, England, pp. 66–96.
10. Pfaff, M. and Bäuerle, E. U.: 1979, *Konjunktur und Ausgabenentwicklung im Gesundheitswesen*. Eine Analyse des Einflusses konjunktureller Kräfte auf das Ausgabenwachstum in der GKV im Zeitraum 1965–1977, Workshop of the Hanns-Seidel-Stiftung: 'Kostendämpfungsgesetz im Lichte der Konjunkturpolitik der Bundesregierung', Gauting 13.10.1979 (Unpublished Working Paper).
11. Pfaff, M., Schneider, M., Kerschreiter, M., and Stumpf, M.: 1981, *Wahltarife in der Krankenversicherung*, Bonn, F.R.G.
12. Pfaff, M., Schneider, M., and Stumpf, M.: 1980, 'Objectives and Forms of Cost Sharing', in: Brandt, A., Horrisberger, B., and Wartburg, W. P. von (eds.), *Cost Sharing in Health Care*, Springer Publishing Co. Berlin, Heidelberg, New York, pp. 117–129.
13. Pfaff, M. and Schneider, M.: 1980, 'Die wirtschaftlichen Auswirkungen der Judikatur im Bereich der Kranken- und Rentenversicherung', *Vierteljahresschrift für Socialrecht* VIII, No. 1, pp. 17–28.

14. Pfaff, M. and Schneider, M.: 1980, 'Unterscheiden sich beitragsgedeckte und nicht-beitragsgedeckte Systeme Sozialer Sicherung hinsichtlich ihrer ökonomischen Voraussetzungen und Wirkungen, insbesondere auch im Hinblick auf die Umverteilung?', in: Zacher, H. F. (ed.), *Die Rolle des Beitrags in der Sozialen Sicherung*, Berlin, pp. 391–423.
15. Pfaff, M. and Bäuerle, E. U.: 1981, 'The Utilization of Public Health Services by Socio-Economic Groups: The Case of the Federal Republic of Germany,' in: Molt, W., Hartmann, H., and Springer, P. (eds.), *Advances in Economic Psychology*, Edition Meyn,.
16. Pfaff, M. *et al*: 1984, 'Ausgewogene Absicherung von Gesundheitsrisiken,' *WIdO-Schriftenreihe 7*, AOK Wissenschaftliches Institut der Ortskrankenkassen, Bonn, pp. 15–176.
17. Sommer, J. H.: 1980, *Kostenkontrolle im Gesundheitswesen: Die nordamerikanischen Erfahrungen mit staatlich regulierenden und wettbewerblich orientierten Ansätzen*, CH-8253, Diessenhofen, pp. 401–420.
18. Thust, W.: 1988, Political-Medical Allocations, *in this volume*, pp. 255–265.
19. Wissenschaftliches Institut der Ortskrankenkassen (ed.): 1980, *Leistung und Finanzierung des Gesundheitswesens in den 80er Jahren: Fragen und Antworten zur künftigen Entwicklung*, WIdO-Materialien No. 8, 3rd ed., Bonn, F.R.G.
20. Wissenschaftliches Institut der Ortskrankenkassen (ed.): 1981, *Leistungssteigerungen im Gesundheitswesen bei Nullwachstum*, WIdO-Materialien No. 14, Bonn.

BETTINA SCHÖNE-SEIFERT

# MEDICAL MICRO-ALLOCATION: IS AND OUGHT

In this paper I will argue that current public concern about the cost-explosion in health care offers a unique chance of reform: a public understanding of the necessity of rationing[1] might be induced as the basis for a transparent *ex ante* pattern of resource allocation for medical services.

## I. INTRODUCTION

Any pluralistic society would subscribe to the ideal of patients autonomously controlling both the quantity and quality of their personal health care[2] out of the full array of biomedicine's technical feasibilities. The various European and American deviations from this medical utopia are about to be further aggravated by increasing financial constraints. Theoretically, medicine with its ever growing provision of diagnostic and therapeutic possibilities is a bottomless pit; and it is simply a matter of prudence to find a stopping point somewhere. The normative question, however, where exactly to set this limit is not deducible from conceptual analysis of health and disease, themselves being value laden.[3]

Although as old as medicine itself, the problem of micro-allocation (which medical care is given to a particular patient?), as the centerpiece of distributive decisions, has only recently gained burning social interest: Facing drastically increasing medical expenditures, most industrialized countries have come to depict the scenario as one of imminent scarcity of medical resources. This has triggered a public debate about the appropriate level of health care, access to which society is obliged to provide to its members.[4]

The first question in this debate is whether indeed society has some such obligation (or, correspondingly, whether individuals have a "right" to health care),[5] rather than viewing provision of care to the poor as a supererogatory act of charity by the wealthier.[6] Although Western welfare states have traditionally rejected the latter libertarian view in practice, it is only recently that the underlying values have been

*Hans-Martin Sass and Robert U. Massey (eds.), Health Care Systems*, pp. 293–311.

subjected to consistent ethical analysis and justification. Though not *per se* a normative justification, the factual acknowledgement of some "right" to health care makes it legitimate for me to assume its importance for future health policy design, and to postpone its justification until the end of this paper.

The term "scarcity," despite its frequent use as a value-free description, presupposes a decision as to medicine's monetary worth, which itself cannot be a freestanding determination, but requires an evaluation of output and of quality.[7] Only transparency of both input and output data can provide the basis for any reasonable health care design under non-utopian conditions.

Balancing the (medical) output against its (financial)[8] input should take place in two steps, asking first for efficiency and second for quality. Obviously, the evaluation of quality is contingent on society's level of wealth and technological know-how. Also, without detailed knowledge about both, a health care system's efficiency and the marginal costs or savings of output changes, any decision to freeze, decrease, or increase its input would be arbitrary with regard to a qualitative evaluation.[9]

Along these lines, it is easy to convict Western health care designers, be they German, Dutch, Swedish, or American, of some arbitrariness; the evaluative centerpiece, i.e., quantification and qualification of medical output data, is imperfect at best. None of our health care systems is consistently based on sound data on the effectiveness of diagnostic and therapeutic procedures, on quantification and comparison of degrees and probabilities of well-being before and after medical interventions, on calculations of false negative and false positive test results, which requires burdensome and meticulous work, and countless controlled clinical trials. How then did our various democratic societies get along without being accused of unjustifiable arbitrariness?

Three possible circumventions to a transparent evaluation based on cost/benefit analysis can be identified:

(1) Pretending indefinite affluence, thus allegedly guaranteeing anybody's access to whatever medical care might be "beneficial." This policy avoids admission that normative decisions are indeed being made on various allocative levels.

(2) Reducing the input/output balance to its efficiency aspect; hence claiming that intended or performed cut-backs of health care expenditures are totally independent of normative output evaluations. How-

ever, even this claim implies the desirability of the current output's *status quo* (which is promised to be left uncompromised).

(3) Denying that a normative consensus on input/output *desiderata* is of any relevance (because there is no "right" to health care), or could ever be rationally achieved, thus leaving their regulation to the ostensibly fair operation of a free market system.

Crudely simplifying the matter, it might be argued that Germany like other Western welfare states has already adoped policy as described in paragraph (1) and is now, under the public cost-containment debate, coming to adopt more and more elements of paragraph (2). Here, where a compulsory health insurance system has provided a high degree of equality of access to medical care according to need, the ethical problem of justly allocating medical resources thus becomes itself a searching for the "decent maximum" that society is obliged to provide.

The United States, on the other hand, with a tradition of a largely libertarian policy (3), has come to be concerned about the corresponding "decent minimum." Here the predominant free market type of health policy has left some 30 million Americans uninsured, for most of whom access even to the minimum of minimal medical care is not guaranteed [27]. Under the pressure of growing costs, Americans have also come to focus on the efficiency aspect, adopting a policy similar to paragraph (2).

All industrialized countries have come to face the same problem of a normative evaluation of medical care in order to design a publicly acceptable allocation system, whether in search of a decent maximum or minimum. These might well turn out to be very different, reflecting differences of values, but they might also converge into one; even then implications for practical policy will be different. However, much time and effort can be saved by realizing that both the ethical problem of justifying a pattern of allocation (and that will mean of rationing) of medical care, and the practical problem of adequate policy design cross the international boundaries of health policy.

## II. CONCEPTUAL CONSIDERATIONS

In order to analyze the complex of allocation procedures it is helpful to adduce Engelhardt's ([18], p. 10) four-level model of allocative decisions:

(1) *High-level macro-allocation*: regarding the total amount of health care expenditures as a share of the gross national product.

(2) *Low-level macro-allocation*: determining the pattern by which the overall expenditure is distributed into medicine's various domains (i.e., preventive medicine, health education, curative, palliative care, etc.).

(3) *High-level micro-allocation*: determining the criteria for the provision of resources to patients.

(4) *Low-level micro-allocation*: regarding an actual patient's reception of resources.

An allocative system is not characterized by the synopsis of decisions made on these four levels, which are interdependent, but rather by the relative weight given to each of them. It is this "meta-level" of allocative decisions that most obviously invites ethical scrutiny.

High-level macro-allocation expresses society's estimation of medicine in comparison with other social goods, provided it is based on a representative evaluation. Theoretically, there are two ways empirically to determine the value society ascribes to medicine by aggregating individual preferences (consumer choices). One is to ask for their willingness to pay, but even in its merely hypothetical form this question, when taken as a measure of medicine's worth, is incompatible with the idea of a right to health care independent of wealth. As has been stressed by proponents of egalitarianism [41], answers to this question would largely reflect the *status quo* of income distribution. Rejecting this measure does not, however, necessarily imply rejection of inequality of incomes, but it affirms the latter's irrelevance for the question at stake.

In accordance with this position, West Germany has based its health cost distribution on the "solidarity principle"[10] instead of free market regulation, thus building an egalitarian island in an otherwise non-egalitarian society. For over 90 percent[11] of the population medical care is almost completely covered[12] by one of the statutory sickness funds.

In such a system a "representative" macro-determination could, alternatively, be achieved by allocating to medicine, *a posteriori*, whatever the aggregate consumer demand. Hence, this "meta-allocative" decision would permit level (4), i.e., allocation of resources to individual patients, to determine by a feed-back mechanism level (3), which in turn would determine levels (2) and, ultimately, (1).

As an alternative to these empirical determinations, macro-allocative decisions could representatively be made *a priori*, thereby reversing the direction of sequential interdependence; level (4) allocations would be determined by overriding distributive decisions. Since, however, any non-arbitrary macro-allocative determination is to be output oriented, level (1) decisions cannot rationally be made without feedback from levels (2) and (3). It turns out then, that, using the above classification, the crucial "meta-allocative" decision is not as to the dominance of macro- (1; 2) or micro- (3; 4) decisions, but to the dominance of the anonymous abstract (1; 2; 3) or the direct personal (4) levels ([20], pp. 5–6).

## III. GERMAN HEALTH CARE ALLOCATION: THE STATUS QUO

With some qualification, e.g., some quantity restrictions in ambulatory care, or in the size of screening programs by the sickness funds, the German health care system is characterized by the feedback allocative dominance of level (4). Certain of retrospective financial coverage for "medically required" interventions by the sickness funds, individual physicians function as allocators vis-à-vis their individual patients, who, however, differ from ideal consumers in ways important enough to undermine the analogy. Having no personal economic interests to restrict expenses caused by their medical care, patients do not balance values on their individual level, but rather tend to exploit the health care system. Moreover, the indirect personal interest to unburden the pool and thereby decrease premiums to the sickness funds is outweighed by the lack of assurance that other patients would do likewise. A further deviation from the ideal consumer analogy is caused by physicians' ability to induce demand for medical services, both by taking advantage of their patients' medical ignorance and by shaping the concept of "medical benefit."[13] To be sure, to some extent patients limit the medical intervention they are willing to undergo because of the costs of time and physical inconvenience. But many expensive medical procedures do not add substantial amounts of time and discomfort costs.

It is therefore not a normative, but a logical conclusion that the aggregate claim on health care services under the current German conditions cannot be taken as an indicator of collective evaluation of medical services' worth.

Usually it is physicians who label an intervention as "benefitting" on the level (4) of direct personal allocation. It is they who largely distribute resources to the areas of diagnosis, therapy, and research, following complex behavioral patterns, compounded of moral intuitions, their understanding of the medical ethos, psychological reasons, and for varying degrees of medical and economic reasons. However, physicians do not openly take on the function of allocation and probably do not even realize that they are performing this task. Literature discussing medical micro-allocation frequently describes it as "tragic," triage-type, paralleling the who-is-to-stay-on-the-lifeboat paradigm (e.g., [10, 32]). The typical hypothetical situation is that of two patients competing for one intensive care bed or for one kidney dialysis machine. And whatever allocative principle (level (3)) is argued for, e.g., lottery; first come, first served; preference for younger age, the whole situation is described as unfortunate, estranging the physician from his true vocation, and due to immoral first order determination.

This factual description of physicians as unwilling allocators, corresponds to the preponderant view of the medical ethos: provide whatever service is potentially of medical benefit, regardless of third-party interests. However, within our open loop system the available remedy for such "tragic" bottlenecks relative to medical need consists in the provision of enough "lifeboats," rather than in convincing the public of the future acceptability of, e.g., lottery allocation. Also, in our system the triage type decision is a rare exception; to focus on it hides the normative character of any other medical microallocative decision (level (4)): Given the *a posteriori* determination of levels (1) through (3), ascribing "benefit" to a certain intervention for a certain patient has the normative implication that its output is worth being paid for by society. And obviously, this decision is not made by physicians *qua* medical experts.

In other health care systems physicians are often justified in accusing macro- or meta-allocations of leading to a lack of "lifeboats": the first, in closed loop systems, as in the U.K., with fixed health budgets that lead to shortcomings potentially for everybody; or the latter, in predominantly free market systems, as in the U.S., that lead to shortcomings for the nonaffluent.

It has been described ([20], p. 9) as an empirical phenomenon, that life-negating decisions on the macro-level are generally perceived as less moral-egregious than those on the micro-level. But in those systems, too, strong mechanisms prevail that tend to hide the life or health negating consequences; there is much evidence [1, 37] that British patients are regularly deceived about their medical "treatability" if they happen to fall outside the category (level (3)) for which treatment resources are available.

A similar effect is produced by what Bobitt and Calabresi ([9], p. 135) have coined as "first-order sufficiency paradox": by unlimited spending for those life-saving treatments that gain much publicity, society affirms its subscription of the pricelessness of human life, despite its rationing of life-saving resources in other areas.

Returning to the German system, office-based physicians who commonly have no hospital privileges are paid on a fee-for-service system, negotiated between the national associations of physicians and the sickness funds, which then pass the costs to the pool of their members. Obviously, the system invites not only patients, but physicians as well, to get out of it as much as possible. And indeed there is empirical evidence that quantity and pattern of medical services provided to a given population depend both on the number of available physicians and the size of the fee [31], thus supporting the "target income hypothesis." Nonetheless, physicians tend to deny the induction of demand, claiming that they provide only those services that are needed for medical or forensic reasons.

German hospital physicians, commonly working on fixed salaries, lack a personal economic interest in increasing medical services. But, much as their free-practicing colleagues, they have no incentive to cut expenses, either. Rather, because of a retrospective per diem reimbursement, hospitals have an economic disincentive to shorten the length of stay of their patients. As a result, German patients are hospitalized longer than in any other industrialized country, another example of the elasticity of "medical requirements".

Hospital costs, currently making up 33% of the health budget [28], are increasing disproportionately. A lion's share of costly diagnoses and treatments, like high technology procedures, chemotherapy, total parenteral nutrition, and terminal care is performed as inpatient care and often sets the standard for subsequent outpatient care.

In the ongoing debate various factors have been identified as cost intensive behavior of (hospital) physicians: ignorance of the costs of procedures, misdiagnosing, psychological difficulties to admit the non-treatability of a patient's disease, duplication of procedures (in part because of the separation between hospital and ambulatory care), use of the experimental therapies outside of controlled clinical studies, administering therapies known to be ineffective on demand of patients. Another crucial factor is the costs of misunderstood academic acknowledgements particularly sought for by younger physicians: proudly and avidly they pursue improbable but "exotic" and expensive differential diagnoses, not only out of ignorance, but so as to document their broad knowledge to colleagues and supervisors – a behavior that will rarely be sanctioned.

Avoidance of these factors could, it is frequently argued, at least postpone the need for rationing, of tragic quality-of-life assessments. Obviously, the argument implies the false premise of a simple dichotomy between *wasted* and *needed* interventions: the former have no beneficial effect for the patient and are to be eliminated and the latter are to be provided without question. However, medical interventions could theoretically be arranged along a continuum of decreasing benefits. And the crucial question is where on this slope to set the cut-off point between the provided and the not-provided (both of which differ only by their degree of benefit). Taking the above dichotomy literally, any diminishingly small potential benefit would qualify a medical intervention as "needed" – an impossible postulate. Operative is here once again the denial of normative elements in daily medical decision making.[14]

Standards of quality evaluation are commonly unreflected in the medical profession. They are implied in the informal protocols of hospitals ([29], pp. 36–37), they are based on personal intuition, they are compared against the objectives and results of customary medical care in a particular subsystem.[15] Choice of any other standard than the utopian one of technical feasibility means that some form of rationing takes place, that somehow probabilities of risks, degrees of disability, prices for life and limb are being accepted, commonly without much reflection, as constituting "good medical practice."

Consider as an example the question of whether screening for occult fecal blood as an indicator of colorectal cancer ought to be included in the mass screening program provided by the sickness funds, as it has indeed been since 1977. According to recent studies the procedure's sensitivity is below 50%, its positive predictive value about 5% [2]. Thus, less than half of all asymptomatic tumors are being identified, 95% of blood positive patients are unnecessarily worried and have to be subjected to further expensive diagnostic procedures to rule out the suspicion of tumor. Nevertheless, the test is valued as unquestionably beneficial by some; as not beneficial, and hence unjustified, by others. This is exactly the kind of allocative problem typical for health planning. And these problems are handled arbitrarily in our historically grown supercomplex system; limitations to the *ad hoc* decisions of physicians (level (4)), rare as they are, are set without discernible decision making structures [25, 26]. This is part of what Calabresi and Bobbit have categorized as the "customary evolutionary approach" [9] of allocative decisions. It implies the avoidance of self-conscious choice, having the advantage for patients, physicians, and politicians of avoiding the social costs of fundamental values in conflict.

In summary, this state of affairs evokes in patients the wrong impression that society until now has provided full coverage of any medical service technologically feasible, however small the marginal benefit, which later would be determined by physicians *qua* medical experts. Therefore, the very idea "suddenly" to ration medical resources provokes a public outcry.

## IV. STATUS DESIDERABILIS

Various cost control remedies that have recently been considered or initiated in Germany have in common a focus on input rather than output data and the neglect of allocative normativity. They differ, however, in their approach to the common goal: cutting waste by modifying level (4) decisions, those made at the physician and patient level. The most far-reaching reform considered for import by some German experts [30] is the DRG (diagnosis-related group) payment, which has rightly been called a revolutionary in the American health care system, where it is about to be adopted by more and more subsystems. Under this policy hospitals receive a predetermined price per patient admission according to one of 468 diagnosis-related groups. This is meant to induce cost-effective behavior in hospitals (and subsequently by their physicians) via financial incentives (gaining by underusing the diagnosis-related price) and disincentives (losing by exceeding the lump payment).

Competition between and decentralization of the sickness funds will probably increase efficiency. By monitoring the services provided and by introducing adequate sanctions, sickness funds could easily prevent physicians from unnecessary duplications, or from using expensive technology without clear indication. But their restriction of physicians will need to be publicly evaluated for its impact on quality. The same is true for the for-profit hospitals restricting their physicians, and even for non-profit hospitals with the potential of surpluses (as will be the case under the future German hospital financing plan). The DRG system further increases these dangers [16, 29, 39].

In ambulatory care, the second leading share of the health budget [28], such a questionable restrictive policy has recently been exemplified by the national association of sickness funds. In some areas they have introduced quantity controls by monitoring physicians' spending patterns over time. Significant overuse of services as compared to the customary standard has to be justified and possibly even paid for by the physician. Attractive as such a measure is at first sight, it might not adequately respect the nature of a particular patient population. Moreover, the customary standard is not necessarily identical with the desired standard.

Still more problematic from an ethical perspective is the suggested introduction of direct financial incentives for physicians to achieve cost containment. Be it in the form of profit sharing between hospitals and physicians, or insurance companies providing efficiency premiums for outpatient care: given the personal identity of provider and winner public control becomes even more difficult and economic interests may undermine the desired quality of care. Where asymmetry of information between physician and patient makes it difficult enough for the latter to recognize therapeutic or diagnostic overuse in the providers' financial interests (fee-for-service outpatient care; per diem reimbursement of hospitals), this is even worse if providers profit from underuse. In the former situation patients could at least potentially exercise control by asking for a second opinion about the proposed intervention. Such a check, however, is lacking in the latter situation, where patients are simply kept ignorant about the potentially beneficial options they are denied.[16]

In addition to these interventions at the supplier-side, suggestions have been made to reduce costs by means of further copayments, deductibles, or health education. There is an ongoing debate about those measures' relative effectiveness [24, 25, 26, 30, 31], which is beyond the scope of this paper. From an ethical point of view it is important to note that all these measures have the sole objective of cost containment and propose to achieve this without compromising the quality of care. Lacking data, however, on where clear-cut waste starts, and lacking both data and consensus on where the marginal benefit is so small as to qualify an intervention as medically unnecessary, third party interests, arbitrary customary standards, or recipients' poverty, possibly without being noticed, take over the setting of quality norms. Introducing more market control into the system hence must be accompanied by safeguards to keep the desired quality of care.

And indeed, concerns about cost containment, but also about the quality of care for Medicare patients under the new DRG system, has led in the U.S. to the establishment of peer review organizations (PROs) for utilization and quality control.[17] Law requires the PROs to be nonaffiliated with either health care facilities or third party payers. PROs' admission and quality objectives, in summary, aim at reducing hospitalization in favor of outpatient care, the number of unnecessary interventions, and the number of readmissions.

To be sure, there are several criteria for waste and frankly bad quality that a PRO group will unanimously identify and, methodological problems aside, apply in the reviewing process of medical charts, e.g., time delay in the initiation of treatment; incompetence of care providers as expressed in mistreating, oversights, wrong causal conclusions; non-availability of valuable services. However, quality parameters often are neither obvious nor agreed upon even within the medical profession: with regard to the length of stay in hospitals, e.g., cutting mere "hotel days" will usually not only increase efficiency but also contribute to the well-being of patients. However, "squeezing out every last non-acute day from each hospitalization" might, as Dans and his coworkers have stressed ([15], p. 1135), endanger good quality because of extreme stress and anxiety during the time of inpatient care. What is needed is, once again, the identification of outcome measures, and a determination of tolerable risks and degrees of compromised well-being. These will have to serve as criteria against which to compare actual medical care. PROs will have to spend much of their activity developing such standards which might then be subjected to public review.

This brings us, finally, to the question of ethical justification of such *ex ante* allocative patterns (level (3)). A common minimal denominator of various secular moral theories is to accept as morally fair what a hypothetical, impartial, and rational judgement would prescribe as categorical. One well-known attempt to interpret these two conditions is Rawls' [33] thought experiment of an "original position" where fictitious contractors have come to an *ex ante* agreement about the principles of justice by which to govern their society-to-be. In addition to being rational, these initial contractors are in possession of the currently available body of knowledge about human nature and society, thereby reflecting the conditions of a particular society. However, they are placed behind a "veil of ignorance" which keeps them from knowing any of their personal future particularities, such as sex, race wealth, and value preferences. Hence, it is this veil that forces the contracting parties to be impartial even if choosing from self-interest, another built-in assumption of the original position. I agree with Hare ([23], p. 25) that this Rawlsian construct is formally analogous to the common justificatory fictions of utilitarianism, the "impartial observer" or the "universal prescriber", but I will stay with the former because of its being so pictorial.

Defining health as "normal species functioning" being "an important component of the opportunity range open to individuals in a society" ([12], p. 41), neo-Rawlsian ethicists like Daniels, Gauthier, or Veatch (e.g., [12, 13, 14, 21, 31]), on whose striking divergences I cannot comment here, have enlarged Rawls' category of "primary goods" so as to embrace health. When arguing along strict Rawlsian lines [41], the distribution of health care has then to be governed by the difference principle, i.e., consistently giving priority to the reduction of suffering of the least well off. However, this conclusion entails the additional assumption that the representative contractors would want to treat health care, despite its convertibility, as independent of wealth, even if the latter were justly distributed. Gauthier [21], not making this assumption, rather concludes the fairness of a correlation between one's access to health services and wealth, supplemented by a general right to basic care resulting from *ex ante* agreement. Detaching the distribution of health care from that of the original primary goods creates even further problems, which should not be explored here, since my own approach does not gain plausibility from its formal association with Rawls' work.

The veil of ignorance, being one adequate translation of the condition of impartiality with regard to health care, means that the original contractors are potential patients, ignorant about their future health status or predispositions to illness, having identical chances to acquire any disease according to its incidence.

Rationality, however, the interpretation of which ultimately accounts for Rawls' whole theory, is characterized by too much conceptual looseness to allow for anything like a pure deduction.[18] It is by reference to "rationality" that most different allocative principles are justified by various authors.[19]

What rationality leads the initial contractors to opt for cannot be considered independent of their motivation, which, I take it, is the desire to maximize their future opportunity for goods that are necessary for the full opportunity range open to all individuals. Identifying those primary goods, including health, and agreeing on their desirability (which corresponds to Rawls' "thin theory of the good")([33], pp. 395–399), the agreed upon principles of justice are not independent of the good. Hence, the whole construct is based on a teleological motivation assumption.

With regard to health care distribution, rational self-interested life planning would let the contractors opt for an income neutral maximum reduction of suffering, because detachment of health and wealth furthers the overall goal, i.e., the maximum aggregate utility.[20]

The two utilitarian standard assumptions, i.e., fairly identical utility curves for everybody, and the law of declining marginal utility, brings the utilitarian close to an egalitarian pattern of allocation: e.g., declining marginal benefits of anti-suffering procedures makes maximization compatible with our moral intuition, that it is more desirable to achieve relief for a single person with severe and painful inflammation of the pancreas than to shorten the duration of the common cold for 100 persons at the same total cost. With the help of this additional assumption of declining marginal utility, the common objection to utilitarianism, that it ignores the ethical differences of various distributive patterns, could be met.

Moreover, the right to health care is a derivative principle which results from the high disutility (weak social values, mistrust, anxiety) of knowing health care to be contingent on wealth, prospective earnings or merits. Thus it is only seemingly a paradox to reject on utilitarian grounds allocation based on prospective productivity ([34], p. 178).

The initial contractors, viewing their prospective lives as wholes, rather than in time-slices, would not allocate resources linearly over time, but devote relatively less to old age with its already diminished life expectancy. As Daniels [13] has convincingly argued, from an impartial *ex ante* perspective this does not amount to discrimination of the elderly, since every initial contractor has the same chance of reaching old age and of suffering from any particular disease in young age.

Some goals, such as freedom from pain, would surely be regarded as age independent, but by and large old age (extension of life expectancy) would receive fewer resources than young age. Ideally, medical interventions would therefore be evaluated according to their: (1) likelihood of success score times (2) resulting approximation to normal functioning (age-relative) score times (3) life expectancy score. Whereas scores (1) and (3) can be determined by medical expertise, score (2) is the catchall parameter of normative evaluation.

Fairness (i.e., impartiality and rationality) of level (3) allocative patterns will not prevent us from suffering and mourning in case we ourselves or somebody we can identify becomes a "victim" of *ex ante* rationing, and does not receive a potentially beneficial service. However, it must be kept in mind that such an anticipatory pattern of rationing, based on rationality and information about the technically possible quality level of medical care, would probably not do away with most quality objectives of current customary care. The British health plan, with its denial of kidney dialysis to anybody over age 65, probably exemplifies an allocative decision (level (3)) not based on an *ex ante* consensus, given this treatment's extremely high potential to restore relatively normal daily functioning even for the elderly. Just because of lacking an *ex ante* consensus, British "victims" of rationing are apparently held in ignorance about their potential treatability. This well documented fact [1, 38] has been used as an argument against *de jure* rationing in general [6, 38]. The latter, it is claimed, can only be achieved at the expense of honesty in the patient-physician encounter, thus undermining patients' autonomy.

But obviously, transparency is a necessary condition for the realization of the stipulated *ex ante* model of allocative decision making. Knowing the game's rules in advance and accepting them as rational, to be a "loser" is viewed as unfortunate but not unfair. As Gibbard argues, we will have to "refine our ways of thinking about risks" ([22], p. 178). After all, realization of known but voluntarily uninsured risks is not limited to medicine, but is intrinsic to rational human life; something we must learn to accept as unavoidable despite its current denial.

Moreover, a level (3) dominance in allocation saves the individual physician from being himself an allocator, a role that, as has frequently been argued in the literature, would neither be ethically justifiable ([4], p. 213), nor appropriate in daily practice [19]. Reference to the traditional role of physicians as their patients' uncompromised agents is sometimes used as an anti-rationing argument by the profession. However, this does not hold at the level of anticipatory rationing, resource availability having always set the limits to such patient agency. Even if a patient happens to fall in a rationing category, there remains enough room for honest high quality care with regard to the patient's other medical needs.

The delineated approach to allocation design might seem disappointing for its lack of content and, moreover, procedurally impossible. However, its objective being a theoretical justification of rationing, it has some practical implications. Being forced by the current public awareness that self-conscious choices must be made, society ought to structure its allocative system transparently, rationally, and impartially. Practically, this will always remain an approximation to the ideal. Not every medical output can be publicly evaluated. The majority of the current customary medical standard will simply and intuitively be accepted as desirable. Nonetheless, the tasks of the future lie in the field of health service research which according to Blanpain [5] has been irrationally underfunded in all industrialized countries.

Quality assessment is to be initiated and publicly subsidized in Germany, possibly drawing on the experiences the United States has had with its peer review organizations. Such output data will be needed both as a basis for safeguard *vis-a-vis* policies that aim at increasing efficiency, and as a basis for public *ex ante* rationing. Ultimately, it is by politicians and their democratic supervision that the fictitious *ex ante* contract situation has to be approximated.

Whatever the content of rationing policies, they ought to be transparent so as to be scrutinized and accepted in the actual "victim situation": although unfortunate, it is one thing to be denied an artificial hip joint at age 80 if this is known to be a consistent[21] allocative rule. But it is another, quite unfair, thing to be denied treatment because inconsistent rationing by physicians or third parties who might conceal moral intuitions and idiosyncrasies, ignorance, or financial interests behind the concept of medical benefit.

In summary, using the four level allocation model, its current hierarchical structure in Germany can be depicted as follows:

$$(1/2) \leftarrow (3) \leftarrow (4).$$

A nondesirable alternative, the closed loop system, looks like this:

$$(1/2) \rightarrow (3) \rightarrow (4).$$

From an ethical perspective the ideal allocative system would be level (3) governed by the political process:

$$(1/2) \leftarrow (3) \rightarrow (4).$$

And while a society with patients autonomously controlling the quality and quantity of their medical care will always remain ____ utopia, autonomous patients freely accepting medical rationing in accordance with a rational, consistent, and transparent *ex ante* allocation might represent the most acceptable solution.

*Universität Göttingen*
*Institut für Geschichte der Medizin*

NOTES

[1] It has been suggested ([10], p. 1414) the terms "allocation" and "rationing" be used interchangeably in the context of social philosophy. However, although not systematically analyzed, it seems to me that frequently "allocation" is used in the very general sense of distribution, whereas "rationing" implies calculated non-provision of desired goods to some persons.

[2] I will use the terms "health care," "medical care," and "medical intervention" synonymously and in their broadest sense. Included are the development and application of all relevant services by physicians and other health professionals, by laboratories and research institutions, and by the various agencies involved in their organization and supervision.

[3] Hence, Daniels' [12] modified biomedical concept of disease as negative deviation from age-relative, species-typical functioning has the built-in value-variable of the "species-typical." Obviously the "species-typical" stands in a dialectic relation to the range of health care services known and available to a society.

[4] The debate concerning the ethical imperative of "equality of health" (see e.g., [8, 11, 14, 21, 41]) has resulted in various interpretations: equality of outcome, of spent resources, of access to care for identical cases – which last I take to be the only convincing one.

[5] Regarding the "right" to health care problem, with both its formal and its substantial aspects, see, e.g., [7, 14, 17, 21].

[6] As a proponent of this libertarian view see, e.g., [18].

[7] As others have before me ([35], p. 582) I shall use the term "quality" as solely concerned with outcome. Hence, a system can be charaterized by both high quality and high inefficiency without a contradiction.

[8] I realize that more than financial costs have to be taken into consideration: social costs resulting from the weakening of stabilizing values; occasional costs such as anxiety and waiting time. But they can ultimately be translated into monetary terms.

[9] Martin J. Bailey in his interesting essay ([3], p. 121) reverses this relation: he emphasizes the importance of ascribing values to various health programs so as to increase the system's efficiency.

[10] Where the healthy pay for the sick and the affluent for the poor. See [36], p. 252.

[11] The remaining 10% are either wealthy enough to be exempted from statutory insurance and commonly buy private health insurance; or their medical care is covered by public assistance.

[12] Exempted from coverage are: certain prescriptions for common diseases; small copayments (DM 2, per drug; and DM 10, for each of the first 14 days of hospitalization, maximally once a year).
[13] To simplify the matter, the empirically small discrepancy between the physician determined "medically beneficial" and its actual use, which turns on the question of shared decision making and informed consent, will be neglected here.
[14] A good deal of the bioethical literature is inevitably "contaminated" by the normative technical ambiguity of the medical evaluative terminology: e.g., Childress ([11], p. 1415) wants to limit physicians' exclusion of patients from treatment to those whose conditions "would certainly prevent *successful* treatment." Correspondingly, Thurow ([40], p. 613) wants them to begin treatment only when "sure that it will make a noticeable *improvement*" (italics mine).
[15] For instance, see Rutstein *et al.* [35] in their pioneer attempt to set up negative indices for the quality of international care, their standard for "unnecessary disease and disability" and "unnecessary untimely deaths" was the current one used in industrialized countries.
[16] I owe this argument to Morreim [29].
[17] For an overview of history, function, and pitfalls of PROs see [15].
[18] Rawls' mode of inference is ambiguous. Predominantly he argues as if the "original position" were the basis for demonstrative arguments. On various occasions, however, Rawls introduces a second justification for his normative principles, i.e., compatibility with our "reflective equilibrium," which is, moreover, not presented as an independent proof.
[19] Just to cite two examples with regard to micro-allocation: Childress argues that "*rational* persons may indeed responsibly choose to use some form of chance because it preserves several of their values better than any other approach to rationing" ([10], p. 1416), and Fletcher, quoted by Childress at the same location, holds that the use of a lottery principle for allocation is "literally irresponsible, a rejection of the burden. Its refusal to be *rational* is a deliberate dehumanization . . ." (italics mine).
[20] Ultimately this is a question, which cannot be further explored here, of the relation between rationality and risk-taking, which decision theory does not unanimously answer.
[21] This touches an important problem, discussion of which exceeds the scope of this paper: i.e., the "right" of the affluent to opt out of a rationed system and to buy extra care, in particular in the case of public research subsidy.

## BIBLIOGRAPHY

1. Aaron, H. J. and Schwartz, W. B.: 1984, *The Painful Prescription: Rationing Hospital Care*, Brookings Institute, Washington D.C..
2. A. T. I. Arzneimittelinformation (ed.): 1985, 'Blutnachweis im Stuhl . . . Darmkrebsfrühdiagnostik mit Haemoccult umstritten', in *Arzneitelelegramm* **10**, p. 80.
3. Bailey, M. J.: 1980, 'Measuring the Benefits of Life-Saving', in Rhoads, S. E. (ed.): *Valuing Life: Public Policy Dilemmas*, Westview Press, Boulder, Colorado, pp. 105–124.
4. Beauchamp, T. L. and Childress, J. F.: 1983, *Principles of Biomedical Ethics*, 2nd edition, Oxford University Press, New York, Oxford.

5. Blanpain, J. E.: 1983, 'Maßnahmen zur Kostendämpfung im Gesundheitswesen in ausgewählten Ländern Europas', *Öffentliches Gesundheitswesen* **45**, 512–517.
6. Brody, B.: 1988, 'The Macro-allocation of Health Care Resources', in this volume, pp. 213–236.
7. Buchanan, A.: 1983, 'The Right to a Decent Minimum of Health Care', in The President's Commission for the Study of Ethical Problems in Medicine, Biomedical and Behavioral Research (ed.): 1983, *Securing Access to Health Care*, Vol. 2, U.S. Government Printing Office, Washington D.C..
8. Buchanan, A.: 1988, 'An Ethical Evaluation of Health Care in the U.S., in this volume, pp. 39–58.
9. Calabresi, G., and Bobbitt, P.: 1978, *Tragic Choices*, Norton & Company, New York.
10. Childress, J. F.: 1982, 'Rationing Medical Treatment', in Reich, W. T. (ed.): *Encyclopedia of Bioethics*, vol. 4, 2nd ed., The Free Press, London, New York, pp. 1414–1419.
11. Childress, F. J.: 1984, 'Right to Health Care in a Democratic Society', in Humber, J. M. and Almeder, R. T. (eds.), *Biomedical Ethics Review*, Human Press, Cliffton, New Jersey, pp. 47–69.
12. Daniels, N.: 1981, 'Health-Care Needs and Distributive Justice', *Philosophy and Public Affairs* **10**, 146–179. Reprinted Cohen, M., Nagel, T., and Scanlon, T. (eds.), 1981, *Medicine and Moral Philosophy*, Princeton University Press, Princeton, New Jersey, pp. 81–114. Page numbers refer to the latter location.
13. Daniels, N.: 1983, 'Am I my Parents' Keeper?' in *Securing Access to Health-Care, loc. cit.*, pp. 265–291.
14. Daniels, N.: 1983, 'Equity to Health Care: Some Conceptual and Ethical Issues', in *Securing Access to Health Care, loc. cit.*, pp. 2–9.
15. Dans, P. E., Weiner, J. P., and Otter, Sharon E.: 1985, 'Peer Review Organisations: Promises and Potential Pitfalls', *New England Journal of Medicine* **313**, 1131–1137.
16. Dolenc, D. A. and Dougherty, C. J.: 1985, 'DRGs: The Counterrevolution in Financing Health Care', *Hastings Center Report* **15**, 19–29.
17. Engelhardt, H. T., Jr.: 1979, 'Rights to Health Care: A Critical Appraisal' *The Journal of Medicine and Philosophy* **4** (2), 113–117.
18. Engelhardt, H. T., Jr.: 1988, 'National Health Care Systems: Conflicting Visions', in this volume, pp. 3–13.
19. Fuchs, V. R.: 1984, 'The Rationing of Medical Care', *The New England Journal of Medicine* **311**, 1572–1573.
20. Gäfgen, G.: 1984, '*Die Ethische Problematik von Allokationsentscheidungen: I Am Beispiel des Ressourceneinsatzes im Gesundheitswesen*', Forschungsstelle für Wirtschafts- und Sozialwissenschaften (ed.), Hochschule St Gallen, Schweiz.
21. Gauthier, D.: 1983, 'Unequal Need: A Problem of Equity in Access to Health Care', in *Securing Access to Health Care, loc cit.*, pp. 179–205.
22. Gibbard, A.: 1983, 'The Prospective Pareto Principle and Equity of Access to Health Care', in *Securing Access to Health Care, loc. cit.*, pp. 153–178.
23. Hare, R. M.: 1976, 'Ethical Theory and Utilitarianism', in Lewis, H. D. (ed.): *Contemporary British Philosophy*, Allen & Unwin, London. Reprinted in Sen, A. and Williams, B. (eds.): 1982, *Utilitarianism and beyond*, Cambridge University Press, Cambridge, London, New York. pp. 23–38. [Page numbers refer to the latter location.]

24. Herder-Dorneich, P.: 1984, 'Zwischen Utopie und Pragmatik', *Bundesarbeitsblatt* **12**, 5–8.
25. Henke, K.-D.: 1983, 'Gesundheitsplanung im Sinne makroökonomischer Ressourcenplanung', *Öffentliches Gesundheitswesen* **45**, 349–361.
26. Henke, K.-D.: 1984, 'Ergebnisorientierung', *Bundesarbeitsblatt* **12**, 11–13.
27. Iglehart, J. K.: 1985, 'Medical Care of the Poor – A Growing Problem', *The New England Journal of Medicine* **313**, 59–63.
28. Kassenärztliche Bundesvereinigung (ed.): 1984, *Grunddaten zur Kassenärztlichen Versorgung in der Bundesrepublik Deutschland*, Deutscher Ärzteverlag, Köln.
29. Morreim, H. E.: 1985, 'The MD and the DRG', *Hastings Center Report* **15**, 30–38.
30. Münnich, F. E.: 1984, 'Mehr Markt', *Bundesarbeitsblatt* **12**, 8–11.
31. Pfaff, M.: 1984, "Internationale Erfahrungen", *Bundesarbeitsblatt* **12**, 13–18.
32. Ramsey, P.: 1970, *The Patient as Person: Explorations in Medical Ethics*, Yale University Press, New Haven, London, England.
33. Rawls, J.: 1971, *A Theory of Justice*, Harvard University Press, Cambridge, England.
34. Rescher, N.: 1969, 'The Allocation of Exotic Medical Life-Saving Therapy', *Ethics* **79**, 173–189.
35. Rutstein, D. D. *et al.*: 1976, 'Measuring the Quality of Medical Care. A Clinical Method', *The New England Journal of Medicine* **294**, 582–588.
36. Schulenburg, Graf von der, J.-M.: 1984, 'Möglichkeiten und Probleme der Steuerung der Nachfrage nach Gesundheitsleistungen', *Mensch, Medizin und Gesellschaft* **4**, 251–259.
37. Schulenburg, Graf von der, J.-M.: 1985, 'Die Ärzteschwemme und ihre Auswirkungen auf die ambulante Versorgung', *WZB discussion papers*, IIM/IP, 85–6.
38. Schwartz, R. and Grubb, A.: 1985, 'Why Britain Can't Afford Informed Consent', *Hastings Center Report* **15**, 19–25.
39. Stern, R. S. and Epstein, A.M.: 1985, 'Institutional Responses to Prospective Payment Based on Diagnosis-Related Groups: Implications for Cost, Quality and Access', *The New England Journal of Medicine* **312**, 621–627.
40. Thurow, L. C.: 1985, 'Medicine versus Economics', *The New England Journal of Medicine* **313**, 611–614.
41. Veatch, R. M.: 1979, 'Justice and Valuing Lifes', in Rhoads, S. E. (ed.), *Valuing Life: Public Policy Dilemmas*, Westview Press, Boulder, Colorado, pp. 147–160.

# SECTION IV

# PREVENTIVE MEDICINE, OCCUPATIONAL HEALTH, AND FUTURE ISSUES

HERBERT B. VIEFHUES

# PREVENTIVE INTERVENTIONISM AND INDIVIDUAL LIBERTY

> It is your right to reproach me with my behavior but it is my right not to care about your reproaches.
> Aristippus (435–360 B.C.)

## I

Prevention is primarily a medical question. But regardless of its possible medical utility it is clearly one of public policy; it is also an ethical question, and by this a question of pragmatic philosophy, because it is related to the question of values.

To exist man needs his fellow men. Living is always living together. The assumptions for acting with others are self respect and respect for the other, a mutual taking seriously of the principle of mutual respect ([7], p. 156). The point of reference for ethics in our Western world cannot, then, be a collective – the group, the society, or the society as organism, but only *the individual in relation to other individuals* [2].

We are living in a community which we may conceive under system theory. We are well aware that the main philosophical defect of all systems theories, however, is their intrinsic unreflectiveness. But we may take it up as a useful model. The different social subsystems have different social functions, e.g.,

| | |
|---|---|
| (a) the Legal System | – the pursuit of "inner" peace; |
| (b) the Economic System | – the production and distribution of goods; |
| (c) the Educational System | – the self-responsible conduct for living; |
| (d) the Social System | – the compensation for social handicaps; and |
| (e) the Health System | – the care of psychophysiological existence. |

According to these different functions, each of these systems is centered about different value-sets and attitudes. The whole system is therefore not a model of concentric circles, encompassing a common

*Hans-Martin Sass and Robert U. Massey (eds.), Health Care Systems*, pp. 315–322.

value as center, but it exists as a pattern of ex-centric yet overlapping circles. In the hierarchy of systems of the community there is the political system, for all systems comprise decision-making structures. It has to enforce the supreme and dominant value of the community in order to balance the particular ethics of the subsystems. Insofar as there is a primacy of politics and of public policy, West Germany and the U.S. are each by nature an internal multiplicity, a complex pluralistic state.

In contrast, our constitutions reflect the self-restrictions of all political power where freedom *from* government is guaranteed *by* government. This especially concerns the protected private life of each citizen. As the system which regulates the general overall formation of public life, the political system restricts the space of action of the subsystems and the validity of their norms. Given the question: "What are the positive ethics of the political system?", one answers: "the guarantee of a viable anarchy" [4]. This also includes the free disposition of the invidual to form his way of life, or, in a word, the self-determination of the individual within the limits required by the freedom of other individuals.

## II

One of these subsystems of our total societal system is the health care system. The center of its ethics is, of course, health. The task of the health care system is to maintain or to restore health. Health maintenance, in general, has the objective of prevention, whereas prophylaxis is the prevention of special diseases.

It is a truism that "prevention is better than cure," for since its beginning medicine was actively involved in preventive intervention (e.g., sanitation). But our focus is a different one.

It is alleged that in former times the political system could maintain an equilibrium within its subsystems so that no single system would dominate the whole system by suppressing the norms and values of other subsystems. For the past fifteen years, the health care system has been accused of "over-medicalizing" society [9]. There is no doubt that medicine, through genetic technology and artificial fertilization, for example, may influence the fate of our grandchildren, or that abortion or intensive care medicine can terminate, restore, or prolong life. The progress of medical research makes for an increasing influence on human existence. The fact that modern medicine is able to reduce mortality and to assure longevity was partly the reason for the "medicalization" of our daily living. This medicalization, however, was necessary in order to establish new ethical agreements. But it must be kept within the boundaries of scientific and professional medicine which form the institutions and core of the health care system.

In the last few decades, a remarkable change took place in the health care system in the Federal Republic of Germany. Side by side with the professional health service a second health system was established in which only a small portion of the professional system participated. The main promoters are non-professional, sometimes even anti-professional, lay organizations with a growing influence in connection with certain circles within the WHO. The leading persons of this new, second health system have transformed the original concept of health into a purely social one in which medical facts have only an ancillary function. This "extra-professional" health care system thus intends to abolish the traditional, professional health system and to replace it with what goes under the slogan, "prevention instead of cure." At the same time, this changing health system attempts to identify its particular values and ethics with the current political system in order to replace the principle of *freedom* with the absolute value of *health* [1, 6].

The main tactic of this second health system – designed to conquer the freedom of the community – is its emphasis on the idea of *preventive intervention*. Because man is doubtless mortal, and illness and injury are part of the *conditio humana*, all prevention has limits in principle. But the idea of preventive intervention has assumed a high status in the mentality of modern man and reaches far beyond its original medical intent. The argument of the non-professionals appears at first glimpse like a medical one, but its intent and ends contradict, in our opinion, those of earlier professional medicine. Certainly there are overlapping interests and fields of action between the professional and the non-professional health system (like self-help groups when medical experts act as "gurus" in the second system). But experience reveals that both parts are acting with an amazing independency in the practical contexts of everyday medicine.

One example of the second health system is the concept of "risk factor medicine" [3]. It is based on social epidemiology, which combines statistical data on illness with biopathological findings and behavioral traits. Risk factor medicine claims to predict accurately illness and death and thus to be able safely to prevent illness by measurements which direct early intervention based on the following considerations: Certain pathophysiological findings are documented in patients with certain diseases (e.g., coronary heart disease); certain factors like overweight, hyperlipemia, hypertension, etc., as well as certain behavioral items (like smoking) are correlated with each other on the bases of statistical data. These procedures not only lead to prognostic possibilities which express the risk to individuals of falling ill, but lead also to an appeal to change one's life style for the prevention of health risks. But I suggest the following: To warn the population of risks to health is a legitimate medical purpose, for early notice of risks to health and illness is very useful. In Germany, however, in contrast to other European countries, "health" is now the chief concern in all public opinion polls. I do not believe in the power of a medical élite, and the medicalization of which I spoke is "not occurring through any increase in political power of physicians" [9]. But preventive intervention is related to a special kind of medicalization in our culture and has a special social etiology. Hence preventive interventionism seems to meet a popular need because "health" is now a supreme value and even competes with personal liberty.

We noticed that (with the exception of the principle of health) there is no shared value-system in our pluralistic society. "Values" and "meaning of life" are today part of each individual's private life, and here the most heterogeneous values have equal risk: Somebody loves his country and somebody his dog! Because of the value of liberty we have the freedom to choose our own self-imposed obligations. To be able to do so there is only one requirement: to live as long and as healthy a life as possible. But with the eclipse of religion from the public (as from private life) the question of the "meaning of life" is not only privatized but is slowly fading from the awareness of individuals and from society as well. Health has thus become, slowly but incessantly, the supreme value and one which possesses a quasi-religious quality and serves to replace the more complex question of the meaning of personal life.

The concept of preventive intervention is thus strongly related to the individual's way of life. Illness is thereby removed from an objective connection to the individual patient; we now no longer "distance" the patient from his own illness by an objectivization which is necessary for scientific medical treatment. Instead of "distance" and objectivization, illness is now personalized: "It is your own fault for becoming ill." This is the new message. Illness becomes a personal failure and the patient is guilty. This not only excludes numerous social factors from illness etiology, but it also causes a regression to the medieval notion of illness as sin. But in contrast to this medieval view, this sin is no longer related to transcendency and to a personal, yet merciful, Lord. The new sinner is guilty of social sins, and society becomes an a-personal and merciless construction. There is no pardon.

Preventive intervention is designed to be a new doctrine of salvation – a health-theology. To quote from a German textbook on social medicine:

> How about the maintenance of health as a moral obligation of the individual? . . . numerous diseases are connected with the wrong life style, which the patient could well change. . . . Therefore we must expect that the patient will change his life style and will live more in compliance with health. . . . It is the misfortune of our present society that the repercussions of the wrong behavior towards health are not even in the consciousness of the masses. . . . Each citizen who is overweight and each excessive smoker is a destructive parasite, a fact which is not yet clear to our society ([5], pp. 271, 235, 253, 274).

This is an aggressive, sectarian expression which is to be found in numerous papers and heard in many lectures. From "wrong life style" to "change your life." He who does not live the healthier life is a "destructive parasite" within our society, an evildoer, malefactor, an outlaw and social sinner. One should compel him to change his ways by force and social ostracism.

This is now the recommendation for a healthier life – to convert to a new and healthier being in a new and healthier society, even to a "Health Society". The interventionists' actions now profoundly interfere with the individual's private life, and open the private life to various manipulative techniques within the second health system. It is astonishing that all this began (to a great extent) in the United States. "Prohibition" completely failed. But this new wave of prohibiting life styles is already leading to almost Orwellian consequences: A smoking campaign in Virginia, for example, has had the result that in some counties smokers will no longer be employed as police officers or firemen. The U.S. Surgeon General, Everett Koop, M.D., hopes to transform the United States into a "smoke-free society" by the year 2000 [8]. Similar campaigns one can find in Germany.

## III

Illness and disease are highly complex phenomena open to explications, interpretations, and explanations from various scientific aspects. If we continue to rely on only one paradigm which works only with simplistic arguments, we shall fall back on merely general statements of everyday life experiences. Although they are not wrong, they are very unspecific and unscientific. In this way we shall surely return to the traditional catalogue of virtues in which *modestia* is at the center of the *sapientia secularis* for "eubiotics." To live by the virtues is morally good, but they are not precise enough to prevent illness and disease which we can now differentiate by means of the very elaborate models and constructions of modern scientific medicine. It is interesting that only sexuality is excluded from the ancient meaning of *modestia*!

The aim of contemporary medical systems is to treat the injured and burdened and to enhance the health and life of humanity. This value is deeply rooted in our civilization with its emphasis upon the personal freedom of each individual. But freedom of action has consequences. These consequences always include certain risks. Today, many wish to have total freedom while at the same time they experience no risks at all. But freedom without risk is not possible, because freedom is fundamentally limited by the freedom of others. Total risk aversion abandons freedom. Total security is total constraint and bondage.

Preventive intervention, therefore, becomes an unseen danger of individual liberty because it interferes with the private expression of the lives of individuals. The slogan – "smoking is a social sin" – expresses the idea of "illness as social guilt." Here mankind is construed as a "mere means" for the collective well being, and is no longer of intrinsic value.

Before preventive interventionism came into the ascendency, medical interventions were inadmissible when they hindered the self-realization of the citizen. In the West German health insurance system persons of a certain base income are statutorily compelled to be members of at least one of the different health insurance plans, but he or she is not forced to be in good health or to shape his or her private life according to the insurer's rules. There are certainly many *voluntary* activities pertinent to primary prevention, health education, and so on; but there is not yet any instrument legally to proscribe one's behavior and choices.

Since the beginning of our sickness insurance system some 100 years ago, the Bismarckian administration was wise in not imposing any obligations on the citizen in order to be eligible for its complete service to the sick, whether the matter be pneumonia or the consequence of a brawl of a notorious drunkard. In addition to this, we have to keep in mind that the German compulsory Sickness Insurance System is a pluralistic one, and based on a certain degree of competition. The state is the guarantor of the service but is not responsible for the service itself. In a nationalized health system it becomes difficult for the political system to be at the same time arbitrator between the norms of the subsystems and one of the subsystems itself. In our Western civilization this role of the state requires a specific qualification of political discrimination, and sensitivity to certain ethical principles.

The Bismarckian insurance scheme, however, has turned out to be a very flexible one. It was founded at a time of transition from the paternalistic, constitutional monarchy, and is still a workable system today within the framework of a modern constructivist constitution. Paternalism is only possible in the context of a monarch who rules absolutely over his subjects. In our constitutional system with its mutual contracts, the *Sozialstaat* (welfare state) is – with its emphasis on the primacy of the principle of freedom – a rational rather than emotional system of mutual-risk associations; there is no place in it for "solidarity" or "fellow feelings."

It is essential today that we begin the discourse on and critique of the domination of scientists – in our case the preventive interventionists – so that the political system may be able to rescue and preserve individual liberty. A patient's illness is to a great extent intrinsic to his private life. Physicians do not treat cases, but persons, and should not wish to prevent all health risks and thus perpetuate a dogmatic theology of health.

*Institut für Sozialmedizin,*
*Ruhr-Universität, Bochum,*
*Federal Republic of Germany*

BIBLIOGRAPHY

1. Baier, H.: 1982, 'Pflicht zur Gesundheit', in Bock, K. D. (ed.), *Risikofaktorenmedizin. Fortschritt oder Irrweg*, Braunschweig und Wiesbaden, pp. 120–136.
2. Blumenberg, H.: 1981, *Wirklichkeiten, in denen wir leben*, Suhrkamp, Frankfurt am Main.
3. Gerhard, U. and Friedrich, H.: 1984, *Risikofaktoren, primäre Prävention und das Problem des richtigen Lebens*, Deutscher Soziologentag, Dortmund.
4. Marcuse, L: 1979, *Meine Geschichte der Philosophie*, Diogenes, Zurich.
5. Schaefer, H. and Blohmke, M.: 1971, *Sozialmedizin*, Thieme, Stuttgart, F.R.G.

6. Scheuch, E.: 1982, 'Das Verhalten der Bevölkerung als Teil des Gesundheitswesens', in Boys, H. *et al.* (eds.), *Gesundheitspolitik zwischen Staat und Selbstverwaltung*, Enke, Stuttgart, F.R.G., pp. 10–132.
7. Tugendhat, E.: 1984, *Probleme der Ethik*, Reclam, Stuttgart, F.R.G.
8. WHO: 1980, *Regional Strategy for Attaining Health for All by the Year 2000*, Regional Office for Europe, Copenhagen, Denmark.
9. Zola, J. K.: 1977, 'On the Way to a Healthier Society', in J. Stolte (ed.), *Veranderde Gezondheitssorg*, Tilburg, The Netherlands, pp. 144–166.

HUBERT BRANDTS

# IMPROVING OCCUPATIONAL HEALTH IN THE FEDERAL REPUBLIC OF GERMANY

## I. INTRODUCTION

According to the constitution of the Federal Republic of Germany, the State is obliged to maintain the physical fitness of its citizens and to provide help for those in need of protection (Constitution art. 2 § 2 and art. 20, § 1). This includes the obligation to create working conditions that meet the requirements of the worker and prevent accidents. In practice, the State fulfills its tasks by a system of institutions, organs, and authorities. The most important institutions of occupational protection and accident prevention are the State factory inspectorate and the carriers of accident insurance, i.e., the industrial injuries insurance institutes. The industrial injuries insurance institutes are self-governing bodies under governmental supervision. They are organized according to industrial branches and proportionally administered by employers and employees. Their tasks can be divided into three main items:

(1) prevention,
(2) rehabilitation,
(3) financial compensation.

Between these theoretically equal tasks an order of precedence exists insofar as successful prevention makes rehabilitation measures unnecessary, and good rehabilitation has a favorable influence on the extent of benefits to be granted. Based on this knowledge, industrial injuries insurance institutes are legally obliged to provide prevention and rehabilitation measures "by all suitable means."

Thus, the first step toward prevention is to avoid injuries. This avoidance is prior to compensation for injury. According to the legal obligation of statutory accident insurance, prevention in its most comprehensive interpretation includes measures and activities for the prevention of occupational accidents and diseases. It aims at maintaining the employees' fitness and thus contributes to realizing the constitutional right to life and physical welfare. Accident prevention, first aid, and occupational medical examinations are, in addition to other important items, the most meaningful sections in the field of occupational protection.

*Hans-Martin Sass and Robert U. Massey (eds.), Health Care Systems,* pp. 323–344.

## II. BACKGROUND

Industrial injuries insurance institutes as carriers of the statutory accident insurance were founded in 1885 (Act relating to accident insurance, introduced in July 1884). Last year the 100th anniversary was celebrated. The origin is even earlier, however. In 1839, a Prussian regulation restricted the work of children and women; in 1878 the Federal German States implemented obligatory factory inspections. The "Gewerbeordnung" (trading regulations) drawn up in 1869 included rules and regulations on work safety from state inspectorates and is, in its new version, still in force. These trading regulations were first solely applicable in the Federation of North German States ("Der Norddeutsche Bund"), but since 1871 applicable everywhere in Germany. In 1891, § 120a was included in trading regulations. This paragraph is still in force and can be regarded as the starting point of occupational protection in companies. According to this paragraph, the employer has to provide workplaces, equipment, machines, and appliances and to organize operations in such a way that employees are protected against dangers to life and health as far as factory operations allow.

The phrase "employer's obligation to provide for the welfare of his employees" became a component of the law in Germany. Among others, this principle was incorporated in the two most important codifications of the German civil law, the Civil Code from 1896, and the Code of Commercial Law from 1897. Section 618 of the Civil Code and § 62 of the Code of Commercial Law show a partly literal accordance with § 120a of the trade regulations.

At first, on introducing the statutory accident insurance system, traumatology had priority in the field of medicine. The first contact between legal accident insurance and developing occupational medicine was made by the introduction of the law of 1925 relating to occupational diseases. New causal relations in the field of legal compensation were developed, and questions regarding general and specific disease prevention were raised. Together with the permanent increase of occupational diseases to be compensated, occupational medical tasks of industrial injuries insurance institutes were enlarged and, in the field of prevention, transferred from particular to general areas. Induced by newly obtained scientific findings regarding occupational medicine which far exceeded the field of occupational diseases, legislature took up this task.

The duties of industrial injuries insurance institutes in the field of occupational medicine comprised not only regulations on occupational diseases: since 1963 these institutions have been responsible for medical checkups where working conditions revealed extraordinary accident and health risks; since 1974 they have been providing general occupational medical care for all industrial branches.

The development of technical and medical occupational protection is characterized by the fact that prevention measures have been forwarded and transferred to the periphery of the endangered area and that the abstract term 'occupational safety' has become more precise.

## III. THE GENERAL ILLNESS PREVENTION SYSTEM

(1) State regulations, a law on the application of chemicals, or a decree relating to the use of working materials – "Chemikaliengesetz," ([5], p. 1718) "Arbeitsstoffverordnung" ([7], p. 144) – govern the production and application of *harmful materials* [13] [14]. Materials covered by the law on chemicals can be marketed only if they have previously been registered by a competent authority. A test certificate facilitating the assessment of whether harmful effects are to be expected is required for registrations.

Moreover, special measures are necessary for carcinogenic materials. The basic requirement for the use of these substances is that concentrations of carcinogenic substances that are indispensable, according to the latest technical standards, are to be restricted to a minimum. Permanent testing has priority with regard to how less harmful or innocuous materials might substitute for harmful substances. If this is possible, the harmless materials have to be substituted. For instance, certain preparations and products containing asbestos may not be manufactured or applied. Their application is either completely prohibited or restricted.

At present, about 50,000 materials dominating our life and our environment are known. Every year, about 500 new materials are added. They have to be regarded as harmful if their effects might lead to the impairment of human beings, animals, or plants, or if they may be released due to particular accident risks, e.g., fires or explosions. The national list of harmful materials which is identical with the list of the European Community comprises, at present, 1,500 substances. This list is constantly being supplemented. Based on animal experiments and experiences with human beings, limits on workplace concentration values have been established for about 400 working materials. Furthermore, 40 substances are regarded as carcinogenic. The importance of harmful materials is demonstrated by the fact that 37 out of 55 occupational diseases recognized by the Federal Republic of Germany are attributed to exposure to those substances.

(2) State regulations (a decree on technical equipment or a decree on the safety of devices and instruments) [Gesetz über technische Arbeitsmittel – Gerätesicherheitsgesetz ([2], p. 717)] prescribe that manufacturers and importers of *technical equipment* provide for their use in accordance with recognized regulations [16]. This is the first "filter" which prevents unsafe technical devices from reaching the operator. This decree does not generally prescribe testing; it is assumed that technical equipment on the market meets this standard. The manufacturer, merchant, or importer may, however, order testing in a recognized test institution. If the equipment meets the requirements it can be marked "GS" (tested safety); this "GS" mark has proven to be very effective for marketing.

If technical equipment is faulty, the competent authority is permitted to prohibit further application.

(3) State regulations demand of all companies (regardless of size or type of production) *minimum requirements* relating to the *safety standard within a factory*, which is intended to support occupational protection and accident prevention, e.g., law for company doctors, safety engineers and other experts on safety work ([3], p. 1885). This law aims at increasing the occupational safety standard and health care in workplace environments by defined operational measures under the employer's responsibility, and does not depend on the efforts of the State factory inspectorate and the technical supervision of industrial injuries insurance institutes [12]. For this purpose, the company must employ the required number of doctors and experts on work safety. With their assistance, a competent application of occupational safety and accident prevention regulations, the competent application of newly ascertained occupational medical and safety techniques, as well as the optimum use of all available means for the benefit of occupational protection and accident prevention, can be achieved [18].

The legislature issued a skeleton law; its actual application belongs to the industrial injuries insurance institutes. The control function assigned to industrial injuries insurance institutes is not limited to assuring an adequate number of company doctors or the application of accident prevention regulations. Also, their function inevitably presupposes vocational continuing education for company doctors, and efforts to assure medical care by utilization of occupational medical institutes outside of the companies. This may be initiated by internal measures or by the promotion of external outpatient institutions and centers. According to the law, company doctors have an extensive variety of tasks, and they are personally and functionally integrated into the company's safety system. In this connection, preventive measures are in the forefront of the medical examinations of employees.

Industrial injuries insurance institutes have no monopoly on the establishment of occupational medical services. Every individual company may apply to any physician who has the required qualification. Nevertheless, industrial injuries insurance institutes have successfully organized an occupatinal medical system, which is legally independent and takes the form of a registered society.

(4) Further State regulations refer to:

(a) the layout of workplaces and working spaces for safety and health (Working Space Decree) ([6], p. 729), [15];

(b) maximum working hours, rest periods, prohibition of work on Sundays and public holidays, and their exceptions (Decree on Working Hours) [7];

(c) protection of youth employment (Law relating to the employment of youth) ([4], p. 965), expectant and nursing mothers (Law relating to the protection of expectant and nursing mothers) ([1], p. 69).

(5) In addition to these State regulations, industrial injuries insurance institutes have used their right to issue autonomous regulations for the prevention of accidents and occupational diseases. They are intended to supplement State regulations and to put them into concrete terms. Moreover, they refer to concrete factors endangering employees and focus on risks specific to the industry. Examples: regulations relating to silo and bunker work, to vehicles, welding and cutting operations and related working procedures, to blasting, protection against mineral dust, diving, laser beams, and noise.

These binding rules, which may prescribe penalties for nonobservance, have equal status with State regulations (e.g., dualism of State laws and regulations and industrial injuries insurance institute accident prevention regulations). Below this network of regulations, a third level exists which includes decrees and rules with reference to practice and according to operational conditions (guidelines, safety regulations, principles, and instructions). In this connection, the relevant regulations of occupational medicine are of great importance. They differ with regard to exactness and with respect to their applicability to general technical, hygienic, and occupational medical development. Examples:

(a) technical regulations on harmful working materials ([8], p.58);

(b) maximum workplace concentrations of harmful substances (lists of threshold limit values) ([8], p. 58);

(c) technical guideline concentrations for carcinogenic substances ([8], p. 58);

(d) regulations issued by industrial injuries insurance institutes on occupational medical checkups [10].

## IV. SPECIAL OCCUPATIONAL MEDICAL BENEFITS PROVISIONS

(1) All regulations on occupational medical benefits are summarized in a basic accident prevention regulation (VBG 100) [9]. Included are exposures and activities which demand occupational medical examinations: "If it can be foreseen that insured persons i) are, during their employment, exposed to chemical, physical, or biological substances or ii) perform dangerous activities . . ., they are subject to occupational medical examinations. Follow-up examinations are to be performed within established periods of time." Table I shows an extract from this document.

The special occupational medical check-up is an examination before one begins employment. The follow-up examination is performed during employment. In the event of exposure to carcinogenic material, follow-up examinations are also required after retirement. These examinations may only be carried out by physicians who are authorized by industrial injuries insurance institutes or other competent authorities. This restriction is intended to guarantee a uniform assessment of all employees. This system has proved to be efficient in the past. If the examination of an employee resulted in the physician's advice against further employment at a certain workplace, permanent further employment is only permitted after the completion of such steps as transfer and treatment by a specialized physician, or by the provision of personal protective equipment. Employer and works council have to be informed about the physician's objections to the workplace in question. Technical precautions in working areas always have priority in preventing exposures.If there is no other way of meeting the physician's requirements, he has several avenues at his disposal, the final one of which is a

TABLE I
Extract of regulations of Industrial Injuries Insurance Institutes

| Chemical, physical, biological influences and endangering activities | Intervals of follow-up examinations (in months) | | Subsequent examinations (in months) |
|---|---|---|---|
| | first follow-up examination | further follow-up examinations | |
| Wearing of breathing protectors | | | |
| persons up to 50 years | 36 | 36 | — |
| persons older than 50 years | | | |
| instrument weight below 5 kg | 24 | 24 | — |
| instrument weight above 5 kg | 12–15 | 12–15 | — |
| Stay in foreign countries under special climate and health conditions | 24–36 | 24–36 | — |
| Benzidine or its salts | 6–9 | 6–12 | ⩽ 60 |
| Benzene | 2 | 3–6 | ⩽ 60 |
| Beryllium or its compounds[a] | ⩽ 60 | ⩽ 60 | ⩽ 60 |
| Lead or its compounds | 1– 2 | 3–12 | — |
| (excluded are: | | | |
| lead alcyls: | 3–6 | 12–24 | — |
| – lead tetraethyl | | | |
| – lead tetraethyl | | | |
| Cadmium and its compounds | 12–18 | 12–24 | — |
| Calcium chromate[a] | 6–9 | 12–24 | ⩽ 60 |
| 1-chlorine-2, 3-epoxypropane (epichlorohydrin) | ⩽ 60 | ⩽ 60 | ⩽ 60 |
| N-chloroformyl-morpholine | ⩽ 60 | ⩽ 60 | ⩽ 60 |
| Chromium III chromate[a] ("chromic-chromate") | | | |
| Chromium VI compounds (excluded are: calcium-chromate, chromium III chromate ("chromic chromate"), strontium chromate, zinc chromate) | 6–9 | 12–24 | |
| Cobalt[a,b] (in the form of inhalable dusts of cobalt metal and cobalt salts of low solubility) | ⩽ 60 | ⩽ 60 | ⩽ 60 |
| Diazomethane | ⩽ 60 | ⩽ 60 | ⩽ 60 |

*Table I (continued)*

| Chemical, physical, biological influences and endangering activities | Intervals of follow-up examinations (in months) | | Subsequent examinations (in months) |
|---|---|---|---|
| | first follow-up examination | further follow-up examinations | |
| 1,2 dibrom-3-chloropropane | ⩽ 60 | ⩽ 60 | ⩽ 60 |
| 1,2 dibromethane | ⩽ 60 | ⩽ 60 | ⩽ 60 |
| 3,3 dichlorobenzidine | 6–9 | 6–12 | ⩽ 60 |
| Dichloromethylether (sym.) | ⩽ 60 | ⩽ 60 | ⩽ 60 |
| Diethyl sulphate | ⩽ 60 | ⩽ 60 | ⩽ 60 |
| Dimethyl carbamide acid chloride | ⩽ 60 | ⩽ 60 | ⩽ 60 |
| 1,2 dimethyl hydrazine | ⩽ 60 | ⩽ 60 | ⩽ 60 |
| N,N-dimethyl nitrosamines | ⩽ 60 | ⩽ 60 | ⩽ 60 |
| Dimethyl sulphate | ⩽ 60 | ⩽ 60 | ⩽ 60 |
| Ethyleneimin | ⩽ 60 | ⩽ 60 | ⩽ 60 |
| Fluor and its inorganic compounds | 12 | 12 | — |

[a] If the material can be inhaled.
[b] Alloys are not included.

prohibition of employment. Even if this venue is not as common in practice as might be concluded from socio-political discussions, the principle that technical measures for exposure prevention always take priority has contributed to a further neutralization of the problem of "prohibition of employment." In general, this prohibition is not permanent but temporary, and can be removed as soon as workplace conditions improve or medical measures for the endangered employee are taken.

The recommendation of the company doctor restricts the employer's scope for decisions and activities. The employer's obligation is so extensive that other employees may only work at the workplace concerned after examination and a determination that employees will be sufficiently protected. Employee or employer may raise an objection against the "certificate" and recommendation of the physician.

(2) "Principles of occupational medical examinations" [10] ensure that – independent of regional or other particulars specific to individual industrial branches – occupational medical checkups are performed in

accordance with the same principles. They guarantee that, in cases of exposure to the same substances and in the case of certain activities, the same criteria for assessment, evaluation, and summary of results are applied. This system of *occupational medical principles* was developed in cooperation with renowned medical experts. Because of their quality, they are generally recognized without being completely legitimized by law. They correspond to the generally recognized regulations of occupational medicine.

(3) All preventive measures aim to eliminate dangers from the beginning. Some dangerous exposures, mostly insidious and prolonged, cannot be controlled; in this case, measures are restricted to the observation of exposure effects and to the determination of the exact time of risk occurrence in order to prevent disease. The *law regulating occupational diseases* takes this aspect into account [19, volume 2].

Occupational diseases are conditions caused by the performance of certain occupational activities and are summarized in a catalogue issued by competent authorities. The legal requirement for inclusion in the list is that the disease was mainly caused by an occupational activity; under the law they have the same status as occupational accidents.

Prevention and rehabilitation are incorporated in the complex system of occupational medical care. The industrial injuries insurance institute has to intervene "with all suitable means" in order to counteract the risk of actual occupational disease, or its recurrence or aggravation. In this case, the whole apparatus of preventive measures is applied, starting from the general or particular effect of risk elimination and considering each available possibility from curative treatment, vocational aid, to financial compensation for having to give up the endangering activity. These measures may be carried out before a decision on disablement is made, which, as already mentioned, must be prevented "with all suitable means." Thus, preventive measures are definitely of prime importance. For example, efforts in the F.R.G. have had the result that in the last 30 years the lifespan of miners affected by silicosis is comparable to that of the total population. Currently, 55 occupational diseases are listed. They comprise diseases of the internal organs, of the nervous system, the skeleton, skin lesions, or impairment of hearing. Not only diseases due to the exposure to a single harmful substance (e.g., lead), but also groups of diseases (e.g., infectious diseases) are included (see Table II).

According to a recommendation of the council of medical experts

TABLE II

| Occupational Disease | | Recognized as Occ. Disease since | Cases Compensated for the First Time | | | | | |
|---|---|---|---|---|---|---|---|---|
| Nr | Term | | 1930[1] | 1950[2] | 1960[2] | 1970[2] | 1980[2] | 1983 |
| 1 | 2 | 3 | 4 | 5 | 6 | 7 | 8 | 9 |
| 1 | Diseases due to Chemical Substances | | | | | | | |
| 11 | Metals and metalloids | | | | | | | |
| 1101 | Lead | 1.7.1925 | 470 | 93 | 112 | 20 | 8 | 6 |
| 1102 | Mercury | 1.7.1925 | 13 | 3 | 3 | 2 | 2 | 1 |
| 1103 | Chromium | 1.8.1952 | — | 6[4]) | 10 | 7 | 4 | 6 |
| 1104 | Cadmium | 1.8.1952 | — | — | 1 | — | — | 2 |
| 1105 | Manganese | 1.1.1929 | 17 | 1 | — | — | 1 | — |
| 1106 | Thallium | 29.4.1961 | — | — | — | — | — | — |
| 1107 | Vanadium | 29.4.1961 | — | — | — | — | — | — |
| 1108 | Arsenic | 1.7.1925 | 22 | — | 1 | 3 | 2 | 1 |
| 1109 | Phosphorus (inorganic) | 1.7.1925 | 1 | 2 | 4 | — | 1 | — |
| 1110 | Beryllium | 1.1.1942 | — | 1 | 1 | 1 | — | — |
| 12 | Asphyxiating gases | | | | | | | |
| 1201 | Carbon monoxide | 1.1.1929 | 48 | 48 | 53 | 15 | 4 | 16 |
| 1202 | Hydrosulphide | 1.1.1929 | 8 | 6 | 2 | 5 | 2 | 1 |
| 13 | Solvents, Pesticides and other Chemical Substances | | | | | | | |
| 1301 | Mucosal changes | 1.4.1937 | — | 9 | 15 | 6 | 16 | 20 |
| 1302 | Halogenated hydrocarbon | 1.4.1937 | — | 6 | 20 | 20 | 6 | 15 |
| 1303 | Benzene | 1.7.1925 | 33 | 16 | 19 | 6 | 10 | 4 |
| 1304 | Nitro- or amino compound of benzene | 1.7.1925 | — | 7 | 2 | 1 | — | 1 |
| 1305 | Carbon bisulphide | 1.7.1925 | 2 | 6 | 3 | 1 | — | — |
| 1306 | Methyl alcohol | 29.4.1961 | — | — | — | — | — | — |
| 1307 | Phosphorus (organic) | 1.7.1925 | — | — | — | — | — | — |
| 1308 | Fluorine | 1.8.1952 | — | — | 1 | — | 3 | 2 |
| 1309 | Nitric acid ester | 1.1.1942 | — | — | — | — | — | 1 |
| 1310 | Alkyl-, aryl- or alcyl aryl oxides | 1.4.1937 | — | — | — | — | 6 | 1 |
| 1311 | Alkyl aryl sulphides | 1.4.1937 | — | — | — | — | — | 1 |
| 1312 | Teeth (acids) | 1.8.1952 | — | — | — | 1 | — | 1 |
| 1313 | Eyes (benzoquinone) | 1.8.1952 | — | — | 1 | 1 | — | — |
| 2 | Diseases due to Physical Influences | | | | | | | |
| 21 | Mechanic influences | | | | | | | |
| 2101 | Tendons | 1.8.1952 | — | — | 13 | 3 | 5 | 12 |
| 2102 | Meniscus lesions | 1.8.1952 | — | — | 1379 | 866 | 450 | 340 |

*Table II (continued)*

| Occupational Disease | | Recognized as Occ. Disease since | Cases Compensated for the First Time | | | | | |
|---|---|---|---|---|---|---|---|---|
| Nr | Term | | 1930[1] | 1950[2] | 1960[2] | 1970[2] | 1980[2] | 1983 |
| 1 | 2 | 3 | 4 | 5 | 6 | 7 | 8 | 9 |
| 2103 | Concussion caused by compressed air tools | 1.1.1929 | 74 | 1210 | 649 | 282 | 205 | 176 |
| 2104 | Circulatory disturbance caused by vibrations | 1.1.1977 | — | — | — | — | 2 | 3 |
| 2105 | Mucous bursa | 1.8.1952 | — | — | 14 | 7 | 1 | 11 |
| 2106 | Pressure paralysis | 1.8.1952 | — | — | 3 | 5 | 5 | 3 |
| 2107 | Spinous process | 1.8.1952 | — | — | 5 | — | — | — |
| 22 | Compressed air | | | | | | | |
| 2201 | Work in compressed air | 1.1.1942 | — | 2 | 3 | 2 | 3 | 1 |
| 23 | Noise | | | | | | | |
| 2301 | Impairment of hearing due to noise | 1.1.1929 | 60 | 15 | 23 | 567 | 2531 | 1453 |
| 24 | Radiation | | | | | | | |
| 2401 | Grey cataract | 1.1.1929 | 48 | 1 | 6 | 2 | 1 | 2 |
| 2402 | Ionising radiation | 1.1.1929 | 6[3] | 1 | 6 | 2 | 3 | 4 |
| 3 | Diseases due to Infections or Parasites and Tropical Diseases | | | | | | | |
| 3101 | Infectious diseases | 1.1.1929 | 77 | 404 | 216 | 361 | 432 | 327 |
| 3102 | Animal-borne diseases | 1.1.1942 | — | 51 | 53 | 10 | 3 | 2 |
| 3103 | Verminosis of miners | 1.7.1925 | 1 | — | — | — | — | — |
| 3104 | Tropical diseases | 1.1.1929 | 11 | — | 13 | 18 | 22 | 9 |
| 4 | Airway and Lung Diseases, Diseases of Pleura Peritoneum | | | | | | | |
| 41 | Diseases due to inorganic dusts | | | | | | | |
| 4101 | Silicosis | 1.1.1929 | 2280 | 6616 | 3791 | 1295 | 1001 | 841 |
| 4102 | Silicotuberculosis | 1.4.1937 | — | 924 | 454 | 227 | 129 | 106 |
| 4103 | Asbestosis | 1.4.1937 | — | 5 | 23 | 63 | 96 | 130 |
| 4104 | Asbestosis with Lung cancer | 1.4.1937 | — | — | — | 2 | 19 | 33 |
| 4105 | Mesothelioma (asbestos) | 1.1.1977 | — | — | — | — | 36 | 74 |
| 4106 | Aluminium | 1.1.1942 | — | 5 | 1 | 1 | — | 1 |
| 4107 | Lung fibrosis | 29.4.1961 | — | — | — | 1 | 1 | 3 |
| 4108 | Phosphate fertilizer | 1.1.1929 | 7 | 1 | 4 | 3 | — | — |
| 42 | Diseases due to organic dusts | | | | | | | |
| 4201 | Farmers' lung | 1.1.1977 | — | — | — | — | — | 1 |
| 4202 | Byssinosis | 1.1.1977 | — | — | — | — | 1 | 5 |

*Table II (continued)*

| Occupational Disease | | Recognized as Occ. Disease since | Cases Compensated for the First Time | | | | | |
|---|---|---|---|---|---|---|---|---|
| Nr | Term | | 1930[1] | 1950[2] | 1960[2] | 1970[2] | 1980[2] | 1983 |
| 1 | 2 | 3 | 4 | 5 | 6 | 7 | 8 | 9 |
| 43 | Obstructive airway diseases | | | | | | | |
| 4301 | Airway diseases (allergic) | 29.4.1961 | — | — | — | 131 | 121 | 139 |
| 4302 | Airway diseases (toxic) | 29.4.1961 | — | — | — | — | 21 | 34 |
| 5 | Skin diseases | | | | | | | |
| 5101 | Skin diseases | 1.4.1937 | — | 150 | 532 | 539 | 397 | 422 |
| 5102 | Cancer of the skin | 1.7.1925 | 77 | 33 | 8 | 17 | 6 | 9 |
| 6 | Diseases due to other causes | | | | | | | |
| 6101 | Nystagmus of miners | 29.4.1961 | — | — | 1 | — | — | — |
| | § 551, paragr. 2 RVO | 1.7.1963 | — | — | — | 1 | 7 | 9 |
| Total occupational diseases | | | 3255 | 9622 | 7445 | 4494 | 5613 | 4223 |

[1] 1930 figures for German Reich
[2] 1950/1983 figures for Federal Republic of Germany
[used with permission.]

established within the Ministry of Labor and Social Affairs, "malignant airway tumors in blast furnace workers after intense long-term exposure to blast furnace gases" have been recognized as an occupational disease. The catalogue will thus be supplemented accordingly.

In 1983, more than 35,000 cases of occupational diseases were reported; about 5,000 of them were compensated. The most frequently occurring disease was impairment of hearing, followed by silicosis and skin diseases. These three groups comprise more than three quarters of all registered occupational diseases (see Figure 1.)

## V. CHARACTERISTICS OF THE SYSTEM

(1) The system in the F.R.G. is characterized by the fact that *prevention has absolute priority*. Prevention of risks is prior to compensation for

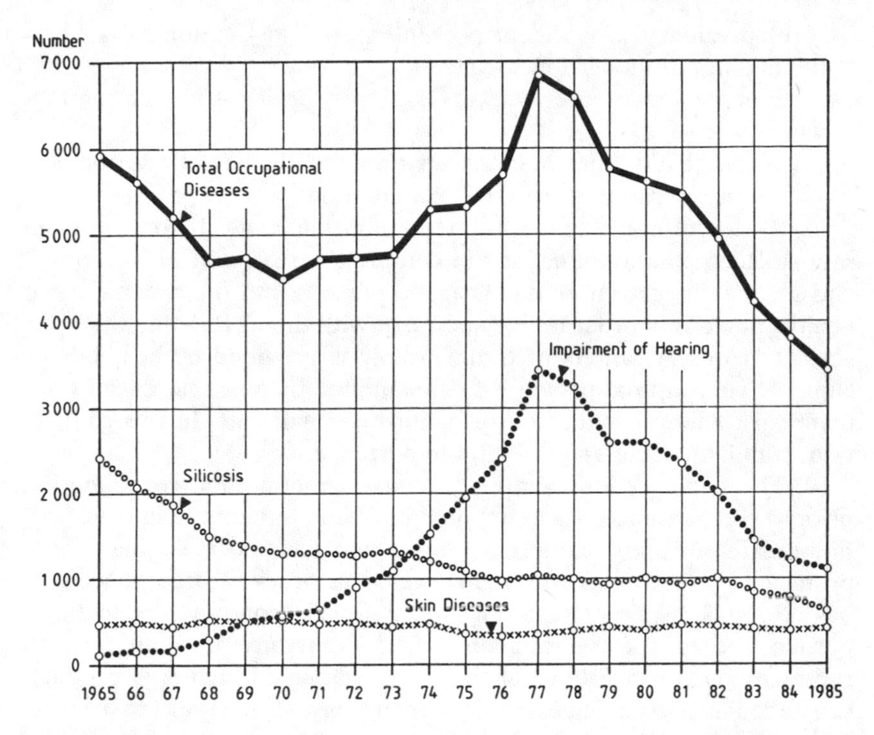


Fig. 1. Occupational diseases compensated for the first time by industrial injuries insurance institutes.

lesions. Prevention comprises the entire strategy of measures and activities directed toward avoidance of occupational accidents and disease risks in their broadest sense. The highest objectives are to maintain the capability to work and the health of all employees.

(2) This development is characterized by the fact that all prevention measures have been transferred to the periphery of the endangered zone. This strategy is called "graded prevention." Here not only classical factory monitoring and advice are concerned but also

(a) monitoring of working materials, recognition of harmful working materials and their prohibition if necessary;

(b) testing of machines and instruments for safety at the construction stage;

(c) safety technique research work in special, modern institutes;

(d) training in the field of occupational safety ;

(e) psychological accident prevention and instruction assisted by the findings of modern medicine;

(f) implementation of workers' rights by accident prevention regulations;

(g) the field of immediate occupational medicine with its complex system of obligatory preventive examinations.

(3) In case of injury or disease, rehabilitation comes before compensation. Rehabilitation aims at the complete restoration or vocational and social reintegration of handicapped persons into the community of healthy persons in order to live and work with them. Rehabilitation for accident cases at centers of traumatology is provided by hospitals or clinics. Every injured person is first examined by an experienced traumatologist who plans the course of further treatment. In this connection, company doctors are of vital importance.

(4) The sequence – prevention prior to rehabilitation, rehabilitation prior to compensation – is based on ethical and humanitarian concerns. For this reason, physical welfare protection has become a basic right protected by the constitution. However, this order of precedence also has economic aspects. Regarding the national economy, the financial burden due to disease, accidents, and premature invalidity, which represents nearly one third of the entire obtainable national product, can be reduced to a great extent by this strategy. With regard to business management, it is also advantageous. According to the contributory system of accident insurance, companies with high accident rates have to pay a higher premium. Thus accident prevention is rewarded.

(5) The occupational medical prevention system is comprehensive. It is assured by a complex system of obligatory preventive examinations, which have to be provided by employers at their own expense, and which are monitored by the State or by the industrial injuries insurance institutes.

(6) The high standard demanded of occupational medicine should be stressed; occupational medicine has become a university subject, and only specially trained physicians are permitted to carry out occupational medical tasks. By the end of 1984, among 192,000 registered physicians, more than 18,000 had specialized in occupational medicine.

(7) The industrial injuries insurance institutes which are responsible for system implementation and monitoring are proportionally represented self-governing bodies. Their structure varies in accordance with each industry. Two outstanding advantages are the consequence:

(a) This organizational form provides the scope necessary to act according to the requirements of *practice* and guarantees that *measures relevant to the corresponding risk* are taken;

(b) thus, working life is organized on a comprehensible scale and on a democratic basis. Furthermore, in accordance with the employees' right of participation and the work council's obligation to act in an advisory capacity, the staffs share responsibility as required by the Labor Management Act in the field of accident prevention, occupational diseases, and health protection.

## VI. BUDGETS AND THE POLICY OF DISTRIBUTED RISKS

The expenses of the preventive system, rehabilitation measures, and required compensation are borne by the employer. This particularly concerns the occupational medical examinations which are directly reimbursed by the companies. Industrial injuries institutes are also paid by employers; the individual employee is not charged directly. The financial contribution to accident insurance is proportional to the risks: Companies with high accident rates have to pay higher fees. Consequently, the prevention of accidents and occupational diseases is financially beneficial.

Although accident insurance is subject to the same cost explosion as other carriers of social insurance – this refers to medicines, doctors' fees, and other benefits – there has been no comparable discussion of expenses. Because of the structure of the system, the increase of expenses parallels that of the increase of the gross national product. This demonstrates the decrease of accidents and occupational disease, and thus reflects successful prevention.

As a consequence of multiple coordinated preventive measures, accidents and occupational diseases decreased in the period between 1960 and 1983 by about 50 percent (reduction of reported cases 52.3 percent; cases compensated for the first time, 43.9 percent; lethal cases 51.9 percent).If accident frequency after 1960 had been constant, compensation benefits of accident insurance would have been higher by about 25 percent, i.e., about 2,000 million DM ($1 billion). According to a study of the main association of industrial injuries insurance institutes, which is based on data from the Federal Office of Occupational Safety and Accident Research, this reduction of accident rates resulted in a saving of more than 30,000 million DM ($15 billion), i.e.,

1.8 percent of the gross national product. In order to avoid misinterpretation, this amount also includes, in addition to accident, disease, and pension insurance benefits, considerable expenses related to business management and the national economy.

## VII. SOME CRITICAL PROBLEMS

The system of preventive measures for the protection of workers, either of a technical or medical nature, also has provoked *criticisms*:

a) The network of rules and regulations that refer to occupational protection is, on the whole, regarded as qualitatively good, but also as too detailed, hair-splitting, fragmented, and obscure; finally, as unfriendly or even inimical to the user. The application of this system of rules and regulations demands too much of small and medium-sized companies. (A draft bill of a new law relating to occupational protection is intended to counteract this criticism.)

b) The results of occupational medical examinations, it is alleged, have a demoralizing and destabilizing effect on a population (socio-psychological argument), and

c) The results of occupational medical examinations tend to unfairly induce a selection process in companies which leads to the premature release of senior employees.

These criticisms engender considerable emotional reaction. A statistical survey will show, however, how ill-founded they really are. In 1983, 1,399,110 occupational medical examinations were performed in the Federal Republic of Germany; only 1.2 percent reflected permanent health impairment and only limited objections were found for 0.9 percent of the examined cases. Premature invalidity, which doubtlessly is on the increase, has basically different causes. Yet today, attention is drawn to risks, mainly through slogans in the media, "Work is injurious to health"; "You are finished at 50." Thus disease-producing factors of the work environment and in life are now being placed in the foreground. Medical findings of other causal factors, like disposition and constitution, bad fortune and aging, and individual habits detrimental to health, are neglected; yet exogenous risk factors in the work environment are overestimated [19].

Another objection to the system of occupational medical measures is, however, more important. It is said that these measures were based on the employee's objectively *unjustified* obligation to accept them [*Dul-*

*dungspflicht*].This objection, too, has a double dimension: (1) a legal one – since these "obligations" are components of the State's system of laws, and (2) a moral one – because the "obligation" to accept these measures in the field of preventive medicine is a moral one.

The legal dimension of this objection leads first to an academic problem. Traditionally, a legal responsibility is involved if a certain behavior is not observed but yet may result in State sanctions if unobserved. This does not apply to the present case: the system of laws in the Federal Republic of Germany does not include sanctions if the employee refuses to take part in occupational medical examinations. The expression "obligation to accept the measures" is therefore misleading. Obviously the critics of occupational precautions and occupational medicine intend something else. The employee who refuses the preventive checkup by a company doctor has to accept certain vocational and economic consequences. He can be assigned to another workplace and may, under certain circumstances, risk being dimissed. The disadvantages which result from his refusal will induce him to participate regularly in occupational medical examinations. Even if the expression "obligation to accept these measures" is not appropriate, it might be appropriate to regard as "obligatory" participation in preventive checkups performed by company doctors.

At the same time, the moral dimension of the problem comes into focus. The indirect effects of legal standards which force the employee to undergo occupational medical examinations affect his moral right to the free and autonomous organization of his way of life. According to critics, this can be even less acceptable as the consequences of omitted occupational medical checkups begin to affect the employee. The employee should decide whether he or she is willing to sacrifice his or her health in biologically, chemically, or physically unfavorable working conditions.

In the case of occupational medical care, the balancing of interests is difficult. The results will in the end be influenced by two points of view: First, company doctors' examinations are, unlike the forced-feeding in the penal system, *only a minor intrusion* into physical integrity. Second, these checkups have the purpose of protecting the community of insured in the social insurance system and are therefore of *legitimate public interest*. In addition, extensive preventive medical measures contribute to a better utilization of the labor force. Consequentially, employees' objections to the "obligation to accept these measures" are not convincing.

Another ethically motivated objection against the kind and extent of occupational medical measures is directed against "biological monitoring." Behind this impressive-sounding term is hidden a familiar element of medical diagnostics: the investigation of human blood and human excretory products: to determine the presence of harmful substances. The core of the argument is the opinion that through "biological monitoring" the worker is "used as a measuring instrument."

But this objection is not convincing either, especially with regard to its ethical premise. In the case of biological monitoring, the employee is not only "used as a measuring instrument," but at the same time is the intended beneficiary of the occupational medical examinations. Most of all, the worker's health and well-being are to be maintained by preventive medical measures. Furthermore, the success of these measures has been confirmed by the reduction of numerous, severe occupational diseases.

Criticism against the system of occupational medicine has also focussed on confidentiality. Mostly jurists have participated in this discussion, and their arguments may be summarized as "health protection *contra* data protection." This area includes the most controversial questions of preventive medicine. How difficult it is to justify the passing on of examination results by a company doctor is demonstrated by the frequent relation to the term "supposed agreement." It is also an appeal to the legislature and courts to clarify or to stabilize the legal situation, thus making an important contribution to legal safety in the Federal Republic of Germany.

Nevertheless, the problem of medical secrecy has not only a judicial but also an ethical component. It is obvious that the passing on of health-related information about a patient involves a moral principle of highest concern. However, the aim of health policy is optimum occupational medical care, which is also an ethical concern. Like all conflicting values, this can only be resolved with great empathy and a thorough analysis of the controversial interests. There should be no question of not maintaining the core of this successful system of occupational medical measures.

Another aspect of activities performed by the industrial injuries insurance institutes raises ethical questions of fundamental importance. Contrary to the problems noted above, not medical but technical measures of occupational protection are concerned. This aspect refers to the acceptance and the *remaining risk* for employees in spite of technical precautions. Frequently, the complete exclusion of health risks will be technically impossible or can only be obtained at enormous expense. The German hard coal mining industry may serve as an example. The permitted limit of quartz dust values would normally, in the course of a working life, exclude the occurrence of silicosis which is to be compensated. Lung disease due to dust levels which are to be categorized below compensation levels cannot be excluded, however, and are accepted under consideration of health policy and technical and economic factors. The ethical relevance of striking a balance here is evident. The health interests of employees always have priority in this case.

## VIII. FUTURE POLICY DIRECTIONS

(1) Our working life is subject to radical changes. Workshops and offices are changing their appearance; demands on employees vary as well. Effects on occupational protection are to be expected. The increased use of new technologies in industrial production as well as in the services sector is of special importance in this connection. For more than ten years, the Federal government has sponsored a program encouraging the promotion of research and development for the "humanization of working life." Hitherto, 994.4 million DM ($449.7 million) have been invested. Studies have so far shown that it is possible in many cases to avoid or eliminate harmful exposures and work environments harmful to health. E.g.,

(a) reduction of extreme exposures to heat, dust, and noise in the hard-coal mining industry;

(b) suppression of the causes of damage to hearing due to noise in, e.g., the metal processing industry;

(c) development of new protection data, guidelines, and minimum requirements for machines, plants, and workshops on the basis of numerous research projects;

(d) promotion of new technologies that meet the requirements of individual workers.

The aim of the program should be a combination of humanization and innovation. The program supports companies, trade unions, and employers' associations in coping with technical changes. The Federal

Republic of Germany will continue with this program. Most of all, the government hopes to have *the results of research translated into practice*. The findings from this program will, it is hoped, by the next decade, as they have in the past, generate multiple improvements in the occupational protection system.

(2) The processing of harmful materials will continue to be important. On a long term basis, the aim is to put enforced regulations into concrete terms and to develop them even further. The Federal Office for Occupational Safety and Accident Research is presently developing a data base on the impact of chemical substances.

(3) According to an extensive "Investigation on the occupational protection system in the Federal Republic of Germany," the government wants to achieve institutional improvement by a *Labor Protection Act*.

(4) The expansion of occupational medical checkups and the extension of occupational medical science inevitably provoke a confrontation with new disease syndromes. This will have consequences for the entire catalogue of occupational diseases. In addition to these monocausal diseases, highly complex diseases induced by multiple factors are more and more in the foreground. For this reason, in addition to the working situation, the entire environment as well as individual physical, psychic, genetic, and social factors have to be considered. Thus, the term "work-related diseases" will be a component in regulations which deal with compensation under statutory accident insurance to be considered by the legislature.

(5) In trade unions which contributed decisively to the establishment of the system in the F.R.G., further developments of the occupational health and safety system are being discussed and promoted, which will have long-term consequences:

(a) an expanded role for industrial injuries insurance institutes;

(b) the systematic registration of injurious factors in workplace environments;

(c) a binding occupational protection program for companies or their industrial branches;

(d) a strengthening of the employee's individual rights, e.g., that the employee may *claim* that certain protective measures be taken by the employer or the employee may *refuse* work which includes risks to his or her health or life.

The future clearly holds excellent prospects for the welfare of workers in the F.R.G.

*Bergbau-Berufsgenossenschaft,*
*Bochum, Federal Republic of Germany*

## BIBLIOGRAPHY

*Laws, Regulations and State Notifications:*

1. 1968, Gesetz zum Schutze der erwerbstätigen Mutter (Mutterschutzgesetz) (BGB1. I, p. 69), (BGB1. I).
2. 1980, Gesetz über technische Arbeitsmittel (Gerätesicherheitsgesetz) (BGB1, p. 717), (BGB1. I).
3. 1976, Arbeitssicherheitsgesetz – Gesetz über Betriebsärzte, Sicherheitsingenieure und andere Fachkräfte für Arbeitssicherheit – (BGBI, I, p. 1885), (BGB1. I).
4. 1976, Gesetz zum Schutze der arbeitenden Jugend (Jugendarbeitsschutzgesetz) (BGB1. I).
5. 1980, Gesetz zum Schutz vor gefährlichen Stoffen (Chemikaliengesetz) (BGB1. I).
6. 1975, Verordnung über Arbeitsstätten (Arbeitsstättenverordnung) (BGB1. I).
7. 1982, Verordnung über gefährliche Arbeitsstoffe (Arbeitsstoffverordnung), (BGB1. I, p. 144).
8. 1984, Technische Regeln für gefährliche Arbeitsstoffe (TRgA 900, betrifft MAK-und TRK-Werte), Bundesarbeitsblatt 10/1984.

*Regulations Issued by Industrial Injuries Insurance Institutes*:

9. 1985 (VBG 100), Verzeichnis der Einzel-Unfallverhütungsvorschriften der gewerblichen Berufsgenossenschaften, Loseblattsammlung, issued by Hauptverband der gewerblichen Berufsgenossenschaften, Selbstverlag, St. Augustin.
10. 1977, Berufsgenossenschaftliche Grundsätze für arbeitsmedizinische Vorsorgeuntersuchungen, Loseblattsammlung, issued by Hauptverband der gewerblichen Berufsgenossenschaften, Gentner Verlag, Stuttgart.

*Comments and Explanations*:

11. Hinrichs, Oswald: 1976, *Verordnung über den Schutz vor Schäden durch ionisierende Strahlen* (Strahlenschutzverordnung), Deutscher Fachschriften-Verlag Braun & Co. KG, Wiesbaden.
12. Krebs, Heinrich: 1984, Arbeitssicherheitsgesetz – *Gesetz über Betriebsärzte, Sicherheitsingenieure und andere Fachkräfte für Arbeitssicherheit*, Loseblattsammlung, Verlag, Schulz, Percha, Kempfenhausen.
13. Nöthlichs, Mathias: 1981, *Chemikaliengesetz, Arbeits-Gesundheits-und Umweltschutzvorschriften*, Loseblattsammlung, Verlag Erich Schmidt, Berlin.
14. Quellmalz Eberhard: 1975, *Verordnung über gefährliche Arbeitsstoffe* (Arbeitsstoffverordnung), WeKa-Verlag, Kissing.

15. Schmatz, Hans und Nöthlichs, M.: 1969, *Arbeitsstättenverordnung*, Loseblattsammlung, Verlag Erich Schmidt, Berlin.
16. Schmatz, Hans und Nöthlichs, M.: 1980, *Gerätesicherheitsgesetz*, Loseblattsammlung, Verlag Erich Schmidt, Berlin.
17. Schmatz, Hans und Nöthlichs, M.: 1981, *Sicherheitstechnik*, Loseblattsammlung, Verlag Erich Schmidt, Berlin.
18. Siller, Ewald und Schliephacke, Jürgen, *sine anno*, *Neues Recht im Arbeitsschutz: Arbeitssicherheitsgesetz*, 2nd ed., (Berufsgenossenschaft der Feinmechanik und Elektrotechnik, Köln).

*Book*

19. Valentin, H. *et al.*: 1979, *Arbeitsmedizin* (2 vols.), 2nd ed., Georg Thieme Verlag, Stuttgart.

H. THOMAS BALLANTINE

# A VIEW FROM A CLINICIAN'S WINDOW

Most of the essays in this volume have been written by individuals who have made distinguished contributions to a further understanding of modern ethical and philosophical issues. Others are erudite in the fields of economics and public health. As one of the few clinicians who has been privileged to be invited to contribute, I will refer to certain of the opinions of my fellow authors, comment on current policy in the public and private sectors as it relates to the delivery of medical care in the United States and, finally, attempt to forecast future developments. In order for the reader to assess clearly the worth of my presentation since it may be biased by my past experiences, a brief biographical sketch follows.

I am a practicing neurosurgeon in the city of Boston who was born in Muskogee, Oklahoma, which was and is somewhere between a town and a city. My father was a family doctor in the true sense of that phrase. Starting at the age of 12, I would make house calls with him and go to his office on many weekends.

In 1933 I entered Johns Hopkins Medical School and worked that next summer as an "assistant" to the struggling young Muskogee doctor who, as City Physician, held clinics for the indigent. From Hopkins I came to the Massachusetts General Hospital and, after World War II, to the University of Michigan for post-graduate training. In 1947, I returned to Boston and ever since have been affiliated with Massachusetts General Hospital and the Harvard Medical School.

All of these experiences prior to 1947 occurred before there was anythingbut a faintvestige of third-party payment formedical services; yet I was unaware of a drastic lack of access to them. Sure, the poor had to queue up for service; yes, they occupied 16 bed wards. But the quality of medical care was as high (or higher!) on the wards of the Massachusetts General as it was in the most expensive part of the hospital. The poor were cared for by the resident staff who were closely supervised by some of the most competent doctors in Boston and access was unfettered.

Physicians were reimbursed on the principle of charging each patient

*Hans-Martin Sass and Robert U. Massey (eds.), Health Care Systems,* pp. 345–351.

according to his means, the so-called "Robin Hood method". Most private hospitals were largely supported by charitable contributions and municipalities built their own.

Although medical care was apparently available to all, demand was to some extent curtailed by cost and availability, particularly in poorer metropolitan areas and rural America. Undoubtedly there were instances of a lack of needed treatment and this came to be identified as a social injustice.

A little over twenty years ago "Equal Opportunity in Our Great Society" became the political rallying cry and led, along with other legislative activities, to the introduction of Medicare and Medicaid for assuring "equal" access to medical care for the elderly and the poor. The rapid growth of health insurance was felt to provide access for other segments of American society.

In all of the developed countries in the Western world this same movement was taking place with, in many instances (most notably in the Scandinavian countries and Great Britain), government control over the provision of and payment for medical services. Care was perceived by those seeking it, as well as the public, as being "free" at the time of purported "need"; the control over demand by the constraints of cost to the individual was lost.

Enoch Powell, Minister of Health in the British Government from July 1962 to October 1965 was among the first to express alarm over the impact of this loss of control over health costs. His writings on this subject have led health economists to speak of "Powell's Law"; i.e., when a service is deemed "free" at its source the demand for it becomes insatiable unless some forms of constraints are instituted. But what are the ethical problems that arise if constraints must be imposed and how might they be solved in a manner which can be considered fair and just? These questions have been under constant recent scrutiny; they were the subject of a report, "Gaining Access to Health Care" by the President's Commission for the Study of Ethical Problems in Medicine and Biomedical and Behavioral Research and were the stimulus for the symposium on which this book is based.

The essayists who have contributed to this volume appear to support certain statements contained in the Declaration of Independence of the United States, particularly that all members of a political entity are "created equal"; i.e., are entitled to equal treatment under law and to equal opportunity for betterment. Furthermore, they seem to agree with the framers of the U.S. Constitution that all members of a political entity have certain "unalienable rights" and among these are the right to "life, liberty and the pursuit of happiness." Implicit in this statement is the notion that the pursuit of good health is one legitimate goal in the overall pursuit of happiness.

In reference to medical care, there again seems to be agreement that a society has an ethical, social, and political obligation to assure that all its members have equitable access to an adequate level of high quality medical care without incurring excessive burdens. This ethical imperative was clearly enunciated by the President's Commission and was perhaps its most important contribution when it pondered the problem of gaining access to health care.

Freedom of choice on the part of consumers and providers of medical care is given great importance in those essays which consider it as highly probable that efforts at cost containment on the part of government and other third party payers will fatally contaminate the ethical obligations of a free society to preserve equitable access and freedom of choice.

In summary, then, there seems to be a consensus that the "ideal" health care delivery system for the developed Western nations would contain four essential ingredients:

(1) Medical care of high (but not necessarily the highest) quality.
(2) Equitable (but not equal) access to that care.
(3) No "excessive burden"s (e.g., financial or geographic).
(4) Preservation of consumer and provider autonomy.

Engelhardt has eloquently delineated the fundamental conflicts or "moral tensions" that arise in attempting to meet these criteria and I will not repeat them here.

But what of the makers of public policy?

Politicians, government bureaucrats, other third party payers, and concerned citizens, all play a role as "public policy makers" and seem to be aware to some extent of the four *desiderata* listed above. Unfortunately, they seem to pay no more than lip service to them.

The concerned citizen, for example, demands high quality care in time of need but seems willing to settle for less in the abstract (particularly for others) if that reduces insurance premium costs. Autonomy of consumer and provider is not considered to be of importance by any of these policy makers. Indeed, the currently popular methods of providing payment for medical care seem designed to reduce autonomy to the lowest level that patients and providers will tolerate. As for the requirement for an access mode which is fair and just, the policy makers in the United States seem to lack an understanding of that concept. But they do know "cost containment," and that is the concept to which all others must bow.

It is ironic that medical care in the United States is being rationed despite Baruch Brody's cogent arguments that rationing is unethical, if not immoral. And it is almost bizarre that such a constraint is to be imposed through price, although much differently from the days before insurance schemes and government involvement dominated the health scene. Let me elaborate:

The Medicare program has progressively increased the deductible, the money that the elderly must pay before becoming eligible for government support; the percentage of coinsurance payments has also increased.

In the private sector an alphabet soup of provider organizations has mushroomed (to mix a metaphor). We have health maintenance organizations (HMOs), preferred provider organizations (PPOs), independent practice associations (IPAs), etc., etc. But they all have one thing in common: they purport to provide for a group of potential patients easy access to high quality medical care with preservation of freedom of choice, all at significantly less premium cost.

An example of one (and, perhaps, one of the better) schemes is "Pru Care" and the quotations which follow are taken from its descriptive brochure. The "Prudential Health Care System" develops a number of "Medical Provider Organizations" (MPOs) in certain designated regions. These organizations will negotiate and receive monthly "capitation-based compensation from the Prudential". Doctors are encouraged to become participating physicians and will have a 20% share, over and above their pre-determined fees, in any financial profit of the organization; but they are also at risk for 20% of any loss that is incurred! "Primary Care Physicians" (defined as specialists in internal medicine, family practice, general practice, pediatrics, or obstetrics/gynecology) are to act as "gatekeepers" who will make access decisions.

Other specialists will also be asked to enter into contracts to deliver services at predetermined fees. There is an implication that these payments may be lower than those currently available in the market place but that such an affiliation "translates into increased patient volume for those who participate."

As to patient autonomy: "PruCare members are able to choose a primary care physician to coordinate their health care needs from physician members in the network; at the same time, a patient will have the freedom to receive health care from providers not designated under the Plan, and be subject to a reduced level of benefits (e.g., deductible and co-insurance). In this way, PruCare members are given the financial incentive to select PruCare-affiliated physicians since this reduces their out-of-pocket costs." One may well ask if this is really freedom of choice.

Think of this scenario: The General Director of the South Side M.P.O. is meeting with his primary care physicians. "Gentlemen," he says, "I have been reviewing our financial position for the past five months of this fiscal year and I regret to tell you that our monthly capitation payments are less than our expenses. The primary causes of this serious situation are a seemingly excessive number of referrals of patients with back problems to orthopedic and neurosurgical specialists, an increase in patient hospitalizations and prolonged length of stay. Now, I am certainly not going to tell you how to care for your patients but I do have the obligation to remind you that if this excess of cost over income continues, you must be prepared to assume 20% of the deficit." Bear in mind that physicians are human, subject to the foibles and temptations of other mortals, (but, perhaps, to a lesser degree) and draw your own conclusions.

All of this will lead in the near term to a three-tiered system for delivering and paying for medical care:

(a) The affluent will have generous insurance benefits that will allow access to high quality care by primary physicians and specialists whom they will be able to choose.

(b) Middle income groups will be encouraged to join patient panels with "gate keepers" who will decide the necessity for access to specialists and/or hospitalization. Patients will be part of a system in which their autonomy and that of their physicians who serve them will be severely limited.

(c) The poor, the near-poor, and those of the elderly who are completely dependent upon government for the payment of medical care expenses, will be herded into corrals of delivery systems from which they cannot escape. Access will be rationed along such financial constraints as "cost-effectiveness on a statistical basis" and "cost-benefit ratios for the population at risk."

This, then, is my scenario for the near term but I am more optimistic about the not-too-distant future. I look forward to a revolt on the part of the middle-income group and the elderly. When enough young mothers are denied access to a pediatrician because some "physician's assistant" decides that such a visit is unnecessary, one of the problems of cut-rate medical care will become apparent. Other problems of equitable access and adequacy will inevitably alert the public to the fact that the burdens incurred when seeking medical care are excessive. Since most of the population is in the middle income or elderly bracket, their demands for change will undoubtedly be met and these changes will filter down to the poor and near-poor.

Then, however, another ethical dilemma will demand attention: the "moral tensions" created by a limitation on the funds available for medical care and the high cost of that care.

A child is in need of a liver transplant, for example. The cost is about $250,000 and the family is on Medicaid. Can society absorb that cost, or rather, should society be required to do so? Our medical care delivery system has already reached this ethical crossroads but society has not provided the system with ethical guideposts to show the way.

The simplest approach is to provide all the funds necessary for any appropriate medical service, but we must also provide for defense, education, police protection, and all the other things that society has decided are necessary for the public good. There is just not enough money for this solution.

Another approach is to accept a financial limitation and attempt to reallocate the available resources, an extremely difficult task in the current ethical climate. We would have to seek answers to a series of thorny philosophical and moral questions such as: "what is life?" and is the quality of life to be considered in deciding whether or not to prolong it? Should "human" life be defined and if so, how? What, exactly, is "personhood"? If it is lost, does human life still exist?

In my view we are spending too much time, energy, and money at the end of life across all age groups and may well be doing the same at its beginning. These are the areas where scarce resources could be reallocated. But this cannot be accomplished without a revision of society's current thinking about the questions listed above, and I am not even sure that all of my fellow authors would agree with this proposal for reallocation.

I am, however, certain that philosophers, ethicists and physicians as educated, informed, and concerned individuals have an obligation to come together to perform the following tasks, among others, for the benefit of society:

(1) To study, contrast, and compare various schemes for medical care delivery in differing governmental, social, and cultural settings.

(2) To consider the advantages and disadvantages of these various schemes.

(3) To discover or create mechanisms to protect the ethical obligation and enhance the opportunity for society to provide equitable access to an adequate level of medical care for all of its constituency without the imposition of excessive burdens while preserving patient and physician autonomy. Without this last essential ingredient all attempts to provide medical care of high quality will prove fruitless.

These endeavors must be aimed toward informing and educating the public since it alone is the true protector of the public good. Moreover, citizens must realize that freedom is not free. Attaining and preserving it demands sacrifice and struggle. A free society can easily become a slave state if its members abuse individual freedom or become apathetic about the concept. The moral tensions involved in designing an ethical system of medical care delivery may be harbingers of future struggles to preserve and protect those rights so highly prized by our founding fathers.

*Massachusetts General Hospital,*
*Boston, Massachusetts*

ROBERT U. MASSEY

# EPILOGUE

In the early decades of the American Republic no one worried very much about medical care; most disease ran its course, doctors by their efforts made little difference, and in most families there was one person, usually the mother, who provided care during times of sickness. There was sometimes a skillful bonesetter in the town, and usually women to help with a delivery. Where all men were equal anyone could be his own doctor. Clearly medical care had not been a major economic consideration in eighteenth-century North America, nor was it for the older nations of Europe. By and large, physicians served the nobility and the wealthier upper classes; everyday care was provided by the mother, using remedies handed down from her mother, or perhaps by some woman on the landlord's or plantation owner's domestic staff, often under the supervision of his wife. Science and technology had not yet given medicine the power and prestige that it has come to have in the twentieth century.

In 1742, Edward Baynard in his "Doctor's Decade" wrote:

> In Ten Words the whole Art is comprised
> For some of the Ten are always advised
> These few Evacuations
> Cure all the Doctor's Patients.
>
> What more they advance,
> Is all done by chance;
> So as to a Cure
> There's none to be sure.
>
> Most other Specificks
> Have no visible Effects,
> But the getting of Fees
> For a Promise of Ease . . . ([6], p. 6).

The development of effective medical care, the means to prevent disease by immunization, the ability to diagnose and to treat with reasonable expectation of success, and the advancing knowledge of

*Hans-Martin Sass and Robert U. Massey (eds.), Health Care Systems,* pp. 353–361.

biological mechanisms have so increased the power, as well as the cost, of medical care that almost all decisions by physicians have come to have social, economic, political, and ethical implications.

Ignorance excused much of the harm done by physicians in the past; technical ignorance now is culpable, and more than common sense and good intentions are needed to understand and manage today's complicated diagnostic and therapeutic procedures. Enough of that knowledge and understanding must somehow be imparted to patients so that they may decide, with the physician, which option to elect. For the physician to act in the patient's best interest, he must seek to understand what those interests are, and to use his special knowledge and skill in ways best suited to the ends upon which they both have agreed.

The machines of modern medicine have costs beyond the means of any single physician and few are portable. Until 1900, the physician could carry most of his tools with him or in his bag: a watch with a second hand, thermometer, stethescope, otoscope, ophthalmoscope, sphygmomanometer, hypodermic needles and syringes, an alcohol lamp, a small apparatus for measuring hemoglobin and performing blood counts, a few surgical instruments, a suture set, a can of ether, and ten or twelve medications, maybe even a microscope, and some slides and stains for simple bacteriologic work. Armed with these and his own clinical skills, a general practitioner could do almost anything modern medicine could do at the turn of the century. Following the discovery of anesthesia and the acceptance of aseptic or antiseptic techniques, even major surgery could be done at home. Arrangements for care were negotiated between physician and patient and the patient's family; except during epidemics or in time of war, the state took little interest in the affair. Whatever happened in the transaction had an insignificant effect on either the economy or the public health.

The exponential growth of biomedical knowledge, the rise of technology, the growth of industry and commerce, the general diffusion of education, the loss of commonly held beliefs, population growth, urbanization, and two world wars have so separated us from our relatively recent past, when most of our values and political systems were formed, that we now grope to define our problems, expecting even less to find satisfactory solutions to them. Writers of epilogues look eagerly for sentences or paragraphs in the collection of papers that will capture and define the issues, reduce them to no more than two or three, and offer solutions that will seem right to everyone on first reading. We recall how the humoral theory of medicine, by referring to the balance of four body fluids, could explain all diseases and provide a rational basis for therapy. Late in the eighteenth century our own Benjamin Rush simplified medicine even more: there is only one disease in all the world, he said, and only one way to treat.

However, that activity of human beings, taking care of each other in matters of their health, is not so easily simplified. The transaction between physician and patient has never been simple, never reducible to three or four principles, even by Benjamin Rush, and is such an important and intimate experience for most of us that we are unlikely to discover a system or organization of care that will satisfy equally our desires for compassion, high quality equity, economy, and freedom.

During the populist movement in the United States in the early nineteenth century, medical licensing laws were repealed; any man could be a doctor. Medicine was only common sense, its apparent complexity artificial. Citizens were free to choose whomever they would to take care of their medical needs. Populism, rugged individualism, the conditions of the frontier, along with a loss of confidence in the heroic medicine of the late eighteenth and early nineteenth centuries, all contributed to breaking the monopoly of regular medicine. Later the development of pharmacology, physiology, pathology, and especially bacteriology in the latter half of the nineteenth century helped to restore public confidence in the effectiveness of medicine and surgery, and licensure laws were returned to the books, the "monopoly" gradually reestablished. Doctors now could claim to base their art upon science, rather than upon some complex system or theory, and were sometimes given credit for the successful outcome even of self-limited illnesses. Any indisposition, no matter how trivial, might be expected to yield to scientific medicine; the medicalization of society had begun. Yet clearly, scientific medicine worked; its success was seen in the decline in perinatal mortality, the improved outcomes in antiseptic surgery, the relief of pain and disability.

However, as Lewis Thomas wrote:

> The great secret, known to internists and learned early in marriage by internists' wives, but still hidden from the general public, is that most things get better by themselves. Most things, in fact, are better by morning.
>
> It is conceivable that we might be able to provide good medical care for everyone needing it, in a new system designed to assure equity, provided we can restrain ourselves, or our computers, from designing a system in which all 200 million of us are assumed to be in constant peril of failed health everyday in our lives. In the same sense that our judicial

system presumes us to be innocent until proved guilty, a medical care system may work best if it starts with the presumption that most people are healthy ([7], p. 85).

But when our good health fails and there is dangerous blood loss from gastrointestinal bleeding, or a life-threatening cardiac arrhythmia, or a retinal detachment, or septic shock following an apparently innocent infection, it won't do to call in the local chiropractor, homeopath, or faith-healer; we may have a choice among surgeons and hospitals, but whoever is chosen will be among those in the monopoly crowd.

Analogies to this are to be found everywhere. If time is of no concern, travel may be by foot, bicycle, horse cart, bus, train, or jetliner. But if we must be in London tomorrow, we may have a choice among airlines, but the pilot will be one of the monopoly, educated and licensed under standards set by his peers and confirmed by the state.

In reading these papers from England, France, East and West Germany, the Netherlands, and the United States, as well as Feshbach's description of the state of medical care in the USSR, we recognize, not surprisingly, how dependent their organizational forms of medical care are upon their cultural and political history; indeed it is easier to understand them from their historical antecedents than from their moral or public policy presuppositions.

When we read of conflicts between physicians and a national health care system, we are not surprised. Physicians and most others involved in the medical care enterprise see themselves standing in a tradition with roots in the Hippocratic principles of beneficence and non-malfeasance:

> I will follow that system of regimen which, according to my ability and judgment, I consider for the benefit of my patients, and abstain from whatever is deleterious and mischievous. I will give no deadly medicine to anyone if asked, nor suggest any such counsel; and in like manner I will not give to a woman a pessary to produce abortion.
>
> Into whatever houses I enter, I will go into them for the benefit of the sick ([1], p. 780).

Physicians deal with their patients one by one. They are, it is said, numerator people. Policy makers, politicians, bureaucrats, hospital presidents, and HMO administrators are, by contrast, denominator people, dealing in populations, health statistics, and votes. Tension between them and physicians is inevitable.

Most of our contributors sought to describe a balance between the interests of the state or community with the needs and wants of individuals. This is at the heart of the philosopher's concern, as it is for the corporate executive, the economist, and the politician. However, for the physician, as physician, each patient, one by one, becomes the center of his world, and, confronted with the patient's pain or illness, all other concerns fade into insignificance; that, he believes, is as it should be, at the heart of his professionalism. Any resolution between these conflicting points of view will be tentative, fragile, and often unfriendly. Add to this the self-interest of the physician in maintaining his social and economic position, and conflict is assured.

This essential conflict is common to all systems, perhaps most intense in those collectivist systems where the state claims to be supreme. Physicians, individualists by the nature of their work, find it difficult to square their primary concern for the care of an individual with the conflicting demands of any group or society. Bound by ethical principles of beneficence, non-malfeasance, and respect for patient autonomy, they are unlikely, when acting as physicians, unless coerced, to be moved by principles of solidarity and economy, choosing rather to be governed by their patients' needs and wants, their responsibility to make clinical decisions, and their freedom to pursue their art. Ultimately the sum of their decisions at the micro-allocational level will determine macro-allocational policy. If called upon, they would seek to formulate policy based upon experience and the scientific analysis of data.

The issues have not qualitatively changed; when modern medicine was only half as old as it is now, the report, "Medical Care and the American People," began:

> The problems of providing satisfactory medical service to all the people of the United States at costs which they can meet is a pressing one. At the present time, many persons do not receive service which is adequate either in quantity or in quality, and the costs of service are inequably distributed. The result is a tremendous amount of preventable physical pain and mental anguish, needless deaths, economic inefficiency, and social waste. Furthermore, these conditions are, as the following pages will show, largely unnecessary....
>
> The most satisfactory solution of the problem of providing adequate scientific medical service to every person according to his needs will be found only when the leaders of the public and of the professions join hands on a basis of mutual understanding, respect, and confidence....
>
> A barrier – in large part economic – stands between practitioners, able and eager to serve, and patients who need the service but are unwilling or unable to pay for it ([9], p.2).

It is instructive for the physician, accustomed as he is to seeing medical care as a sequence of unique situations, to be required to examine the principles upon which he apparently has been making his decisions, and to fit these into some larger philosophical, political, or economic whole. It is threatening, after being comfortable with one vantage point, to be led to another and asked to re-examine and describe the scene from this unfamiliar and novel point of view. Threatening or not, only in this way do his perceptions come closer to "reality" as the world sees it. Yet, returning and re-entering that other world which is peculiarly his own, peopled with the anxious, the fearful, the hurting, the sick, the injured, the depressed, and the dying, he wonders how he will use his new-found wisdom.

> "Beyond the Wild Wood comes the Wide World," said the Rat. "And that's something that doesn't matter, either to you or me. I've never been there, and I'm never going, nor you either, if you've got any sense at all. Don't ever refer to it again, please. Now then! Here's our backwater at last, where we're going to lunch" ([4], p. 8).

Asked for their assessment of health care systems in the late twentieth century, physicians might at first complain of insensitive bureaucrats and greedy lawyers, but then assert, as they have in the past, that the chief problem is ignorance, their own and others'; those political masters responsible for macro-allocational decisions should increase their investments in basic research. Gerald Weissmann wrote in an earlier volume in this series:

> There is ample support for the idea that 'science for its own sake,' the esthetic view, yields utilitarian benefits. Research oriented not to disease, but into the nature of soil fungi, gave us streptomycin, inquiry into the nature of cells in culture led to the Salk vaccine, and studies of the cell cycle in onion root tips have eventuated in the rational treatment of leukemia ([8], p. 106).

The major health problems of the world, malaria, schistosomiasis, pulmonary and cardiovascular disease, malignancy, dementia, AIDS, drug and alcohol abuse, and a host of others equally important to those who suffer and die of them, will not be resolved by changing health care systems, as important as that may be, or by reorganizing society along collectivist or capitalistic lines. We do not have the knowledge to deal effectively with these problems, yet in the United States we spend less than two percent of our "macro-allocation" for health on research; that is 0.2 per cent or less of our GNP, and declining.

Increasing our knowledge base and ingeniously redesigning the system of organizing and financing medical care will be of limited value without assuring the high intellectual and ethical standards of those who are to provide it. Neither economic incentives nor regulations assure virtuous behavior in physicans and all those others engaged in the health care enterprise; yet the morality of the system depends finally upon the virtues of those who give the care.

We all have our list of desirable qualities we would hope to find in our own doctor; years ago, Ivan Bennett listed these qualities which he had remembered from his professor of surgery, Samuel Harvey of Yale: integrity, intelligence, capacity for hard work, common sense or good judgment, and the faculty for ascertaining the truth. I might add compassion and empathy, although these may be encompassed somewhere in the other five. The essential difficulty that a physician finds in working in accordance with any preordained system is illustrated in this quotation from Alexis Carrell with which Bennett preceded his listing of the qualities needed by a physician:

> He needs sound judgment, great physical endurance, and ceaseless activity. He is set a task very different from that of a man of science. The latter can confine himself (almost) entirely to the world of symbols. Physicians, on the contrary, have to face both concrete reality and scientific abstractions. Their mind must simultaneously grasp the phenomena and their symbols; search into organs or consciousness, and enter, with each individual, a different world. They are asked to realize the impossible feat of building up a science of the particular. Of course, they might use the expedient of indiscriminately applying their scientific knowledge to each patient as, for instance, a salesman trying to fit the same ready-made coat to people of different sizes. But they do not really fulfill their duty unless they discover the peculiarities of each patient. Their success depends not only on their knowledge, but also on their ability to grasp the characteristics which make each human an individual ([2], pp. 342–346).

Academic or university medical centers are those complex organizations in every nation where health professionals are educated, where biomedical research is conducted, and where the most serious and difficult medical problems are handled. Their share of the macro-allocation of national resources must be sufficient to assure that this work is well done, quite irrespective of the organization or financing of health care.

In these papers the contributors have spoken of the quality of care, a just and decent minimum, basic and equitable care for all, or of priorities in the allocation of care. Yet no one has devised a satisfactory means of measuring quality or even the effectiveness of care in population groups; mostly it has been the examination of process, assuming that if the process meets a minimal standard, quality is assured. Seventy-five years ago, E. A. Codman wrote a landmark article, "The

Product of a Hospital," and urged that "results of treatment obtained at different institutions . . . be made out and published by each hospital in a uniform manner so that comparison will be possible" ([3], p. 494). While clinical investigators do measure outcomes of specific forms of therapy by carefully devised prospective schemes, outcome measurements comparing institutions, regions, and entire nations have only begun to be devised. Without these data, reorganizing the provision of health services is largely a theoretical exercise in the redrawing of tables of organization. If there must be organizational tampering, and there will be, physicians would urge the experts to devise systems which would minimize conflicts of interest, and avoid incentives to do too much or too little; in fact, they would plead that profit making (or cost saving, which is often the same thing) not be made the goal of medical care.

Finally, the allocation of resources to foster health and prevent disease and disability by public education and environmental improvement has already gained broad support. As the population grows older, and therefore, on average, less adaptable to environmental stress, our policy makers must be given incentives to support prevention of disease and disability as well as medical care.

There is a caution needed here, a reminder that health may not be the highest good. Buchanan spoke of the irrationality and the dubious ethics in "pursuing health to the exclusion of all other goods." Health is a means, but not the sole means to a good life; we may recall examples from art and literature in which illness and disability may have contributed more to our humanity than glowing good health.

Robert S. Morison, concerned with the overriding importance we have come to attach to health, commented:

> Perhaps the most serious result is the way the preoccupation with health occludes the search for other ends. Even though one suspects that other ends may never be found, the renunciation of the search leaves us with nothing to do but order up another angiogram to assess the heart's desire, or to prescribe the most recent tranquilizer to still the remnants of existential angst ([5], p. 63).

Lewis Thomas, you will recall, urged restraint in these matters, suggesting that it might be better if we all could learn the great secret known to most doctors and doctor's wives: "Most things, in fact, are better by morning." Somewhere else he implied that all this concern about health may be unhealthy.

*School of Medicine,*
*University of Connecticut Health Center,*
*Farmington, Connecticut, U.S.A.*

## BIBLIOGRAPHY

1. Adams, F. (trans.): 1939, *The Genuine Works of Hippocrates*, The William & Wilkins Company, Baltimore, Maryland.
2. Bennett, I. L.: 1973, 'Trends and Objectives in Medical Education', in *The Education of Tomorrow's Physicians*, The New York Academy of Medicine, New York, pp. 342–346, reprinted from the *Bulletin of the New York Academy of Medicine* **49** (4) (April 1973), 337–348.
3. Codman, E. A.: 1914, 'The Product of a Hospital', *Surgery, Gynecology and Obstetrics* **18**, 491–496.
4. Grahame, K.: 1980, *The Wind in the Willows*, Ariel Books, Holt, Rinehart and Winston, New York.
5. Morison, R. S.: 1980, 'A Further Note of Visions', *Daedalus* **109**, 1: 55–64.
6. Shryock, R. H.: 1960, *Medicine and Society in America 1660–1860*, New York University Press, New York.
7. Thomas, L: 1974, *The Lives of a Cell*, The Viking Press, New York.
8. Weissman, G.: 1982, 'The Need to Know: Utilitarian and Esthetic Values of Biomedical Science', in W. B. Bondeson, *et. al.* (eds.), *Philosophy and New Knowledge in the Biomedical Sciences*, *Philosophy and Medicine*, Vol. 10, Reidel, Dordrecht, Holland, pp. 105–112.
9. Wilbur, R. L.: 1932, (Chairman, The Committee on the Costs of Medical Care) *Medical Care for the American People*, The University of Chicago Press, Chicago, Illinois.

# NOTES ON CONTRIBUTORS

H. Thomas Ballantine, M.D., is Professor Emeritus, Harvard Medical School and Massachusetts General Hospital, Boston, Massachusetts, U.S.A.

Hubert Brandts, J. D., is Director, Bergbauberufsgenossenschaft, Bochum, F.R.G.

Baruch A. Brody, Ph.D., is Leon Jaworski Professor of Biomedical Ethics, and Director of the Center for Ethics, Medicine, and Public Issues, Baylor College of Medicine, Houston, Texas, U.S.A.

Allen E. Buchanan, Ph.D., is Professor of Philosophy, University of Arizona, Tucson, Arizona, U.S.A.

Corinna Delkeskamp-Hayes, Ph.D., Buchbergstrasse 17, D-6463 Freigericht 1, F.R.G.

H. Tristram Engelhardt, Jr., Ph.D., M.D., is Professor of Medicine and Community Medicine, and Member of the Center for Ethics, Medicine, and Public Issues, Baylor College of Medicine, Houston, Texas, U.S.A.

Murray Feshbach, Ph.D., is Research Professor of Demography, Georgetown University, and President of the American Association for the Advancement of Slavic Studies, Washington, D.C., U.S.A.

Jean-François Lacronique, M.D., is Director of the Hospital and Medical Department, Institut Curie, Paris, France.

Robert U. Massey, M.D., is Professor of Medicine and Community Medicine and Health Care, Dean Emeritus, School of Medicine, University of Connecticut Health Center, Farmington, Connecticut, U.S.A.

Ruth Mattheis, M.D., Senat für Gesundheit und Soziales, Berlin, F.R.G.

Alan Maynard, Ph.D., is Professor of Economics and Director of the Center for Health Economics, University of York, Great Britain.

Martin Pfaff, Ph.D., is Professor of Economics, University of Augsburg, and Director of the Internatioal Institute for Empirical Social Economics (INFES), Augsburg, F.R.G.

Michael A. Rie, M.D., is Assistant Professor of Anesthesia, Harvard Medical School and Massachusetts General Hospital, Boston, Massachuesetts, U.S.A.

Hans-Martin Sass, Ph.D. is Professor of Philosophy, and Director, Zentrum für medizinische Ethik, Ruhr-Universität Bochum, F.R.G., and Senior Research Fellow, Kennedy Institute of Ethics, Georgetown University, Washington, D.C., U.S.A.

Bettina Schöne-Seifert, M.A., M.D., Department of Pediatrics, Universitätskliniken, Göttingen, F.R.G.

J.-Matthias Graf von der Schulenburg, Ph.D., is Professor of Economics, Maximilians Universität, München, and Senior Research Fellow, International Institute of Management, Wissenschaftszentrum Berlin, F.R.G.

Stuart F. Spicker, Ph.D., is Professor of Community Medicine and Health Care (philosophy), School of Medicine, University of Connecticut Health Center, Farmington, Connecticut, U.S.A.

Wolfdieter Thust, Ph.D., is Head, Division of Economics and Statistics, Kassenärztliche Bundesvereinigung, Köln, F.R.G.

Herbert B. Viefhues, M.D., D.M.S.A. (Edin.), is Professor of Social Medicine, Director Zentrum für medizinische Ethik, Bochum, F.R.G.

Maurice M.A. de Wachter, Ph.D., is Director of the Institute for Bioethics, Maastricht, the Netherlands.

# INDEX

*The Philosophy and Medicine Book Series*

*Editors*

H. Tristram Engelhardt, Jr. and Stuart F. Spicker

1. **Evaluation and Explanation in the Biomedical Sciences**
   1975, vi + 240 pp. ISBN 90–277–0553–4
2. **Philosophical Dimensions of the Neuro-Medical Sciences**
   1976, vi + 274 pp. ISBN 90–277–0672–7
3. **Philosophical Medical Ethics: Its Nature and Significance**
   1977, vi + 252 pp. ISBN 90–277–0772–3
4. **Mental Health: Philosophical Perspectives**
   1978, xxii + 302 pp. ISBN 90–277–0828–2
5. **Mental Illness: Law and Public Policy**
   1980, xvii + 254 pp. ISBN 90–277–1057–0
6. **Clinical Judgment: A Critical Appraisal**
   1979, xxvi + 278 pp. ISBN 90–277–0952–1
7. **Organism, Medicine, and Metaphysics**
   *Essays in Honor of Hans Jonas on his 75th Birthday, May 10, 1978*
   1978, xxvii + 330 pp. ISBN 90–277–0823–1
8. **Justice and Health Care**
   1981, xiv + 238 pp. ISBN 90–277–1207–7 (HB)/90–277–1251–4 (PB)
9. **The Law-Medicine Relation: A Philosophical Exploration**
   1981, xxx + 292 pp. ISBN 90–277–1217–4
10. **New Knowledge in the Biomedical Sciences**
    1982, xviii + 244 pp. ISBN 90–277–1319–7
11. **Beneficence and Health Care**
    1982, xvi + 264 pp. ISBN 90–277–1377–4
12. **Responsibility in Health Care**
    1982, xxiii + 285 pp. ISBN 90–277–1417–7
13. **Abortion and the Status of the Fetus**
    1983, xxxii + 349 pp. ISBN 90–277–1493–2
14. **The Clinical Encounter**
    1983, xvi + 309 pp. ISBN 90–277–1593–9
15. **Ethics and Mental Retardation**
    1984, xvi + 254 pp. ISBN 90–277–1630–7
16. **Health, Disease, and Causal Explanations in Medicine**
    1984, xxx + 250 pp. ISBN 90–277–1660–9
17. **Virtue and Medicine**
    *Explorations in the Character of Medicine*
    1985, xx + 363 pp. ISBN 90–277–1808–3
18. **Medical Ethics in Antiquity**
    *Philosophical Perspectives on Abortion and Euthanasia*
    1985, xxvi + 242 pp. ISBN 90–277–1825–3 (HB)/90–277–1915–2 (PB)
19. **Ethics and Critical Care Medicine**
    1985, xxii + 236 pp. ISBN 90–277–1820–2
20. **Theology and Bioethics**
    *Exploring the Foundations and Frontiers*
    1985, xxiv + 314 pp. ISBN 90–277–1857–1